Analytics and Knowledge Management

Data Analytics Applications

Series Editor: Jay Liebowitz

PUBLISHED

Actionable Intelligence for Healthcare
by Jay Liebowitz and Amanda Dawson
ISBN: 978-1-4987-6665-4

Analytics and Knowledge Management
by Suliman Hawamdeh and Hsia-Ching Chang
ISBN 978-1-1386-3026-0

Big Data Analytics in Cybersecurity
by Onur Savas and Julia Deng
ISBN: 978-1-4987-7212-9

Big Data and Analytics Applications in Government:
Current Practices and Future Opportunities
by Gregory Richards
ISBN: 978-1-4987-6434-6

Big Data in the Arts and Humanities: Theory and Practice
by Giovanni Schiuma and Daniela Carlucci
ISBN 978-1-4987-6585-5

Data Analytics Applications in Education
by Jan Vanthienen and Kristoff De Witte
ISBN: 978-1-4987-6927-3

Data Analytics Applications in Latin America and Emerging Economies
by Eduardo Rodriguez
ISBN: 978-1-4987-6276-2

Data Analytics for Smart Cities
by Amir Alavi and William G. Buttlar
ISBN 978-1-138-30877-0

Data-Driven Law: Data Analytics and the New Legal Services
by Edward J. Walters
ISBN 978-1-4987-6665-4

Intuition, Trust, and Analytics
by Jay Liebowitz, Joanna Paliszkiewicz, and Jerzy Gołuchowski
ISBN: 978-1-138-71912-5

Research Analytics: Boosting University Productivity and Competitiveness
through Scientometrics
by Francisco J. Cantú-Ortiz
ISBN: 978-1-4987-6126-0

Sport Business Analytics: Using Data to Increase Revenue and
Improve Operational Efficiency
by C. Keith Harrison and Scott Bukstein
ISBN: 978-1-4987-8542-6

Analytics and Knowledge Management

Edited by
Suliman Hawamdeh
Hsia-Ching Chang

CRC Press
Taylor & Francis Group
Boca Raton London New York

CRC Press is an imprint of the
Taylor & Francis Group, an **informa** business

AN AUERBACH BOOK

CRC Press
Taylor & Francis Group
6000 Broken Sound Parkway NW, Suite 300
Boca Raton, FL 33487-2742

© 2018 by Taylor & Francis Group, LLC
CRC Press is an imprint of Taylor & Francis Group, an Informa business

No claim to original U.S. Government works

Printed on acid-free paper

International Standard Book Number-13: 978-1-1386-3026-0 (Hardback)

Visit the Taylor & Francis Web site at
http://www.taylorandfrancis.com

and the CRC Press Web site at
http://www.crcpress.com

Contents

Preface

The terms data analytics, Big Data, and data science have gained popularity in recent years for a number of good reasons. The most obvious reason is the exponential growth in digital information and the challenge of managing large sets of data. Big Data forms a challenge and opportunity at the same time. It is a challenge if not managed properly and the organization does not make the needed investment in the knowledge infrastructure recognizing the value of data as an organizational asset. Knowledge infrastructure is made of several components including intellectual capital (human capital, social capital, intellectual property, and content), physical capital, and financial capital. What makes Big Data an opportunity is the prospects of knowledge discovery from Big Data and the value of such knowledge in enhancing an organization's competitive advantage through improved products and services, as well as enhanced decision-making processes.

Given the cost associated with managing Big Data, organizations must adopt a knowledge management strategy in which Big Data is viewed as a key organizational asset. This also includes making the necessary investment in data science and data analytics tools and technologies. Knowledge management places a higher emphasis on people and human capital as a key to realizing the concept of the knowledge-based economy. This means any knowledge management strategy must include a plan to educate and enhance the capacity of those working with Big Data and knowledge discovery.

The shift toward the knowledge economy and the realization of the importance of data as an organizational asset within the context of knowledge management has given rise to the emerging fields of data science and data analytics. The White House's "Data to Knowledge to Action" initiative in 2013 aimed at building Big Data partnerships with academia, industries, and public sectors. This initiative led to the National Science Foundation (NSF) increasing the nation's data science capacity by investing in human capital and infrastructure development. The Big Data to Knowledge (BD2K) initiative by the National Institutes of Health (NIH) in 2012 and Google's Knowledge Graph were also aimed at building big data capacity. Such capabilities will be based on well-established knowledge infrastructures made of a network of individuals, organizations, routines, shared norms, and practices. Building knowledge infrastructures requires human interactions through connecting people, organizations, and practices to facilitate knowledge discovery from Big Data. However, relatively few organizations have developed data governance and knowledge management plans for handling and managing Big Data and big data analytics projects.

Knowledge management is an interdisciplinary approach to dealing with all aspects of knowledge processes and practices. Many of these activities are critical to the notion of managing data and creating information and knowledge infrastructure in the long run. We all understand the importance of information organization and data management for data (big or small) to be useful. The term "garbage in, garbage out" is used to describe poor data and information organization practices. The long-term preservation of big data as part of the knowledge retention process is crucial to the knowledge discovery process.

The process of transforming data into actionable knowledge is a complex process that requires the use of powerful machines and advanced analytics techniques. Analytics, on the other hand, is the examination, interpretation, and discovery of meaningful patterns, trends, and knowledge from data and textual information. It provides the basis for knowledge discovery and completes the cycle in which knowledge management and knowledge utilization happen. This book examines the role of analytics in knowledge management and the integration of Big Data theories, methods, and techniques into the organizational knowledge management framework. The peer-reviewed chapters included in this book provide an insight into theories, models, techniques, applications, and case studies in the use of analytics in organizations. The following are summaries of each chapter.

Chapter 1, "Knowledge Management for Action-Oriented Analytics" by Edwards and Rodriguez traces the evolution of analytics, compares several practical concepts (analytics, business analytics, business intelligence, and Big Data), and associates those concepts with knowledge management. They recommend that organizations develop knowledge focuses on data quality, application domain, selecting analytics techniques, and on how to take actions based on patterns and insights derived from analytics. Such actions rest on leveraging people, processes, and technology. They discuss the classifications of analytics from the dual perspectives of types of analytics techniques (descriptive, predictive, and prescriptive analytics) and business management (strategic, managerial, operational, customer-facing, or scientific analytics). From a global and cross-industry perspective, they also provide examples that demonstrate how each individual analytics type corresponds to one or multiple knowledge focuses.

Chapter 2, "Data Analytics Process: An Application Case on Predicting Student Attrition" by Delen discusses a significant and coincidental resemblance between the most popular data mining methodology, Cross-Industry Standard Process for Data Mining (CRISP-DM), which addresses analytics processes, and the Six Sigma-based DMAIC (define, measure, analyze, improve, and control) methodology that not only improves organizational performances but supports knowledge management processes. To exemplify the CRISP-DM approach, Delen introduces a specific case of data analytics in higher education regarding predicting student success and student attrition.

Chapter 3, "Transforming Knowledge Sharing in Twitter-Based Communities Using Social Media Analytics" by Evangelopoulos, Shakeri, and Bennett showcases eight vignettes of social media analytics and illustrates how different dimensions (such as time, location, topic, and opinion) can contribute to the knowledge base of Twitter-based communities. Incorporating derived facts and derived dimensions, they devise a knowledge base data warehouse (KBDW) schema for presenting transformed explicit knowledge in Twitter-based communities.

Chapter 4, "Data Analytics for Deriving Knowledge from User Feedback" by Chahal and Kapur is a practice-oriented contribution. To make analytics of user feedback data meaningful, they suggest connecting data analytics with data management and knowledge management as a three-step approach. They select the case of the Indian government's demonetization drive in 2016 and describe how the government can better understand public perceptions across the country by means of opinion mining and data mining of Twitter data.

Chapter 5, "Relating Big Data and Data Science to the Wider Concept of Knowledge Management" by Stark and Hawamdeh outlines the concept of data science, a core component being taught within master's programs that focuses on data as well as the skill set desired by current employers who seek to hire data scientists. The chapter reviews some of the current data science and data analytics tools being used to store, share, mine, model, and visualize Big Data. The chapter also reviews some of the case study applications and their relevance to the concept of managing data, small and big.

Chapter 6, "Fundamentals of Data Science for Future Data Scientists" by Chen, Ayala, Alsmadi, and Wang considers data science as an interdisciplinary field revolving around data and puts data science in the context of knowledge management processes. Through literature review and the analysis of 298 job postings, they identify the desired knowledge and skills for data scientists. Additionally, they reviewed existing data science programs in the United States and provided suggestions for integrated curriculum design.

Chapter 7, "Social Media Analytics" by Chong and Chang starts with the evolution of social media and analytics. Besides introducing the definitions and processes of social media analytics from different perspectives, the chapter focuses on identifying the main techniques and tools used for social media analytics.

Chapter 8, "Transactional Value Analytics in Organizational Development" by Stary introduces an approach of value network analysis (VNA) and suggests how organizations can use different proven methods (repertory grid, critical incident technique, and storytelling) to elicit the tacit knowledge from stakeholder transactions for value management. Stary interviewed 14 active analysts who had diverse backgrounds and experiences in practical value management for international organizations in Germany and Austria. Those analysts appeared to reach the consensus that the repertory grid technique combining qualitative with quantitative insights fits better in analyzing stakeholder transactions. In addition, Stary validated that the repertory grid technique works effectively on constructing holomaps for collective sense-making during VNA.

Chapter 9, "Data Visualization Practices and Principles" by Kim and Schuler demonstrates a hypothetical scenario: The Knowledge Management Corporation (KMC) has been working with the fictional More Fun Toys company to assist in understanding market trends and customer purchasing patterns over time. They illustrate various data visualization tools that can support answering different types of business questions to improve profits and customer relationships. The authors showed that data visualization has a critical role in the advancement of modern data analytics, particularly in the field of business intelligence and analytics.

Chapter 10, "Analytics Using Machine Learning-Guided Simulations with Application to Healthcare Scenarios" by Elbattah and Molloy explores the potential opportunities of using data-driven machine learning to create new knowledge, thus informing simulation models for better decision making. Like deep learning, emerging machine learning techniques can greatly reduce human biases during knowledge elicitation process. Elbattah and Molloy use the dataset of elder care from Health Service Executive (HSE) of Ireland to depict an analytics use case combining machine learning and simulation methods in elderly patient discharge planning.

Chapter 11, "Intangible Dynamics: Knowledge Assets in the Context of Big Data and Business Intelligence" by Erickson and Rothberg reframes the conceptualization of intangible assets where the four-layer hierarchy framework starts with data and information, moves from explicit knowledge to tacit knowledge, and ultimately to insight and intelligence. In the proposed framework, they compare and contrast the four layers in terms of definition, domain, source, exchange, range, management, metrics, and indicates.

Chapter 12, "Analyzing Data and Words—Guiding Principles and Lessons Learned" by Bedford exemplifies the approach to intelligent use of analytics and thus enhancing knowledge management methods. The need for analysis to support knowledge management goes beyond what has traditionally been available. The chapter examines quantitative and qualitative analytics and their use including the methods, tools, and technologies available. The technologies needed to do qualitative and quantitative analysis should be built on a strong foundational understanding of languages including the ability to interpret any content given to them at a primitive language level.

Chapter 13, "Data Analytics for Cyber Threat Intelligence" by Chi, Scarlett, and Martin provides an overview of data analysis uses in the field of cyber security and cyber threat intelligence, including insider threat. The chapter examines the process of collecting and organizing data as well as reviewing various tools for text analysis and data analytics and discusses dealing with collections of large datasets and a great deal of diverse data types from legacy system to social networks platforms.

Editors

Suliman Hawamdeh is a professor in the Department of Information Science in the College of Information at the University of North Texas. He is the director of the Information Science PhD program, one of the largest interdisciplinary information science PhD programs in the country. He is the editor in chief of the *Journal of Information and Knowledge Management* (JIKM) and the editor of a book series on innovation and knowledge management published by World Scientific. Dr. Hawamdeh founded and directed a number of academic programs including the first Master of Science in Knowledge Management in Asia at the School of Communication and Information at Nanyang Technological University in Singapore. Dr. Hawamdeh has extensive industrial experience. He was the Managing Director of ITC Information Technology Consultant Ltd, a company that developed and marketed a line of software development products. He worked as a consultant to several organizations including NEC, Institute of Southeast Asian Studies, Petronas, and Shell. Dr. Hawamdeh has authored and edited several books on knowledge management including *Information and Knowledge Society* published by McGraw Hill and *Knowledge Management: Cultivating the Knowledge Professionals* Published by Chandos Publishing, as well as a number of edited and co-edited books published by World Scientific.

Hsia-Ching Chang is an assistant professor in the Department of Information Science, College of Information at the University of North Texas. She received her PhD in informatics and MS in information science from the University at Albany, State University of New York as well as her MA in public policy from the National Taipei University in Taiwan. Her research interests concentrate on data analytics, social media, cybersecurity, knowledge mapping, scientometrics, information architecture, and information interaction.

Contributors

Duha Alsmadi is a PhD candidate in information science at the College of Information at the University of North Texas. She holds a master's degree in computer information systems. Her areas of interest include cloud computing, e-learning, technology adoption, latent semantic analysis, as well as quantitative research methods. She is currently working on her dissertation on the topic of information sharing and storage behavior via cloud computing services.

Brenda Reyes Ayala is a PhD candidate in information science, College of Information at the University of North Texas. She holds a master's degree in human–computer interaction. Her areas of interest include data science, information retrieval, machine learning, and web preservation. She is currently working on her dissertation on the topic of information quality in web archives.

Denise A. D. Bedford is currently an adjunct professor at Georgetown University's Communication Culture and Technology program, adjunct faculty at the Schulich School of Business, York University, a visiting scholar at the University of Coventry, and a Distinguished Practitioner and Virtual Fellow with the U.S. Department of State. She teaches a range of graduate-level courses in knowledge management, enterprise architecture, and data sciences. Her current research interests include knowledge architectures and knowledge engineering, knowledge economies and knowledge cities, intellectual capital management, knowledge sharing behaviors, semantic analysis and text analytics, communities of practice, business architecture, document engineering and content architectures, multilingual architecture, and search system design and architectures. Dr. Bedford retired from the World Bank in 2010 where she was the Senior Information Officer. In 2010, Dr. Bedford accepted the position of Goodyear Professor of Knowledge Management at Kent State University. She retired from this role in 2015. Dr. Bedford has also worked for Intel Corporation, NASA, University of California Systemwide Administration, and Stanford University. Her educational background includes a BA triple major in History, in Russian Language and Literature, and in German Language and Literature from the University of Michigan; an MA in Russian History also from the University of Michigan; an MS in Librarianship from Western Michigan University, and a PhD in Information Science from University of California, Berkeley. She is a certified enterprise architect. She currently serves on several conference and journal editorial boards, and serves as an associate editor of the *Journal of Knowledge Management*.

Andrea R. Bennett is a doctoral student in the Department of Marketing and Logistics in the College of Business at the University of North Texas. She received her master's degree in public administration from the Institute of Public and Nonprofit Studies at Georgia Southern University. Her research interests include social media analytics, brand co-creation, and public-private partnerships.

Kuljit Kaur Chahal is an assistant professor in the Department of Computer Science, Guru Nanak Dev University. Her research interests are in open source software, distributed systems, data analytics, and machine learning. Chahal received a PhD in computer science from Guru Nanak Dev University, Amritsar. She is a POSSE graduate. Contact her at kuljitchahal.cse@gndu.ac.in.

Hsia-Ching Chang is an assistant professor in the Department of Information Science, College of Information at the University of North Texas. She received her PhD in informatics and MS in information science from the University at Albany, State University of New York as well as her MA in public policy from the National Taipei University in Taiwan. Her research interests concentrate on data analytics, social media, cybersecurity, knowledge mapping, scientometrics, information architecture, and information interaction.

Jiangping Chen is a professor in information science in the College of Information at the University of North Texas. She earned her PhD in information transfer from the School of Information Studies at Syracuse University, New York. Dr. Chen conducts research in data science, digital libraries, multilingual information access, and information systems. She is the editor in chief of *The Electronic Library*, an SSCI-indexed academic journal by Emerald Publishing Limited.

Hongmei Chi is an associate professor of Computer & Information Sciences at the Florida A&M University, Tallahassee, Florida. She currently is the director of the FAMU Center for Cyber Security and teaches graduate and undergraduate courses in data mining and cyber security and conducts research in the fields of Big Data and applied security. She has published articles related to data science, parallel computing, and cyber security research and education. Her web page is http://people.sc.fsu.edu/~hcc8471/.

Miyoung Chong is a doctoral candidate at the Department of Information Science in the College of Information at the University of North Texas. Her research interests include open data, computational linguistics, new media theory, and media analysis including social media, film, and mass media.

Dursun Delen is the holder of the William S. Spears Endowed Chair in Business Administration and Patterson Foundation Endowed Chair in Health Analytics, Director of Research for the Center for Health Systems Innovation, and Professor of Management Science and Information Systems in the Spears School of Business at Oklahoma State University (OSU). He received his PhD in industrial engineering and management from OSU in 1997. He authored and co-authored seven books and textbooks in the area of business analytics, decision support systems, business intelligence, data mining and text mining and his research and teaching interests are in business analytics, data and text mining, decision support systems, knowledge management, business intelligence, and enterprise modeling.

John S. Edwards is emeritus professor and professor of knowledge management at Aston Business School, Birmingham, UK. He holds MA and PhD degrees from Cambridge University. His interest has always been in how people can and do (or do not) use models and systems to help them do things. At present, his principal research interests include how knowledge affects risk management; investigating knowledge management strategy and its implementation; and the synergy between knowledge management, analytics, and Big Data. He has written more than 70 peer-reviewed research papers and three books on these topics. He is a consulting editor of the journal *Knowledge Management Research & Practice.*

Mahmoud Elbattah is a PhD candidate at the National University of Ireland, Galway. He is a holder of the Hardiman Scholarship from the College of Engineering and Informatics. His research interests are in the fields of simulation modeling and machine learning. His e-mail address is m.elbattah1@nuigalway.ie.

G. Scott Erickson is a professor of Marketing in the School of Business at Ithaca College, Ithaca, NY where he has also served as department chair and Interim Associate Dean. He holds a PhD from Lehigh University, master's degrees from Thunderbird and Southern Methodist University, and a BA from Haverford College. He has published widely on Big Data, intellectual capital, and business analytics. He spent the 2016–2017 academic year studying knowledge networks related to sustainability as a Fulbright National Science Foundation Arctic Scholar at Akureyri University, Iceland. His latest book, *New Methods in Marketing Research and Analysis* was published by Edward Elgar in 2017.

Nicholas Evangelopoulos is a professor of business analytics at the University of North Texas, where he teaches business statistics, data mining, and Big Data analytics. He received his MS in computer science from the University of Kansas and his PhD in business from Washington State University. His current research interests include text analytics, decision analysis, and applied statistics. His articles appear in *MIS Quarterly*, *Communications of the ACM*, *Decision Sciences*, *Decision Support Systems*, *Communications in Statistics*, and many others. His consulting experience includes corporate projects in text analytics, predictive modeling, customer attrition modeling, and policy evaluation. In 2013, he received the UNT College of Business Senior Faculty Research Award, and in 2016 he was named the 2016–2017 UNT College of Business Professional Development Institute Fellow.

Suliman Hawamdeh is a professor and department chair of the Department of Information Science in the College of Information at the University of North Texas. He is the director of the Information Science PhD program, one of the largest interdisciplinary information science PhD programs in the country. He is the editor in chief of the *Journal of Information and Knowledge Management* (JIKM) and the editor of a book series on innovation and knowledge management published by World Scientific. Dr. Hawamdeh founded and directed a number of academic programs including the first Master of Science in Knowledge Management in Asia at the School of Communication and Information at Nanyang Technological University in Singapore. Dr. Hawamdeh has extensive industrial experience. He was the Managing Director of ITC Information Technology Consultant Ltd, a company that developed and marketed a line of software development products. He worked as a consultant to several organizations including NEC, Institute of Southeast Asian Studies, Petronas, and Shell. Dr. Hawamdeh has authored and edited several books on knowledge management including *Information and Knowledge Society* published by McGraw Hill and *Knowledge Management: Cultivating the Knowledge Professionals* Published by Chandos Publishing, as well as a number of edited and co-edited books published by World Scientific.

Salil Vishnu Kapur is currently pursuing a master's degree in computer science specializing in data science at Dalhousie University, Halifax, Nova Scotia. During this research work, he was working with Capgemini in Pune, India as a senior analyst in the Big Data analytics domain. Kapur received a B.Tech in computer science and engineering from Guru Nanak Dev University. In his undergraduate years, he pursued research on service-oriented computing and published research in the ICSOC and APSEC conferences. Contact him at salilvishnukapur@gmail.com.

Jeonghyun Kim is an associate professor at the Department of Information Science in the College of Information at the University of North Texas. She received her PhD from the School of Communication and Information at Rutgers University. Her areas of research interest range from data curation, digital libraries, data visualization, convergence issues surrounding libraries, archives, and museums to human computer interaction. She has published numerous papers in journals, including the *Journal of the American Society for Information Science and Technology* and the *International Journal of Digital Curation*. She teaches "data visualization and communication" for the master's degree in data science program at the Department of Information Science in the College of Information at the University of North Texas.

Angela R. Martin is an intelligence specialist who works for New Generation Warfare Study Group at USA TRADOC ARCIC as a Cyber Security Strategic Planning and Development Analyst. She has 10 years of experience in the Department of Defense in various roles to include human intelligence collector, senior interrogator, general military intelligence analyst, human terrain specialist, research manager, knowledge management and information technology supervisor, core curriculum instructor and senior all source socio-cultural dynamics analyst.

Owen Molloy is a lecturer in information technology at the National University of Ireland, Galway. His research interests are in the areas of business process modeling and simulation, software engineering, healthcare processes, and information systems. His email address is owen.molloy@nuigalway.ie.

Eduardo Rodriguez PhD, MSc., MBA Eduardo is the sentry endowed chair in Business Analytics University of Wisconsin-Stevens Point. In his work he has created the Analytics Stream of the MBA at University of Fredericton, Fredericton, Canada, analytics adjunct professor at Telfer School of Management at Ottawa University, Ottawa, Kansas, corporate faculty of the MSc in Analytics at Harrisburg University of Science and Technology, Harrisburg, Pennsylvania, senior associate-faculty of the Center for Dynamic Leadership Models in Global Business at http://nz.linkedin.com/company/the-leadership-alliance-inc.?trk=ppro_cprof. The Leadership Alliance Inc. Toronto Canada, and principal at IQAnalytics Inc. Research Centre and Consulting Firm in Ottawa, Canada. He has been a visiting scholar Chongqing University, Chongqing, China, and EAFIT University, Medellín, Colombia, for the Master of Risk Management.

Eduardo has extensive experience in analytics, knowledge, and risk management mainly in the insurance and banking industry. He has been the knowledge management advisor and quantitative analyst at Export Development Canada (EDC) in Ottawa, regional director of PRMIA (Professional Risk Managers International Association) in Ottawa, vice-president in Marketing and Planning for Insurance Companies and Banks in Colombia. Moreover, he has a worked as a part-time professor at Andes University and CESA in Colombia, author of six books in analytics, and a reviewer of several journals and with publications in peer-reviewed journals and conferences. He created and Chair the Analytics Think-Tank, organized and Chair of the International Conference in Analytics ICAS, member of academic committees for conferences in Knowledge Management and international lecturer in the analytics field.

Eduardo holds a PhD from Aston Business School, Aston University, Birmingham, UK, and MSc in Mathematics from Concordia University, Montreal, Canada, Certification of the Advanced Management Program McGill University,

Montreal, Canada, and an MBA and bachelor in Mathematics from Los Andes University, Bogotá, Colombia. His main research interest is in the field of Analytics and Knowledge Management applied to Enterprise Risk Management.

Helen N. Rothberg is a professor of strategy in the School of Management at Marist College, Poughkeepsie, NY. She holds a PhD and MPhil from City University Graduate Center, and an MBA from Baruch College, City University of New York. She is on the faculty of the Fuld-Gilad-Herring Academy of Competitive Intelligence and is a principal of HNR Associates. She has published extensively on topics including competitive intelligence and knowledge management. Her latest book, *The Perfect Mix: Everything I Know About Leadership I Learned as a Bartender* was published by Simon and Schuster in 2017.

Carol Y. Scarlett founded Axion Technologies, in January of 2016, to produce TRNG hardware based on experimentation she was performing. During her seven years at Florida A&M University (FAMU), Tallahassee, Florida, she, an associate professor in physics, has continued research in the areas of astronomy and particle physics. She has over 30 peer-reviewed publications and holds 3 patents on a novel technology for material identification and TRNG devices. Through her work, she developed a technique to utilize magnetic birefringence and microscopy to extend material identification to an unprecedented level of sensitivity. As well, she holds two patents on methods to achieve enhanced optical noise for purposes of encryption and simulations. She maintains active collaborations with Brookhaven National Laboratory (BNL) and Lawrence Livermore National Laboratory (LLNL).

Eric R. Schuler received his PhD in experimental psychology from the University of North Texas, Denton, Texas. He has worked as a teaching fellow in the Department of Psychology and as the research consultant for Department of Information Science's Information Research and Analysis Lab. His research interests include: refining best-practice quantitative techniques based on Monte Carlo statistical simulations, measurement development and validation, and how belief systems shift after a traumatic event. When Eric is not running R-code, he enjoys playing Dungeons and Dragons.

Shadi Shakeri is a doctoral candidate in the Department of Information Science in the College of Information at the University of North Texas. She received her MS in library and information science from Kent State University. Prior to pursuing her master's degree, she worked in industry as a metadata specialist and system analyst in Iran. Her research interests include data and network analytics, knowledge discovery, knowledge management, data modeling, crisis informatics, and metadata designs.

Hillary Stark, with a background in business and marketing, focuses her research towards nutrition marketing efforts, specifically the information relied upon by individuals when purposefully making more healthful eating choices. She is excited and encouraged by the present age of Big Data, as the tools made available to analyze robust amounts of information, from conversations across social media platforms to documents submitted on behalf of passing new dietary legislation, can assist academics, health-care professionals, manufacturers, and the individuals themselves to be more in control of their actions through this newly acquired knowledge. She is currently finishing her PhD studies in Information Science at the University of North Texas, Denton, Texas, and enjoys running and volunteering in her community in her free time.

Christian Stary received his diploma degree in computer science from the Vienna University of Technology, Austria, in 1984; his PhD degree in usability engineering, and also his Habilitation degree from the Vienna University of Technology, Austria, in 1988 and 1993, respectively. He is currently a full Professor of Business Information Systems with the University of Linz. His current research interests include the area of interactive distributed systems, with a strong focus on method-driven learning and explication technologies for personal capacity building and organizational development.

Guonan Wang is an adjunct professor in the Department of Information Science in the College of Information at the University of North Texas. She earned her PhD in physics from McMaster University.

Chapter 1

Knowledge Management for Action-Oriented Analytics

John S. Edwards and Eduardo Rodriguez

Contents

Introduction

Analytics, Big Data, and especially their combination as "Big Data analytics" or "Big Data and analytics" (BDA) continue to be among the hottest current topics in applied information systems. With the exception of artificial intelligence methods such as those focused on deep learning, the emphasis on analytics is now moving away from the purely technical aspects to the strategic, managerial, and longer-term issues: not a moment too soon, some people would say.

In this chapter, we take a strategic perspective. How can an organization use analytics to help it operate more successfully? Or even to help it become the organization that those who run it would like it to be? The need for more thought about this is certainly evident, but the starting point is far from clear.

Our choice of starting point is to build on previous work on the relationship between analytics, Big Data, and knowledge management. This is represented as an action-oriented model connecting data, analytics techniques, knowledge, and patterns and insights. These elements will be used to categorize and analyze examples of analytics and big data projects, comparing the United Kingdom with other countries.

A particular focus is the difference between one-off projects, even if they lead to ongoing results, such as sensor monitoring, and ongoing activities or a series of projects. In the latter, our model adds influences of knowledge on data (what to collect and standardization of meaning), knowledge on techniques (new techniques developed in the light of past examples), and action on knowledge (learning from skilled practice).

Links will also be made to the higher-level issues that drive analytics efforts - or at least should do. These include the definition of problems, goals, and objectives, and the measurement of organizational and project performance. This raises the question of the relationship between knowledge management and strategic risk management.

Looking at the analytics "landscape," we see that opinions, and even reported facts, differ fundamentally about the extent to which organizations are already using analytics. For example, several "white papers" attempt to describe the extent and nature of the use of analytics by employing variants of the well-known five-stage software engineering capability maturity model (CMM) (Paulk, Curtis, Chrissis, & Weber, 1993).

A report commissioned from the UK magazine *Computing* by Sopra Steria uses the same names as the CMM stages for respondents to describe their organization's approach to data and analytics (Sopra Steria, 2016). The results were: initial 18%, repeatable 18%, defined 19%, managed 33%, and optimized 12%.

International Data Corporation (IDC) has a very similar five-stage model for "BDA competency and maturity"—ad hoc, opportunistic, repeatable, manageable, and optimized—but their report gives no figures (Fearnley, 2015). MHR Analytics (MHR Analytics, 2017), a UK consultancy group focusing on human resource management, has developed a five-stage model for what they call the "data journey," comprising the following: unaware 33%, opportunistic 7%, standards led 18%, enterprise 19%, and transformational 30%. However, the fact that these percentages add up to 107% does call their accuracy into question, even if only in the proofreading of the report. Nevertheless, the impression given from these three analyses is that at least half of the organizations responding are well on the way to making good use of analytics and Big Data.

On the more skeptical side, the American Productivity and Quality Center (APQC) reported in 2016 that four-fifths of organizations had not yet begun to take advantage of Big Data (Sims, 2016). This does seem to be closer to our own experience, although research by the McKinsey Global Institute (Henke et al., 2016) found considerable differences between sectors, with retail more advanced than most. As for the trend in Big Data and analytics use, again it may not be as rapid as some articles—usually, those not quoting data—imply. A report by the Economist Intelligence Unit (Moustakerski, 2015) suggests about a 10% movement toward the strategic use of data between 2011 and 2015.

One reason for this difference of opinion/fact is the type of decisions for which analytics and Big Data are being used. Are they the most significant decisions about the strategic direction of the business, characterized by the Economist Intelligence Unit in a report for PwC (Witchalls, 2014) as "big" decisions, or are they everyday decisions? Later in the chapter we will use a classification by Chambers and Dinsmore (2015) to help understand the effect of these differences.

The other crucial element is the extent to which the organization's managers accept, and plan for the use of, analytics; indeed, it might not be an exaggeration to phrase that as the extent to which they *believe in* analytics. The *Computing* survey (Sopra Steria, 2016) found that only 25% of respondents' organizations had a well-defined strategy and—not surprisingly in the light of that figure—that only 10% of analytics projects were "always" driven by business outcomes, with 44% being driven "most of the time."

We see knowledge and its management as the connection between the organization and its strategy on the one hand, and the availability of data and the use of analytics on the other. It is well-recognized that organizations with a culture of knowledge sharing are more successful than those without such a culture (Argote, 2012). A key element in the use of BDA is the presence of a data-driven culture. A data-driven, or data-oriented, culture refers to "a pattern of behaviors and practices by a group of people who share a belief that having, understanding and using certain kinds of data and information plays a critical role in the success of their organization" (Kiron, Ferguson, and Prentice, 2013, p.18). This is the link to the "belief" in analytics that we mentioned earlier.

Nevertheless, belief by itself is not enough for effective use of analytics. A balance needs to be struck between the "having" and "using" elements of the definition, and the "understanding" element. Spender (2007) has found that there are three different types of understanding, although he prefers to call it organizational knowing. These are data, meaning, and skilled practice. To use the distinction identified by Polanyi (1966), some of this understanding and knowing is explicit (codifiable in language), and some is tacit (not readily expressible). Crucially, for skilled practice, the tacit knowledge dominates: even having good data and high-quality technical analysis is not enough on its own.

Another element in our thinking is that the development of the relationship between knowledge and analytics is in permanent evolution. Each time that an analytics solution is obtained, it is then time to start the search for a new analytics solution based on the knowledge acquired from the previous analytics solution implementation. Data changes, and that modifies the outcomes of the models, but at the same time the models are changing, the problems are better defined, and the scope is clearer, based on the experience in successive implementations. In addition, there is a clear challenge in using the freshest data that it is possible to access. In strategic risk, for example, every second the changes in the stock prices are potentially modifying some results in models; in credit risk, there is a need to use a permanent data flow that will potentially modify the outcomes for the decision-making process (granting loans).

When we consider analytics, meaning and skilled practice refer both to the analytics and to the target domain where the analytics techniques are being applied. Skilled practice and meaning also have a two-way relationship with knowledge; they influence it and they are influenced by it.

This two-way interaction of meaning and skilled practice is illustrated by the process of management control systems creation. In one direction the target domain is the development of systems to assure the implementation of strategy in organizations. The analytics domain concerns how to implement the knowledge from accounting, cost control, and so on to develop performance indicators and build up a measurement system for the organization's performance at any level. Nucor Corporation (Anthony and Govindarajan, 2007) is an example where the organization started implementing knowledge management as a management control system, and evolved to develop capabilities to use data in its strategic development and strategy implementation and in particular to the use of business analytics (Hawley, 2016) for improving supply chain management practice, a core process in the business. Thus a better understanding of the requirements of management control systems led a transition from a general analytics view to specific analytics application in a key area of Nucor Corporation's strategy.

Figure 1.1 shows how knowledge, and the various elements involved in an analytics study, fit together when just a single study is considered. The knowledge falls into one or more of four categories:

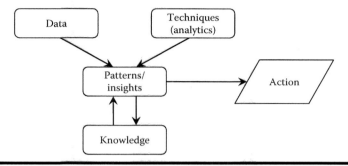

Figure 1.1 The interactions between data, analytics, and human knowledge in a single study. (From Edwards, J.S., and Rodriguez, E., *Proc. Comp. Sci.*, 99, 36–49, 2016.)

- Knowledge about the domain, whether that is medical diagnosis, retail sales, bank customer loyalty, etc.
- Knowledge about the data: are the sources reliable and accurate, are the meanings clear and well defined, etc.
- Knowledge about the analytics techniques being used
- Knowledge about taking action: how to make real change happen; this can be further subdivided into the aspects of people, processes, and technology (Edwards, 2009)

We see analytics acting to supplement gut feel and intuition in the development of solutions to business process problems. Knowledge management supports the development of actions in organizations looking to reduce the storage of knowledge that is never used. Knowledge that is not related to the process of finding solutions is not adding value to the organization. The same is true of data: data without analytics to create knowledge to be used in business solutions cannot provide value. The Economist Intelligence Unit/PwC report mentioned earlier (Witchalls, 2014) indicates that several factors in big decisions can affect the use of analytics. Analytics requires preparation time to create a process that adds value in the organization. Several decisions are made in a short time, forcing a decision based on reactions. The necessary preparation time to avoid reactive solutions or solutions and decisions enforced without good knowledge can be made available through a systematic analytics monitoring system and continuous feedback regarding decisions made and the value added by the analytics.

In the remainder of this chapter, we first offer some relevant definitions, then look at the history of analytics, and move into the related issue of terminology. This enables us to categorize the examples that we consider in the main body of the chapter. We then conclude by considering likely future influences of the political and business environments.

Categorizing Analytics Projects

There is still no generally agreed definition of analytics, although data and models feature in all those we have seen. For example, Davenport and Harris (2007, p.7) define analytics as "the extensive use of data, statistical and quantitative analysis, explanatory and predictive models, and fact-based management to drive decisions and actions." By contrast, the Institute for Operations Research and the Management Sciences (INFORMS) explains that "analytics facilitates realization of business objectives through reporting of data to analyze trends, creating predictive models for forecasting and optimizing business processes for enhanced performance." (Robinson, Levis, and Bennett, 2010).

We prefer those which are more action-oriented; that include a commitment to action, or an expectation of it, rather than just a hope, or even no mention of action at all, and so we use the Davenport and Harris definition.

Classification by Technical Type

Despite the lack of agreement on an overall definition, somewhat strangely there is a much broader agreement about the different types of analytics. By far the most commonly-seen is one that also comes (albeit rarely with proper credit) from the work commissioned from Cap Gemini by INFORMS (Robinson et al., 2010), which identifies three types: descriptive, predictive, and prescriptive analytics.

Descriptive analytics is "the use of data to figure out what happened in the past. Descriptive analytics prepares and analyzes historical data and identifies patterns from samples for reporting of trends" (Robinson et al., 2010). Specific techniques under this heading include data modeling, visualization and regression analysis, and newer techniques such as sentiment and affect analysis, web analytics, and graph mining. Subsequently, Gartner Research extended this taxonomy (Chandler, Hostmann, Rayner, & Herschel, 2011), restricting the category of descriptive analytics to the "what happened?" part, and separating that from the "why?", which they called diagnostic analytics, but this distinction does not seem to have caught on more widely than in Gartner's own materials.

Predictive analytics "uses data to find out what could happen in the future... Predictive analytics predicts future probabilities and trends and finds relationships in data not readily apparent with traditional analysis" (Robinson et al., 2010). Specific techniques here include predictive statistical modeling, the whole range of forecasting methods based on data, machine learning, and some forms of neural network analysis.

Prescriptive analytics "uses data to prescribe the best course of action to increase the chances of realizing the best outcome. Prescriptive analytics evaluates and determines new ways to operate, targets business objectives, and balances all constraints" (Robinson et al., 2010). Techniques such as optimization and simulation are in this group, along with others such as case-based reasoning, and neural networks that aim to recommend a particular course of action. Robinson et al. (2010) add that "Most of us would probably agree that in fact most operations research (OR) techniques reside in this space."

Robinson et al. (2010) see the three types as a distinct hierarchy, commenting that "Businesses, as they strive to become more analytically mature, have indicated a goal to move up the analytics hierarchy to optimize their business or operational processes. They see the prescriptive use of analytics as a differentiating factor for their business that will allow them to break away from the competition." This is somewhat more contentious, much as its appeal to those pushing analytics services and relevant IT products is obvious. INFORMS commissioned the 2010 work to help decide how it addressed the rise of analytics as a hot topic in business. Haight and Park (2015), in a report for Blue Hill Research about the Internet of Things (IoT), also see descriptive, predictive, and prescriptive analytics as the stages in a maturity model. In their case, the ultimate goal is complete automation, that is, to "provide specific recommendations based on live streaming data to an employee or to automatically initiate a process" (p.4).

To understand this issue better, we need to look at a classification of analytics along a different dimension.

Classification from a Business Perspective

Chambers and Dinsmore (2015) identified five types of analytics viewed from the perspective of the business: strategic, managerial, operational, customer-facing, and scientific. We use this classification because of our interest in connecting analytics knowledge with actions across the organization and across multiple projects.

Strategic analytics "typically provide high value and are operationalized through an offline decision or process." (Chambers and Dinsmore, 2015, p.66). These are the key part of what the PwC report (Witchalls, 2014) calls the big decisions. However, a big decision can have components of all five types. Consider the case of a new product line development. The decision requires the strategic evaluation of the business environment, and simultaneously the connection of the opportunity with the capacity of the organization to take the opportunity and the acceptance by, and expected value and impact on the stakeholders. The new product development is at the same time a result of innovation that can be the outcome of scientific analytics.

Managerial analytics, by contrast, "typically provide value through midterm planning. The implementation of this type of analytics is often executed through semi- or fully-automated processes." (Chambers and Dinsmore, 2015, p.66).

Chambers and Dinsmore state that strategic analytics usually lead to action on a timescale of one to three years, whereas managerial analytics are implemented in three months to a year, but we think that this is a little misleading since timescales vary so much between industry sectors. Nuclear power generation, for example, has a vastly longer timescale than retail fashion.

Operational analytics are "embedded into the front-line processes for an organization and are executed as part of the day-to-day operations. Operational analytics ranges from real-time (now) to short-time horizons (today or this week)." (Chambers and Dinsmore, 2015, p.66). No doubt about the timescales for this type.

Customer-facing analytics "typically provide value by providing insights about customers. They also tend to range from real-time to short-time horizons." (Chambers and Dinsmore, 2015, p.67). The difference between this type and the operational category is more clear cut in manufacturing than in some service industry sectors. In the latter, sometimes all that matters is what the customer thinks of the service.

Scientific analytics "add new knowledge—typically in the form of new intellectual property—to an organization. The frequency may be periodic (every year) or occasional (once every several years)." (Chambers and Dinsmore, 2015, p.67). This type of analytics is specifically related to intellectual capital, and is rather different from the managerial, operational, and customer-facing types, in that it is related to building capacity or capability rather than to operations directly.

We will now begin to use these categories in considering the history and terminology of analytics.

How Analytics Developed

Despite its recent "hot topic" status, analytics in organizations is far from a new phenomenon (Holsapple, Lee-Post, and Pakath, 2014; Kirby and Capey, 1998). Its roots can be traced back to at least the middle of the nineteenth century, with arguably the best-known early examples being from public health rather than business or industry. These include John Snow's analysis of a cholera outbreak in London in 1854, and Florence Nightingale's statistical work on causes of patient mortality in British Army hospitals, also begun in the 1850s. Both of these examples consisted of descriptive analytics, and both meet our action-oriented requirement. Snow's work led to the immediate action of removing the handle of a suspect street water pump so that it could not be used, and in the longer term to confirming his theory that cholera was transmitted through contaminated water. Nightingale's work revolutionized approaches to sanitation in the British Army (during the Crimean War, poor sanitation was responsible for the death of more British soldiers than combat), and subsequently she revolutionized British public health as well.

Reported examples of recognizable use of analytics in business or industry can definitely be found from early in the twentieth century. We do not wish to spend much space here debating whether or not Taylorism counts as analytics. However, it is noteworthy that Kirby and Capey (1998), writing well before the recent upsurge in interest in analytics, would surely have said that it wasn't, describing Taylorism as "grounded in a semi-intuitive feel for the practical aspects of manufacturing processes rather than in the application of mathematical and statistical analysis"

(Kirby and Capey, 1998, p.308). In the United Kingdom, there is the clear example of the Shirley Research Institute, analyzing large amounts of data from the cotton textiles industry in the 1920s (Kirby and Capey, 1998). In the United States, management consultants such as Arthur D. Little, McKinsey & Company, and Booz Allen Hamilton were well-known by the 1930s, and some of their work used analytical approaches.

What we would now call analytics really began to gain traction in the 1950s, spurred on by military uses in the 1940s. In the United Kingdom, J. Lyons & Co. built the world's first business computer in 1952 to carry out analytics work on food sales in their chain of cafés. In the United States, UPS initiated its first corporate analytics group in 1954 (Davenport and Dyché, 2013). For more early examples, see Kirby and Capey (1998) for the United Kingdom and Holsapple, Lee-Post, and Pakath (2014) for the United States.

Improvements in information technology in the twenty-first century have led to the current boom in interest in analytics. Some authors have attempted to identify different generations of analytics. Chen, Chiang, and Storey (2012) argued that Business Intelligence and Analytics (BI&A), as they call it, evolved from BI&A 1.0 (database management system-based structured content) to BI&A 2.0 (web-based unstructured content) and BI&A 3.0 (mobile and sensor-based content). Davenport's list, also of three generations, describes an evolution from Analytics 1.0 (the era of business intelligence) to Analytics 2.0 (the era of Big Data), and moving toward Analytics 3.0 (the era of data-enriched offerings) (Davenport, 2013; Davenport and Dyché, 2013). Davenport and Dyché (2013) actually date the three eras, with Analytics 1.0 running to 2009, Analytics 2.0 from 2005 (so there is an overlap) to 2012, and Analytics 3.0 from 2013.

Analytics, Operations Research/Management Science, and Business Intelligence

As a result of this history of gradual evolution, there has been little standardization of terminology, as those two lists of generations demonstrate. The reaction of operations research/management science professionals to the growth of analytics has split into two camps. Some have embraced the term as what they have been doing all along. This can be seen in a widely-quoted example of how Coca-Cola ensures orange juice products stick to the "product recipe" even though the characteristics of the available ingredients vary (Stanford, 2013). This was publicized as an analytics example, but the original article correctly describes it more specifically as a technique; an algorithm (one of the many optimization techniques that come under the prescriptive analytics category). It is a type of algorithm that has been used in operations research/management science for decades. Only its complexity and speed of operation differ from what could have been seen in the 1960s. Indeed, one of this chapter's authors worked on an application of a similar recipe algorithm for a brand of cat food more than 30 years ago. Again, this was prescriptive analytics in action. Both this and the Coca-Cola orange juice example are operational analytics, although in the cat food example the recalculation was only done every few days. This was because of the purchasing cycle for the ingredients rather than technological limitations; calculating the results (using what was then called a mini-computer) took less than ten minutes.

On the other side of the divide, some operations research/management science professionals seem to be ignoring the analytics "bandwagon" entirely. For example, the study into improving paper production in Finland by Mezei, Brunelli, and Carlsson (2016) uses detailed data and machine operators' knowledge to fine-tune the production machinery. This could appropriately have been described as an example of predictive analytics, but neither the terms analytics nor Big Data appear anywhere in their paper.

The problem with the bandwagon effect is that analytics has now reached the point where consultancies and other players seek to differentiate themselves from others in the market by using somewhat different terminology, and academics try to make sense of the various labels. Early in the "boom" period, Davenport and Harris (2007) described analytics as a subset of business intelligence. We have already seen that Chen, Chiang, and Storey (2012) couple the two terms together. Business intelligence, derived from the military sense of that term, and so meaning information rather than cleverness, also has a long history. Holsapple, Lee-Post, and Pakath (2014) point out that it was first coined at IBM in the 1950s (Luhn, 1958), although it is often credited incorrectly to the Gartner Group in the 1990s (see e.g., Watson & Wixom, 2007), much as Gartner helped to popularize the term. Summing up the confusion, Watson & Wixom (2007) observed that "BI [Business Intelligence] is now widely used, especially in the world of practice, to describe analytic applications."

The MHR report (MHR Analytics, 2017) is a good illustration of how this confusion of terminology continues. It has data analytics in its subtitle, but begins with the statement "Business Intelligence (BI) is not a new topic," and refers most frequently to "BI and [data] analytics." Just to add to the mix, the *Computing* report (Sopra Steria, 2016) concentrates on the phrase "applied data and analytics."

Overview of Analytics Examples

Rather than seeking to resolve these issues of detailed terminology, we will regard any study or project that fits the definitions of descriptive, predictive, or prescriptive analytics, as defined above, as being an example of analytics.

We will then look at the examples on two dimensions. One is the Chambers and Dinsmore (2015) analytics categorization from the perspective of the business: strategic, managerial, operational, customer-facing, or scientific. The other is the focus of the knowledge: domain, data, techniques, or action (Table 1.1). Each example can fit only one analytics type, but may involve more than one knowledge focus.

Table 1.1 Examples Categorized by Analytics Type and Knowledge Focus

Analytics type	Knowledge Focus			
	Domain	*Data*	*Techniques*	*Action (People, Process, Technology)*
Strategic	Tesco portfolio approach U.S. Air Force	UK banks	Luxottica	Uber Hong Kong Efficency Unit
Managerial	Christie hospital Caterpillar Marine Warehouse Group Marks & Spencer Schneider National	Gelderse Vallei hospital Target	Warehouse Group Schneider National UPS	Ford Motor Company UPS
Operational	Ocado warehouse	eBay eHarmony	eBay Equifax Geneia Blue Cross-Nexidia	Coca-Cola
Customer-facing	Ocado e-mail HEQCO	Bank of America	SEB Dunnhumby contribution to Tesco loyalty cards IPSoft contribution to SEB	Tesco loyalty cards Ocado e-mail SEB Colruyt
Scientific			Panama Papers	

To explain this further, let us look again at the two UK public health examples from the nineteenth century. Although a "business" perspective does not strictly apply, the categories of Chambers and Dinsmore can still be used. It is clear that Snow's first study was operational analytics, but later work became strategic analytics as more studies were carried out. In the first study, the crucial knowledge elements were knowledge of techniques and then knowledge of action. Snow needed to develop a new technique—an early form of data visualization—to identify the patterns in the data. Having identified the offending pump, Snow knew that just putting a notice on it would not stop people using it, especially as many people in London at that time could not read, and so the only way to stop them from drinking the contaminated water was to disable the pump: the people element of action was crucial. His later strategic work contributed to knowledge of the domain—the theory of how diseases are transmitted.

Nightingale's work also relied on knowledge of the people element of action, specifically how to influence army generals and politicians to adopt her ideas, especially given that they were being proposed by a woman. Similarly, her work went on to represent a major contribution to knowledge in the domain of public sanitation.

Single-Project Examples

Strategic Analytics

Let's begin the section with another historical example. When retail banks in the United Kingdom began to automate their processes in the late 1950s, based on magnetic ink character recognition technology, the key piece of data was the account number. As the use of computers expanded through the 1960s and 1970s, databases continued to be structured with the account number as the primary key. However, this was no help in identifying which accounts belonged to the same customer. Integrating data to make the customer centrally involved considerable rewriting of the banks' systems, and only the twin drivers of concerns about Year 2000 (Y2K) problems and the rise of online banking finally persuaded UK banks to make the necessary investment to do it properly. The whole process of switching the data focus from the account to the person took about 20 years in the United Kingdom, and arguably is still not entirely complete. This is an example of knowledge about data linked to strategic analytics, which was simply not possible (even for descriptive analytics) with the old account-centric computer systems.

For the next example, we present Tesco, the United Kingdom's number one supermarket, and one of the world's largest retailers. Tesco has always been an analytics leader among UK retailers, even before the term came into regular use. For example, it pioneered the use of geographic information systems in United Kingdom retailing in the 1980s, to analyze the "catchment area" of a supermarket, aided by one valuable strategic insight. Tesco managers realized that deciding where to locate a new store and evaluating the performance of an existing store were basically variants of the same question: if there is a store located here, how well should it perform? To gather the data and do the predictive analytics, they set up the Tesco Site Research Unit. This enabled them to take a portfolio approach to opening new stores (and acquiring stores from competitors) instead of considering each potential new store decision in isolation. This was a strategic use of analytics resulting from knowledge about the domain and was one of the elements that enabled Tesco to become the United Kingdom's number one supermarket, a position it has retained ever since. We shall return to that latter point later in the chapter.

Another example is a project by the RAND Corporation (Chenoweth, Moore, Cox, Mele, & Sollinger, 2012), developing supplier relationship management (SRM) for the U.S. Air Force. Chenoweth et al. (2012) observed "As is the case with all military services, the U.S. Air Force (USAF) is under pressure to reduce the costs of its logistics operations while simultaneously improving their performance." The USAF addressed the problem on a premise of managing a good relationship with suppliers based on the development of analytics capacity in the SRM system. The scope of this project made it another strategic use of analytics based on domain knowledge. The analytical capabilities were concentrated on the construction of appropriate supplier scorecards that are associated with identification of improvement opportunities in areas ranging from maintenance to resource planning systems. According to Teradata, the main benefits for the USAF are concentrated on data and analytics tools use "...to streamline workflows, increase efficiencies and productivity, track and manage assets, and replace scheduled/time-based maintenance with condition-based maintenance (CBM) – saving at least U.S.$1.5 million in one year."

Returning to retail, the case of Luxottica Retail North America is also a strategic view of analytics, this time based on knowledge of techniques. Luxottica sells luxury and sports eyewear through multiple channels. The project consisted of the integration of data and applications from internal and external sources. Its purpose was to support the marketing strategy associated with data integration of multiple distribution channels and the appropriate use of resources in marketing development. According to IBM "...Luxottica gained a 360-degree view of its customers and can now fine tune its marketing efforts to ensure customers are targeted with products they actually want to buy." (Pittman, 2016).

For an example of a focus on knowledge about technology leading to disruptive innovation at the strategic level, we turn to Uber. The understanding of costs related to taxi operations and the analysis of customers' needs led to the development of a substitute taxi service based on network and data analytics, where the technology (phone applications) was crucial to effective action. Berger, Chen, and Frey (2017) pointed out, "Unlike a traditional taxi business, Uber does not own any cars; instead, it provides a matching platform for passengers and self-employed drivers and profits by taking a cut from each ride." The use of technology in understanding customers' needs, transportation operation, and development is leading better customer services and at the same time improvements in the organization's performance.

Innovation is not the only engine to connect strategic analytics and actions. The Hong Kong Efficiency Unit (Hong Kong Efficiency Unit, 2017) illustrates strategic analytics and all three aspects of knowledge about action: people, process, and technology. To provide a good service to Hong Kong's citizens, the Efficiency Unit created a system for managing the answers to the contacts that people have with the government. The use of analytics on data relating to prior contacts has provided a capacity to anticipate complaints and offer a 24 × 7 service providing rapid answers to the issues that have been reported.

Managerial Analytics

Moving to the slightly shorter timescales and lower-level decisions of managerial analytics, we return to retail (remember we said it was one of the most advanced sectors?), but cross to the other side of the globe and The Warehouse Group, New Zealand's largest retailer. Warehouse Stationery, part of the group, brought in a count mechanism, which was based on thermal imaging, so they could understand how many people were coming into a store. This was put alongside the transaction data from the tills so that stores could understand how their conversion worked, and higher-level managers could target stores that needed special emphasis. Kevin Rowland, Group Business Intelligence Manager, commented "we very quickly managed to create a series of dashboards that showed by hour, by day, to store level, what was going on in terms of counts through the door and the conversion rate of that… at the end of it, the GM of operations stood up and said, 'I could kiss you,'" (Ashok, 2014). In this case, the descriptive analytics project relied on knowledge of techniques in data visualization, and in turn contributed to knowledge of the domain-patterns of conversion rates in the stores.

Expanding our scope to a worldwide example, the Ford Motor Company operates the Global Warranty Measurement System. This is now available in the form of an application, enabling Ford dealers to carry out descriptive analytics on warranty repair costs, claims, and performance. It also enables Ford to gauge if dealers will meet Ford's expectations on their performance, so moving into the predictive category. Cohen and Kotorov (2016) quote Jim Lollar, Ford's Business Systems Manager of Global Warranty Operations as saying: "We have always had data on actual costs but we weren't giving it to dealers" (p.107). The crucial knowledge focus here was knowledge of action-process; how to deliver the data in a way (an application) that was convenient to use. One of the most important contributions of the Global Warranty Measurement System is its effect on the dealerships' performance management practice. Given the volume of data, the capabilities to generate reports, and the governance of the data, it has been possible to develop metrics that provide benchmarks as an indication of how the dealerships are operating. For example, it is possible to predict and manage the cost per vehicle serviced; repairs per 1,000 vehicles serviced; and cost per repair (Cohen and Kotorov, 2016). Dealerships can thus adjust their business processes to obtain better results.

Returning to the United Kingdom, "the Christie" is a specialized cancer hospital in the National Health System (NHS) (the UK public healthcare system). The hospital makes extensive use of statistical process control (SPC) in clinical processes (McKenna, 2016b) to identify processes needing investigation or improvement. SPC is not at all new as a technique for descriptive or predictive analytics, dating back to the work of Shewhart in the 1920s, but the technology now available vastly improves the speed of producing the control charts that originally had to be drawn by hand. Areas of use at the Christie include monitoring patients' length of stay, bed availability, and discharge times, and in the future, it is planned to monitor the cost of pharmacy dispensing. In all these examples we find the analytics work contributing to knowledge of the domain.

Marks and Spencer (M&S) is another of the United Kingdom's best-known retailers, concentrating on clothing and food. They have used some form of descriptive analytics to examine their sales for as long as current employees can remember. Originally this was done informally in each store but later was centralized. Current technology enables the analysis of much larger datasets, and lets the end users do it themselves. This descriptive analytics leads to improved knowledge of the domain, as illustrated by the following two examples from Goodwin (2014).

In one case, new clothing range launches were proving less successful than hoped. Descriptive analytics revealed that the new clothes were available in stores in the right total quantities but the wrong mixture of sizes. This is partly because in the United Kingdom, as in many other countries, changes in diet mean that people tend to be different in size and shape from their parents a generation ago. This also led to supplier availability performance being shown as poor, because people complained that a particular style was unavailable in their size, when in fact the suppliers were delivering exactly what M&S had ordered, and on time. The second example is from the other side of M&S's business, food: smaller stores were finding that they could not exploit new food products because the minimum order they could place (determined by the tray size) was too large. So M&S halved the tray size, and increased sales. Both examples show analytics as contributing to knowledge about the domain.

Moving away from retail, our next example of managerial analytics is from the logistics sector. Schneider National is one of North America's largest trucking firms, and is expanding the use of sensor data from its trucks. This includes "a process where the sensor data, along with other factors, goes into a model that predicts which drivers may be at greater risk of a safety incident. The use of predictive analytics produces a score that initiates a pre-emptive conversation with the driver and leads to less safety-related incidents." (Davenport and Dyché, 2013). Here, the input of knowledge about the techniques in use contributes to developing knowledge about the domain; for example, frequent hard braking is often an indicator of a riskier driver.

Another logistics example comes from UPS, the world's largest package delivery organization. They also have telematics sensors in their fleet of over 46,000 vehicles (Davenport and Dyché, 2013). The data recorded from their trucks includes, for example, their speed, direction, braking, and drive train performance. UPS has embarked on a higher-level use of managerial analytics, which is clearly OR on a big scale—predictive analytics. "The data is not only used to monitor daily performance, but to drive a major redesign of UPS drivers' route structures. This initiative, called On-Road Integrated Optimization and Navigation (ORION), is arguably the world's largest operations research project." (Davenport and Dyché, 2013, p.4) Davenport and Dyché (2013) also report that ORION had "already led to savings in 2011 of more than 8.4 million gallons of fuel by cutting 85 million miles off of daily routes." This is an example of using both knowledge of suitable OR techniques, to take advantage of real online map data, and knowledge of how to put them into action, the technology to carry out the calculations quickly enough.

The next example uses somewhat similar technology—sensors in transportation—but this time we see analytics that makes a contribution to knowledge of the domain. It comes from Caterpillar Marine (Marr, 2017). Caterpillar's Marine Division provides services to shipping fleet owners for whom fuel usage is a crucial factor affecting profitability. One of Caterpillar's customers knew that hull contamination (corrosion, barnacles, seaweed, etc.) must affect fuel consumption, but had previously had no way of quantifying its effect. "Data collected from ship-board sensors as the fleet performed maneuvers under a variety of circumstances and conditions—cleaned and uncleaned—was used to identify the correlation between the amount of money spent on cleaning, and performance improvements." (Marr, 2017)

The outcome of the predictive analytics was the conclusion that ship hulls should be cleaned much more frequently; roughly every six months as opposed to every two years. Potential savings from these changes would amount to several hundred thousand U.S. dollars per year (Marr, 2017).

We complete this section of managerial analytics examples with two examples where the knowledge focus is on the data, both related to healthcare. The first is from the healthcare sector in the Netherlands (SAS Institute, 2016b). A key performance indicator for surgery in public hospitals in the Netherlands, as in the United Kingdom, is the length of time patients must wait before being admitted for surgery. The Gelderse Vallei hospital's waiting time for hernia surgery had jumped. Rik Eding, Data Specialist and Information Analyst, explains. "When we looked closer, it was because two patients postponed their operations due to holidays. If we left these two cases out of consideration, our waiting period had actually decreased" (SAS Institute, 2016b, p.13). Understanding outliers like this in the data is a crucial element of descriptive analytics. Perhaps postponements by the patient should not be included in a calculation of the average waiting time, but that would depend on the reason. A postponement because the patient's condition had worsened surely should be included. Standardization of data is necessary for reliable analytics results, but it is not always easy to do it appropriately.

Wider issues still can arise. North American retailer Target offers a cautionary tale (Hill, 2012). They analyzed purchase histories of women who had entered themselves on Target's baby registry, and found patterns so clear that they could estimate not just if a customer was pregnant, but roughly when her baby was due. They went on to send coupons for baby items to customers according to their pregnancy scores. (It seems highly unlikely that this behavior would have been legal in most European countries.) Unfortunately, they did not restrict this to customers on the baby registry. This hit the press when an angry father went to the Target store manager in Minneapolis to complain about the coupons that were being sent to his daughter, who was still in high school. Now, the store manager had little understanding of the analytics process behind this marketing campaign, and was very surprised to discover what was happening. It did transpire a few days later that the girl was indeed pregnant. So, it was a technically sound use of predictive analytics based on knowledge of data, but Target had not had the sense to restrict it only to women already on the baby registry, and it earned them a great deal of dubious publicity. Hence we have classified this under the managerial category. It is not clear if Target still uses the pregnancy scores in its marketing.

Operational Analytics

The Coca-Cola orange juice case mentioned earlier is an example of operational analytics, with the knowledge focus being on action—the technology needed to carry out the predictive analytics fast enough.

Another brand name that is well-known worldwide gives us an operational analytics example where the knowledge focus is about the data and the techniques. eBay uses machine learning to translate listings of items for sale into different languages, thus reaching more potential buyers (Burns, 2016). This is not a straightforward use of natural language processing (NLP), because the eBay listings that form the data have specific characteristics. For example, does the English word "apple" in a listing refer to the fruit (which should be translated) or the technology company or brand (which should not)? Similarly, should an often-used acronym like NIB, which stands for "new in box," be translated or not? Some non-English speakers will still recognize the meaning of NIB even if they do not know what the acronym actually stands for. Therefore, eBay has had to develop its own statistical algorithms to supplement more common NLP techniques, thus contributing to knowledge of techniques (Burns, 2016). Machine learning for NLP does not fit as well into the descriptive/predictive/prescriptive classification as most techniques, but is most often regarded as predictive—the translation effectively serves as a forecast of what the best translation might be.

Equifax, the credit scoring agency, has also developed its own techniques (the knowledge focus) for key operational decisions (Hall, Phan, & Whitson, 2016). Automated credit lending decisions have typically been based on logistic regression models. However, Equifax has now developed its own Neural Decision

Technology (NDT), based on artificial neural networks with simple constraints, which it is claimed delivers measurably more accurate results and produces "the mandatory reason codes that explain the logic behind a credit lending decision" (Hall et al., 2016). It also requires less human input, since "logistic regression models often require segmentation schemes to achieve the best results" (Hall et al., 2016). This is predictive analytics, but the dividing line between predictive and prescriptive analytics often comes down simply to the use that managers in the organization decide to make of the results: do they review them (which makes it predictive), or always do what the software recommends (which would be prescriptive)?

Geneia, a US healthcare organization, is another operational analytics example where the principal knowledge focus is on techniques, supplemented by domain knowledge (Hall et al., 2016). Geneia uses well-tried statistical and machine learning techniques, but has developed them using its own knowledge of the market and regulatory privacy requirements. These comprise "statistical and machine learning techniques—including principal component analysis—to assimilate claims, clinical, consumer, actuarial, demographic, and physiologic data…[b]inary logistic regression, a straightforward classification technique, helps identify predictors and develop risk-scoring algorithms to identify where sick patients fall on a wellness spectrum and the most appropriate remediation" (Hall et al., 2016, p.20).

That example related to relatively traditional databases and analytics methods. However, even in healthcare the use of nonstructured data for providing or improving services is expanding. Blue Cross reported an application of speech recognition for improving patients' experience. They use Nexidia solutions, so clearly the knowledge focus is on techniques. The purpose has been the development of an effective and efficient voice of the customer (VoC) by using data available from the contacts and interactions between customers and the company with the purpose of improving customer service. The analytics system allows not only a better quantification of the issues to solve based on the data collected but also supported the solution of problems, opening the coordination of areas and employees involved in a customer's solution. This coordination allows a reduction of contact with employees and provides better service (Nexidia, 2014). This example also shows how the analytics type and knowledge focus can differ even when the broad application area is the same; contrast this Blue Cross case with the Hong Kong Efficiency Unit case above, and the Ocado case in the next section.

Customer-Facing Analytics

Gaining insights into customers and their behavior is one of the most active areas of analytics at present, perhaps because of the perception that this has led to the success of companies like Amazon.

For our first example in this section, we look to Tesco again. In the 1990s, Tesco was the first UK retailer to make a success of using customer data from loyalty cards, in collaboration with market intelligence company Dunnhumby. Tesco had to issue its own cards, because UK data protection laws, unlike those of some other countries, do not permit retailers to identify customers from their credit card data. The crucial knowledge focus here for Tesco was on action—the people aspect of knowing what benefits (discount coupons, etc.) customers needed to make it worth their while to obtain a loyalty card and remember to use it. This illustrates our point about action-orientation very well. Without these insights, the predictive analytics (knowledge about techniques from Dunnhumby) would not have achieved anything for Tesco.

One of Tesco's UK competitors is Ocado. Ocado has no physical stores. As an online-only grocer, most of their customer contact is via email. Coping with the volume of emails can be a particular problem at the holiday season or during freak weather conditions that delay deliveries. Ocado is now trialing machine learning to help respond to the emails it receives. There are two success criteria for this use of predictive analytics: "does the model behave in the same way a human would?" and "is it helping to speed up resolution times?" (Donnelly, 2016). It is evident that the first of these criteria relates to knowledge of the domain, and the second to action-oriented knowledge about the process.

Tesco still uses its loyalty cards, but these days delivers most of the benefits to customers online, including discount coupons. However, Colruyt, one of the largest supermarket chains in Belgium, combines old and new technology. Colruyt still sends out paper coupons for discounts, but by using predictive analytics, Colruyt can "calculate purchasing probabilities based on past customer behavior, as well as on household and demographic information stored in its database. Of the 400 products on promotion during a given two-week period, each household receives four pages of coupons specifically tailored to it, down from 32 pages previously." (Alford, 2016). Again, this revolves around knowledge of action, both people (the customers and how they behave) and the processes (obviously Colruyt's consumers are still happy with paper coupons).

Another European country gives us a very different example of predictive analytics. Skandinaviska Enskilda Banken (SEB) is a large full-service Swedish bank. Toward the end of 2016, SEB introduced Amelia to its customer service operations. Amelia is machine learning software; a "chatbot" that can understand text and respond to customer queries (Flinders, 2016). Knowledge about action is again central here, with a people focus. Flinders (2016) describes the key issues as what people will put up with, and different people wanting different things, since the main channel for customer queries at present is the phone, rather than text-based chat. Domain knowledge may turn out to be relevant (as in the eBay case earlier), and also the knowledge of language processing techniques contributed by IPSoft, for whom SEB represents the first users of Amelia in a language other than English.

In a different sector, student retention and attrition are topics of great concern to education authorities. Domain knowledge of the causes and factors affecting the levels of final graduation indicators is a very important area of analytics for them.

Lopez-Rabson and McCloy (2013) carried out a descriptive analytics study of colleges in Toronto, Canada, for the Higher Education Quality Council of Ontario. Their study discovered factors that provide insights to education institutions to develop strategies and tactics to support students' success, so it is indeed customer-facing even though it is in the strategic knowledge category.

Closing this section, we look at Bank of America for an example of predictive analytics where knowledge of the data is the main driver (Davenport and Dyché, 2013), although the business goal, as with all customer-facing analytics, is understanding the customer. The developments come from combining customer data across all channels. The predictive models establish that "primary relationship customers may have a credit card, or a mortgage loan that could benefit from refinancing at a competitor. When the customer comes online, calls a call center, or visits a branch, that information is available to the online app, or the sales associate to present the offer" (Davenport and Dyché, 2013, p.16) There is also a minor action focus on process knowledge here, in that having combined the data from all channels, the offer is also made available across all channels, and an incomplete application (e.g., online) can be followed up through a different channel (e.g., an e-mail offering to set up a face-to-face appointment).

Scientific Analytics

Our chapter devotes less space to scientific analytics than the other categories, since by definition, the purpose of work in this category is to add to an organization's intellectual capital. The close links between intellectual capital and knowledge management should mean that the case for the relevance of knowledge to action does not need to be made. Nevertheless, one example is particularly instructive. The so-called Panama Papers (McKenna, 2016a) represented a success for descriptive analytics using the new technique of graph databases. The papers comprised a huge set of documents (11.5 million) from the Panama-based legal services firm Mossack Fonseca, which specializes in offshore organizations, including those in countries often regarded as tax havens. The new technique made it possible for news organizations such as the International Consortium of Investigative Journalists (ICIJ), the BBC, UK newspaper the Guardian, and German newspaper the Süddeutsche Zeitung to analyze connections in the data in ways that would have been near-impossible before. So the knowledge of the technique turned the Panama Papers into a resource they would not otherwise have been, especially for the ICIJ.

Multiple Project Examples

It is not difficult to find many other examples of successful analytics use online in blogs and press releases. Documented successful application is far less common in the academic literature, with the typical 2–3 year time lag between data collection

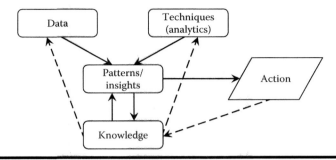

Figure 1.2   The interactions between data, analytics, and human knowledge over several studies. (From Edwards, J.S., and Rodriguez, E., *Proc. Comp. Sci.*, 99, 36–49, 2016.)

and publication. However, even in blogs and press releases, relatively few articles look at the evolution of analytics in an organization, and the way that the organization progresses from one project to another. We have already hinted at some examples of this, and in this section we consider them in more detail.

The essential relationships between data, analytics techniques, patterns/insights, knowledge, and action remain the same across a series of projects or studies, but we expect to see three additional influences, as shown by the dashed arrows in Figure 1.2. These are all types of learning: influences of knowledge on data (what to collect and standardization of meaning); of knowledge on techniques (new analytics techniques developed in the light of past examples); and of action on knowledge (learning from skilled practice).

Beginning again with Tesco, many of the analytical activities are still the same as when they first went into analytics decades ago, like sales forecasting. But there, for example, their priority is now real-time analytics: doing the same things, but doing them much faster (Marr, 2016). This involves both newer technology, like Teradata, Hadoop, and a data lake model, and also changing business processes. So we see the contribution of knowledge of action—processes and technology—with Tesco learning a very great deal from their existing skilled practice, and trying to preserve their lead over competitors in this respect. The actual examples can be managerial, operational, or customer-facing, depending on the detail. However, especially at the managerial level, at present it can take 9–10 months to progress from data to action (Marr, 2016). This is not unusual: the findings of Halper (2015) were that just the stage of putting an agreed predictive model into action can take several months. She found that only 31% of respondents said this process took a month or less, with 14% responding nine months or more, and the mean and modal response being three to five months.

Other examples of learning from skilled practice include Ocado and SEB. Before the customer email project, Ocado already used machine learning in its warehouse operations, co-locating items that are frequently bought together to reduce packing and shipping time (Donnelly, 2016). Similarly, SEB first used Amelia to answer queries coming to its internal IT help desk (Flinders, 2016) before rolling out their customer-facing system. Both of these were operational examples of predictive analytics, whereas the later projects were customer-facing. We thus see that the analytics type and the knowledge focus can change between related studies or projects, without affecting the possibility of transferring learning—of action influencing knowledge.

Another change of type is the plan for UPS's ORION to include online map and traffic data, so that it will reconfigure a driver's pickups and drop offs in real time, moving from managerial to operational (Davenport and Dyché, 2013). However, as far as we know, this part of the system has not yet been rolled out.

In both the Warehouse Group and Marks & Spencer, the new knowledge covers the whole domain of using analytics. Kevin Rowland of the Warehouse Group again: "I would actually design something in the beginning and give it to the business, and ask them what they think, as opposed to going to the business with a blank page and asking them what they want. Now they understand what the tool can do but back then they didn't." (Ashok, 2014). Equally, Paul Williams, M&S Head of Enterprise Analytics, observes: "People are asking questions they wouldn't have asked before. And asking one question breeds curiosity, often leading to another 10, 20, or 30 questions." (TIBCO Software, 2015).

In terms of learning about the data, the UK banking example has relied crucially on standardization of meaning, by shifting from account number to a customer identifier, even when that leads to counter-intuitive structures (in database terms) such as two "unique" customer identifiers for each joint account. As for Caterpillar Marine, James Stascavage, Intelligence Technology Manager, concluded: "I think the best lesson we learned is that you can't collect too much information…the days of data storage being expensive are gone…if you don't collect the data, keep the data and analyse the data, you might not ever find the relationships." (Marr, 2017)

Equifax is a prime example of developing new techniques in the light of past examples. The NDT's increased accuracy could also lead to credit lending in a broader portion of the market, such as new-to-credit consumers, than previously possible (Hall et al., 2016).

The history of the online dating website eHarmony shows the evolution of data knowledge and the operational analytics solutions that were developed from it. eHarmony started at the end of the last century with the original use of questionnaire-style variables and data to predict matching and compatibility for good relationships among couples. Given the need to develop trust among the service users, eHarmony evolved to use social media data to manage (some of) the dating uncertainty. Nevertheless, the business these days is affected not only by the customers' use but also by the business model. As Piskorski, Halaburda, and Smith (2008) pointed out "eHarmony's CEO needs to decide how to react to imitations of its business model, encroachment by competing models and ascendance of free substitutes." Thus, as we said earlier, all five types of analytics may turn out to be linked.

Future Developments

In this section we present three main aspects of analytics and knowledge in the action-oriented view: organization environment evolution, political environment evolution, and transforming the analytics workflow into models that can be embedded into business processes and evolve according to new data input.

Organizational Environment

One of the key difficulties in organizational innovation revealed by the business process re-engineering movement in the 1990s (Davenport and Short, 1990; Hammer, 1990) was the presence of organizational silos: vertical divisions that acted as barriers to better organizational communication and hence performance.

Over 25 years later, silos still rule in far too many organizations. Henke et al. (2016) point out that the old organizational silos that often still exist have been joined by data silos: different systems that cannot be connected to each other. They give examples of the barriers presented by these silos in three sectors: banking, manufacturing, and smaller specialized retailers. A survey carried out by Cap Gemini Consulting in November 2014 (reported by Burns, 2015, p.11) backs this up. They found that "data scattered in silos across various units," cited by 46% of respondents, was the greatest challenge in making better use of Big Data.

Virgo (2017) understands these issues better than most, because of his particular interest in IT in national and local government, observing that there seems to have been little improvement generally in the past 10 years. He comments on a successful example of smart city analytics from New York (unfortunately his article gives no more details) that "the breakthrough had been to get departments to share sensitive information on problems they were shy of discussing with others." Virgo goes on to identify the three main reasons for this shyness as:

1. Fear that others will learn just how poor (quality, accuracy, etc.) our own data is
2. The inability to agree on common terminology
3. Fear that others will abuse "our" data

These are specific examples of what Alvesson and Spicer (2012) have labeled a stupidity-based theory of organizations.

Organizational silos come under the people heading of action issues, and data silos under the technology (and people) headings. However, taking action to move forward can also raise difficulties of changing processes. In Equifax, the company's operations are entirely structured around the existing analytics techniques. This is an instance of the well-known problem that first emerged in the 1970s: it is usually easier to computerize a manual process than to switch from one computer system to another.

The last big silo, especially for customer-facing analytics, is the one surrounding personal data. How much of our personal data are we willing to share with organizations? A report for the Future Foundation/SAS about UK attitudes (SAS Institute, 2016a) finds that younger people are much more willing to share personal data than older people when there is some direct benefit in it for them, but less willing when the benefits are for society as a whole (e.g., sharing energy use data to reduce the risk of brownouts or blackouts).

Despite these barriers, some organizations have succeeded in analytics and will endeavor to continue to do so. Marr (2016) reports four key challenges for Tesco in the future. Below we relate them to our discussion:

- Continuing to gain a better understanding of the changing nature of consumer behavior. This is learning from skilled practice.
- Creating efficiencies in their logistics and distribution chains, to keep down costs and minimize environmental impact. This is also learning, but with more emphasis on new techniques.
- Facing up to the challenge of emerging business models that compete with their own (such as Ocado, for example). Tesco's business model is still based around their network of stores despite a large and successful online presence: Tesco is regularly in the top four UK online retailers across all sectors. This is clearly the main strategic challenge, but Tesco has used analytics to help with strategy before, as we saw in their portfolio approach to new store development.
- Reducing the amount of food that goes to waste at their stores. This is a new direction for Tesco's operational analytics.

Political Environment

As the use of analytics, especially predictive and prescriptive systems based on deep machine learning, becomes more widespread, the regulatory and governance environment will undoubtedly need to change to cope with it. This is not a new issue. The practice of "redlining" in the banking and insurance industries, whereby loans or insurance are refused to anyone in a particular geographic area (originally shown by a red line on a map) has been outlawed by many laws, legal rulings, and judgments over the past 50 years. Similarly, the European Court of Justice ruled in 2011 that it was against European Union equality laws to use gender in calculating car driver insurance premiums. A machine learning system that "reinvented" redlining or gender discrimination would put its owners in a difficult legal position.

As a consequence, the UK Labour Party (currently the main opposition) is calling for the algorithms used in big data analytics to be subject to regulation (Burton, 2016). This would not be a long step from current policy: the UK National Audit

Office already has a detailed framework for the evaluation of models used in government departments (White & Jordan, 2016). However, enforcement in the private sector would present a new challenge.

With a worldwide perspective, many feel that artificial intelligence (AI) needs regulation, resulting in developments such as the Asilomar AI Principles (Future of Life Institute, 2017), endorsed by Elon Musk and Stephen Hawking amongst others. Two of the principles are very relevant to our discussions:

#7 Failure Transparency: If an AI system causes harm, it should be possible to ascertain why.

#8 Judicial Transparency: Any involvement by an autonomous system in judicial decision-making should provide a satisfactory explanation auditable by a competent human authority.

Judicial use of analytics would certainly be a big step forward from credit scoring, a chatbot, or even a self-driving car.

Analytics Workflow Embedded in Business Processes

In the search to improve knowledge management for action-oriented analytics, one factor that is crucial is to keep updating the results based on the permanent feed of data used. The new data and the changes in the outcomes will lead to a better design of intelligent systems in analytics. Big Data leads the need to create adaptive systems given the potential of creating new knowledge, learning, and modifying actions in an ongoing changing business environment, as shown above in Figure 1.2. The approach to the analytics process is a continuous set of predefined and sequential steps that requires adaptive development according to the ongoing feedback (Rodriguez, 2017). Adaptive means that the analytics system is in a permanent state of reviewing the feedback that new data, tests, and actions produce. The tools that the analytic system possesses to create Predictive Model Markup Language (PMML) (Pechter, 2009) outcomes are a key component of the analytics system to generate actions in business processes. This means that, for example, predictive analytics at any level (strategic, managerial, etc.) will have a great deal to contribute when new data arrive, and the systems will be able to recalculate and develop new outcomes to manage the decision-making process. Adaptive models will require organizational changes because what is true now (belief) will not necessarily be true tomorrow. In risk management, it is crucial to understand the variations that are expected, in particular, strategic risk and credit risk. In strategic risk, opportunities can be important today because of diverse factors that change permanently and modify the expected worth of big decisions, while credit risk can have new variables in the profiles to define good or bad prospects, thus modifying managerial decisions.

Conclusion

We have explained why we believe that knowledge management is the key to successfully putting analytics into action. We discussed many examples, looking at data, the analytics techniques, and the people, process, and technology issues of implementing the system.

Despite barriers such as silos, and the potential legal issues of relying too much on a machine learning system that cannot explain its decisions in terms that a human can understand, the use of analytics will surely continue to expand rapidly. The millennial generation seems to be very much at home with it.

However, at the heart of any analytics project there has to be data. Sometimes the data will pose problems that cannot be overcome. For example, predicting how long it will take to develop computer software has been an issue for over 40 years. Sadly, the data required to drive a predictive model giving an accurate (within 10%) forecast is simply not available at the time when the estimate is needed for business purposes (Edwards and Moores, 1994). Alternatively, the data may not be collected, even now. As we saw, Hall et al. (2016) reported Equifax as claiming that the new technique delivers "measurably more accurate" results than their older one. This begs the question of what data, if any, they had on loans that were declined. It is almost a logical impossibility to know that a loan that was declined should in fact have been offered.

Thus an increasing strategic risk stems from poor understanding of what to do with the knowledge created from data. Knowledge creation through analytics can be costly in both the search for solutions and in their implementation. Companies can lose this investment because they are insufficiently prepared to use the new knowledge (in our terms, lack of domain knowledge). Another strategic risk is to consider data as a resource and not as an asset. If data are an asset, adequate governance and care should be provided. The beauty of data and analytics development is that in many cases organizations will be able to do more with the same resources. Scope economies will take a predominant place in benefits generation instead of continuing only with the paradigm of scale economies and growth. Several companies are returning to their core business to use their resources in a better way to create wealth (GE is one example); data are a particular resource to use to manage uncertainty and risk.

Turning to the strategic risks of data analytics, these are mainly associated with the creation of an analytics process that can support accuracy in predictions and prescriptions, and possibly most important the generation of meaning and understanding for the organization's management. And of course, it all depends on the accuracy of the data…

So, even in the age of Big Data, analytics, and deep machine learning, an adage that is as old as the use of computers holds true—"garbage in, garbage out"!

References

Alford, J. (2016). *Keep them coming back: Your guide to building customer loyalty with analytics.* Cary, NC: SAS Institute.

Alvesson, M., & Spicer, A. (2012). A stupidity-based theory of organizations. *Journal of Management Studies, 49*(7), 1194–1220. doi:10.1111/j.1467-6486.2012.01072.x.

Anthony, R., & Govindarajan, V. (2007). *Management control systems.* New York, NY: McGraw-Hill.

Argote, L. (2012). *Organizational learning: Creating, retaining and transferring knowledge.* New York, NY: Springer Science and Business Media.

Ashok, S. M. (2014). Case study: Warehouse Group does more with data. *Computerworld New Zealand.* Retrieved January 30, 2018 from http://www.computerworld.co.nz/article/541672/case_study_warehouse_group_does_more_data/

Berger, T., Chen, C., & Frey, C. B. (January 23, 2017). Drivers of disruption? Estimating the Uber effect. Retrieved March, 11, 2017, from http://www.oxfordmartin.ox.ac.uk/downloads/academic/Uber_Drivers_of_Disruption.pdf

Burns, E. (2015). Big data challenges include what info to use—and what not to. The Key to Maximizing the Value of Data (pp. 7–16): Information Builders/SearchBusinessAnalytics. Retrieved January 30, 2018 from http://docs.media.bitpipe.com/io_12x/io_122156/item_1108017/InformationBuilders_sBusinessAnalytics_IO%23122156_Eguide_032315_LI%231108017.pdf.

Burns, E. (2016). Ebay uses machine learning techniques to translate listings. Retrieved January 30, 2018 from http://searchbusinessanalytics.techtarget.com/feature/EBay-uses-machine-learning-techniques-to-translate-listings?utm_content=control&utm_medium=EM&asrc=EM_ERU_72462481&utm_campaign=20170210_ERU%20Transmission%20for%2002/10/2017%20(UserUniverse:%202298243)&utm_source=ERU&src=5608542

Burton, G. (2016). Labour Party to call for regulation of technology companies' algorithms. *Computing.* Retrieved January 30, 2018 from http://www.computing.co.uk/ctg/news/3001432/labour-party-to-call-for-regulation-of-technology-companies-algorithms

Chambers, M., & Dinsmore, T. W. (2015). *Advanced analytics methodologies: Driving business value with analytics.* Upper Saddle River, NJ: Pearson Education.

Chandler, N., Hostmann, B., Rayner, N., & Herschel, G. (2011). *Gartner's business analytics framework.* Stamford, CT: Gartner.

Chen, H., Chiang, R. H. L., & Storey, V. C. (2012). Business intelligence and analytics: From big data to big impact. *MIS Quarterly, 36*(4), 1165–1188.

Chenoweth, M., Moore, N., Cox, A., Mele, J., & Sollinger, J. (2012). *Best practices in supplier relationship management and their early implementation in the Air Force materiel command.* Santa Monica, CA: The RAND Corporation.

Cohen, G., & Kotorov, R. (2016). *Organizational intelligence: How smart companies use information to become more competitive and profitable.* New York, NY: Information Builders.

Davenport, T. H. (2013). Analytics 3.0. *Harvard Business Review, 91*(12), 64–72.

Davenport, T. H., & Dyché, J. (2013). Big data in big companies: International Institute for Analytics. Retrieved January 30, 2018 from http://www.sas.com/resources/asset/Big-Data-in-Big-Companies.pdf

Davenport, T. H., & Harris, J. G. (2007). *Competing on analytics: The new science of winning.* Boston, MA: Harvard Business School Review Press.

Davenport, T. H., & Short, J. E. (1990). The new industrial engineering: Information technology and business process redesign. *Sloan Management Review 31*(4), 11–27.

Donnelly, C. (2016). Machine learning helps Ocado's customer services team wrap up email overload. *Computer Weekly.* 29 November 2016, 10–12.

Edwards, J. S. (2009). Business processes and knowledge management. In M. Khosrow-Pour (Ed.), *Encyclopedia of information science and technology* (2nd ed., Vol. I, pp. 471–476). Hershey, PA: IGI Global.

Edwards, J. S., & Moores, T. T. (1994). A conflict between the use of estimating and planning tools in the management of information systems. *European Journal of Information Systems, 3*(2), 139–147.

Edwards, J. S., & Rodriguez, E. (2016). Using knowledge management to give context to analytics and big data and reduce strategic risk. *Procedia Computer Science, 99,* 36–49. doi:10.1016/j.procs.2016.09.099.

Fearnley, B. (2015). *IDC MaturityScape: Big data and analytics in financial services.* Framingham, MA: IDC Research Retrieved January 30, 2018 from https://www.idc.com/getdoc.jsp?containerId=US40619515.

Flinders, K. (2016). Case study: How Swedish bank prepared robot for customer services. In B. Glick (Ed.), *Artificial intelligence in the enterprise: Your guide to the latest thinking in AI and machine learning* (pp. 22–26): Computer Weekly. Retrieved January 30, 2018 from http://www.computerweekly.com/ehandbook/Focus-Artificial-intelligence-in-the-enterprise.

Future of Life Institute. (2017). Asilomar AI principles. Retrieved February 11, 2017, from https://futureoflife.org/ai-principles/

Goodwin, B. (2014). M&S turns to predictive analytics to keep shelves stocked over Christmas. *Computer Weekly.* Retrieved January 30, 2018 from http://www.computerweekly.com/news/2240236043/MS-turns-to-predictive-analytics-to-keep-shelves-stocked-over-Christmas

Haight, J., & Park, H. (2015). *IoT analytics in practice.* Boston, MA: Blue Hill Research.

Hall, P., Phan, W., & Whitson, K. (2016). *The evolution of analytics: Opportunities and challenges for machine learning in business.* Sebastopol, CA: O'Reilly Media.

Halper, F. (2015). *Operationalizing and embedding analytics for action.* Renton, WA: TDWI Research.

Hammer, M. (1990). Re-engineering work: Don't automate, obliterate. *Harvard Business Review, 68*(4), 104–112.

Hawley, D. (2016). Implementing business analytics within the supply chain: Success and fault factors. *Electronic Journal of Information Systems Evaluation, 19*(2), 112–120.

Henke, N., Bughin, J., Chui, M., Manyika, J., Saleh, T., Wiseman, B., & Sethupathy, G. (2016). The age of analytics: Competing in a data-driven world: McKinsey Global Institute. Retrieved January 30, 2018 from http://www.mckinsey.com/business-functions/mckinsey-analytics/our-insights/the-age-of-analytics-competing-in-a-data-driven-world

Hill, K. (2012). How Target figured out a teen girl was pregnant before her father did. Retrieved January 30, 2018 from https://www.forbes.com/sites/kashmirhill/2012/02/16/how-target-figured-out-a-teen-girl-was-pregnant-before-her-father-did/#5f0167436668

Holsapple, C., Lee-Post, A., & Pakath, R. (2014). A unified foundation for business analytics. *Decision Support Systems, 64,* 130–141.

Hong Kong Efficiency Unit. (2017). Efficiency unit. Retrieved March 12, 2017, from http://www.eu.gov.hk/en/index.html

Kirby, M. W., & Capey, R. (1998). The origins and diffusion of operational research in the UK. *Journal of the Operational Research Society, 49*(4), 307–326. doi:10.1057/palgrave.jors.2600558.

Kiron, D., Ferguson, R. B., & Prentice, P. K. (2013). From value to vision: Reimagining the possible with data analytics. *MIT Sloan Management Review, 54*(3), 1–19.

Lopez-Rabson, T., & McCloy, U. (2013). *Understanding student attrition in the six greater Toronto area (GTA) colleges.* Toronto, Canada: Higher Education Quality Council of Ontario.

Luhn, H. P. (1958). A business intelligence system. *IBM Journal of Research and Development, 2*(4), 314–319.

Marr, B. (2016). Big data at Tesco: Real time analytics at the UK grocery retail giant. Retrieved January 30, 2018 from http://www.forbes.com/sites/bernardmarr/2016/11/17/big-data-at-tesco-real-time-analytics-at-the-uk-grocery-retail-giant/#1c4d5cd4519a

Marr, B. (2017). IoT and big data at Caterpillar: How predictive maintenance saves millions of dollars. Retrieved January 30, 2018 from https://www.forbes.com/sites/bernardmarr/2017/02/07/iot-and-big-data-at-caterpillar-how-predictive-maintenance-saves-millions-of-dollars/#7a18a9867240

McKenna, B. (2016a). Case study: Panama papers revealed by graph database visualisation software. In B. McKenna (Ed.), *Business intelligence in the world of big data* (pp. 22–25): Computer Weekly. Retrieved January 30, 2018 from http://www.computerweekly.com/ehandbook/IT-Project-Business-intelligence-in-the-world-of-big-data.

McKenna, B. (2016b). Case study: The Christie speeds up SPC charts to improve clinical processes. In B. McKenna (Ed.), *Business intelligence in the world of big data* (pp. 14–17): Computer Weekly. Retrieved January 30, 2018 from http://www.computerweekly.com/ehandbook/IT-Project-Business-intelligence-in-the-world-of-big-data.

Mezei, J., Brunelli, M., & Carlsson, C. (2016). A fuzzy approach to using expert knowledge for tuning paper machines. *Journal of the Operational Research Society, Advance Online Publication.* doi:10.1057/s41274-016-0105-3

MHR Analytics. (2017). *Plotting the data journey in the boardroom: The state of data analytics 2017.* Nottingham, UK: MHR Analytics Retrieved January 30, 2018 from http://www.mhr.co.uk/analytics

Moustakerski, P. (2015). *Big data evolution: Forging new corporate capabilities for the long term.* London, UK: Economist Intelligence Unit.

Nexidia. (2014). *Nexidia and Blue Cross and Blue Shield of North Carolina–voice of the customer (VoC) analytics to increase clarity and ease of use for customers.* Mountain View, CA: Frost & Sullivan. Retrieved January 30, 2018 from http://www.nexidia.com/media/2222/fs_casestudy-bcbsnc_final.pdf

Paulk, M. C., Curtis, B., Chrissis, M. B., & Weber, C. V. (1993). Capability maturity model, version 1.1. *IEEE Software, 10*(4), 18–27.

Pechter, R. (2009). What's PMML and what's new in PMML 4.0? *ACM SIGKDD Explorations Newsletter, 11*(1), 19–25.

Piskorski, M. J., Halaburda, H., & Smith, T. (2008). *eHarmony.* Cambridge, MA: Harvard Business School. Retrieved January 30, 2018 from http://www.hbs.edu/faculty/Pages/item.aspx?num=36554

Pittman, D. (2016). Big data, analytics and the retail industry: Luxottica. Retrieved March 11, 2017, from http://www.ibmbigdatahub.com/presentation/big-data-analytics-and-retail-industry-luxottica

Polanyi, M. (1966). *The tacit dimension.* Garden City, NY: Doubleday.

Robinson, A., Levis, J., & Bennett, G. (2010). INFORMS news: INFORMS to officially join analytics movement. *OR/MS Today, 37*(5). Retrieved January 30, 2018 from https:// www.informs.org/ORMS-Today/Public-Articles/October-Volume-37-Number-5/ INFORMS-News-INFORMS-to-Officially-Join-Analytics-Movement

Rodriguez, E. (Ed.). (2017). *The analytics process: Strategic and tactical steps.* Boca Raton, FL: CRC Press.

SAS Institute. (2016a). Analytics for the future: The new "data generation": Future Foundation and SAS Institute. 1483298UK0916. Retrieved January 30, 2018 from http://docs.media.bitpipe.com/io_13x/io_135639/item_1493324/future-of-analytics-report-future-foundation%281%29.pdf

SAS Institute. (2016b). *How any size organization can supersize results with data visualization.* Cary, NC: SAS Institute. Retrieved January 30, 2018 from https://www.sas.com/en_us/whitepapers/any-size-business-can-supersize-results-with-data-visualization-108199.html

Sims, M. (2016). *Big data and analytics: Once a luxury, now a necessity.* Houston, TX: American Productivity and Quality Center. Retrieved January 30, 2018 from https://www.apqc.org/knowledge-base/documents/big-data-and-analytics-once-luxury-now-necessity

Sopra Steria (2016). How to become an analytics powered enterprise: Key insights from the early adopters: Sponsored by Sopra Steria. Retrieved January 30, 2018 from https:// www.soprasteria.com/docs/librariesprovider41/White-Papers/sopra-steria-computing-data-analytics-wp-nov2016.pdf?status=Temp&sfvrsn=0.8211430164259691

Spender, J. C. (2007). Thinking, talking, and doing: How the knowledge-based view might clarify technology's relationship with organizations. *International Journal of Technology Management, 38*(1/2), 178–196.

Stanford, D. (2013). Coke engineers its orange juice with an algorithm. *Bloomberg News.* Retrieved April 24, 2017 from https://www.bloomberg.com/news/articles/2013-01-31/ coke-engineers-its-orange-juice-with-an-algorithm

TIBCO Software. (2015). *Marks and Spencer empowers business analysts with Spotfire.* Palo Alto, CA: TIBCO Software. Retrieved January 30, 2018 from http://spotfire.tibco.com/assets/blt3773b773ca2292eb/ss-marks-and-spencer.pdf

Virgo, P. (2017). Lost in the global cesspit of big data. *Computer Weekly.* Retrieved January 30, 2018 from http://www.computerweekly.com/blog/When-IT-Meets-Politics/ Lost-in-the-global-cesspit-of-Big-Data.

Watson, H., & Wixom, B. (2007). The current state of business intelligence. *IEEE Computer, 40*(9), 96–99.

White, E., & Jordan, T. (2016). *National Audit Office framework to review models.* (pp. 11018–002). London, UK.

Witchalls, C. (2014). Gut & gigabytes: Capitalising on the art & science in decision making: PwC. Retrieved April 24, 2017 from http://www.pwc.com/gx/en/issues/ data-and-analytics/big-decisions-survey/assets/big-decisions2014.pdf

Chapter 2

Data Analytics Process: An Application Case on Predicting Student Attrition

Dursun Delen

Contents

Introduction to Data Analytics Processes

Despite the popularity of data analytics in this age of Big Data, there is still not a proven, standardized, and universally accepted process to conduct data analytics projects. As has been the case in any previous computational paradigms, discovering or creating knowledge from large and diverse data repositories (i.e., Big Data) started as trial-and-error type experimental initiatives. Many have looked at the problem from a different, relatively narrow perspective, trying to characterize what does and doesn't work for their specific problem and purpose. For quite some time, data analytics projects were carried out as experimental or trial-and-error endeavors. However, to systematically conduct data analytics, a standardized process needed to be developed and regularly followed and practiced. Based on the best practices, data analytics researchers and practitioners have proposed several processes (i.e., workflows, in the form of simple step-by-step methodologies) to maximize the possibility of success in conducting data analytics projects. These efforts—mostly modeled after the data mining and knowledge discovery methodologies—resulted in a few standardized processes, of which the most popular ones are described in the following subsections.

Knowledge Discovery in Databases Process

One of the earliest and perhaps the first data mining (i.e., predictive analytics) process is proposed by Fayyad et al. (1996) under the name of knowledge discovery in databases (KDD) methodology. In their seminal work, Fayyad and his colleagues proposed KDD as a methodology within which data mining is believed to be a single step or task where the novel patterns are identified and extracted from data. They proposed KDD as a comprehensive end-to-end process that encompasses

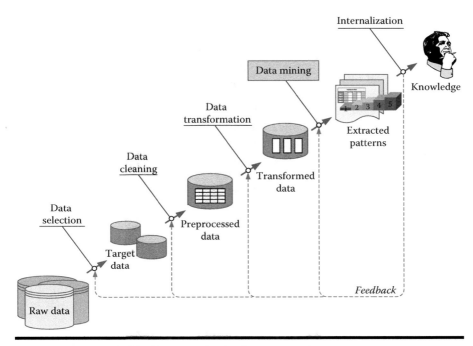

Figure 2.1   Knowledge discovery in databases (KDD) process.

many individual steps or tasks to convert data into knowledge (i.e., actionable insight). A pictorial representation of the KDD process is given in Figure 2.1.

In Figure 2.1, the processing steps are shown as directed arrows with callout labels, and the result of each step is shown as a graphical image representing the artifact. As shown therein, the input to the KDD process is a collection of data coming from organizational databases and/or other external mostly structured data sources. These data sources are often combined in a centralized data repository called a data warehouse. A data warehouse enables the KDD process to be implemented effectively and efficiently because it provides a single source for data to be mined. Once the data are consolidated in a unified data warehouse, the problem-specific data is extracted and prepared for further processing. As the data are usually in a raw, incomplete, and dirty state, a through preprocessing need to be conducted before the modeling can take place. Once the data are preprocessed and transformed into a form for modeling, a variety of modeling techniques are applied to the data to convert it into patterns, correlations, and predictive models. Once the discovered patterns are validated, they need to be interpreted and internalized so that they can be converted into actionable information (i.e., knowledge). One important part of this process is the feedback loop that allows the process flow to redirect backward, from any step to any other previous steps, for rework and readjustments.

Cross-Industry Standard Process for Data Mining

Another standardized analytics process, arguably the most popular one, is called Cross-Industry Standard Process for Data Mining (CRISP-DM), which was proposed in the mid to late 1990s by a European consortium of companies. It was intended to serve as a nonproprietary standard methodology for data mining projects (Chapman et al., 2000). Figure 2.2 illustrates this six step standardized process, which starts with a good understanding of the business problem and the need or objective for the data mining project (i.e., the application domain) and ends with the deployment of the solution that satisfied the specific business need that started the data mining project in the first place (Delen, 2015). Even though these steps are shown sequentially in the graphical representation, there usually is a great deal of backtracking and feedback. Because the data mining is driven by experience and experimentation, depending on the problem situation and the skills, knowledge, and experience of the analyst, the whole process can be very iterative (i.e., one that would require going back and forth through the steps quite a few times) and time demanding. Because a latter step is built on the outcome of the immediate predecessor, one should pay extra attention to the earlier steps in order not to put the whole study on an incorrect path from the onset. What follows is a relatively brief description of each step in CRISP-DM methodology.

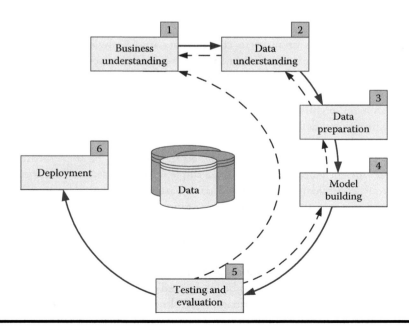

Figure 2.2   CRISP-DM data mining process.

Step 1—Business Understanding

The key to success in any analytics project is to know what the study is meant for (i.e., the purpose or the objective). Answering this question begins with a thorough understanding of the managerial need for new knowledge and an explicit specification of the business objective regarding the study to be conducted. Specific goals such as "What are the common characteristics of the customers we have lost to our competitors recently?" or "What are typical profiles of our customers, and how much value does each of them provide to us?" are needed. Then a project plan for finding such knowledge is developed that specifies the people responsible for collecting the data, analyzing the data, and reporting the findings. At this early stage, a budget to support the study should also be established, at least at a high level with rough numbers. For example, in a customer segmentation model, developed for a retail catalog business, the identification of a business purpose meant identifying the type of customer that would be expected to yield a profitable return. The same or similar analysis would also be useful for credit card distributors. For business purposes, grocery stores often try to identify which items tend to be purchased together so it can be used for affinity positioning within the store, or to intelligently guide promotional campaigns. Data mining has many useful business applications applied to many business problems and opportunities. Intimately knowing the business purpose is very critical to achieving success.

Step 2—Data Understanding

There is a need to make a perfect match between the business problem and the data being used to address it. That is, an analytics study is specific to addressing a well-defined business task, and different business tasks require different sets of data. Following the business understanding, the next activity in the data mining process is to identify the relevant data from many available data sources. There are several key points that must be considered in the data identification and selection process. First and foremost, the analyst should be clear and concise about the description of the data mining task so that the most relevant data can be identified. For example, a retail data mining project may seek to identify spending behaviors of female shoppers who purchase seasonal clothes based on their demographics, credit card transactions, and socioeconomic attributes. Furthermore, the analyst should build an intimate understanding of the data sources (e.g., where the relevant data are stored and in what form; what the process of collecting the data is—automated versus manual; who the collectors of the data are and how often the data are updated) and the variables (e.g., what the most relevant variables are, if there are any synonymous or homonymous variables, if the variables are independent of each other—do they stand as a complete information source without overlapping or conflicting information).

To better understand the data, the data scientist often uses a variety of statistical and graphical tools and techniques, such as simple statistical descriptors or summaries of each variable (e.g., for numeric variables the average, minimum, maximum, median, and standard deviation are among the calculated measures, whereas for categorical variables, the mode and frequency tables are calculated), correlation analysis, scatterplots, histograms, box plots, and cross tabulation. A through process of identification and selection of data sources and the most relevant variables can make it more straightforward for downstream algorithms to quickly and accurately discover useful knowledge patterns.

Data sources for data selection can vary. Normally, data sources for business applications include demographic data (such as income, education, number of people in a household, and age), sociographic data (such as hobby, club membership, and entertainment), and transactional data (such as sales record, credit card spending, and issued checks), among others. Regardless of the sources, data can be categorized as quantitative and qualitative. Quantitative data are measured using numeric values. It can be discrete (such as integers) or continuous (such as real numbers). Qualitative data, also known as categorical data, contain both nominal and ordinal data. Nominal data have finite nonordered values (e.g., gender data, which has two values: male and female). Ordinal data have finite ordered values. For example, customer credit ratings are considered ordinal data because the ratings can be excellent, fair, and bad. Quantitative data can be readily represented by some sort of probability distribution. A probability distribution describes how the data are dispersed and shaped. For instance, normally distributed data are symmetric and is commonly referred to as being a bell-shaped curve. Qualitative data may be coded to numbers and then described by frequency distributions. Once the relevant data are selected according to the analytics project objectives, another critical task—data preparation and preprocessing—would be conducted.

Step 3—Data Preparation

The purpose of data preparation (or more commonly called data preprocessing) is to take the data identified in the previous step and prepare it for analytics methods and algorithms. Compared to the other steps in CRISP-DM, data preprocessing consumes the most amount of time and effort—most believe that this step accounts for more than 80% of the total time spent on an analytics project. The reason for such an enormous effort spent on this step is the fact that real-world data are generally incomplete (lacking attribute values, lacking certain attributes of interest, or containing only aggregate data), noisy (containing errors or outliers), and inconsistent (containing discrepancies in codes or names).

Some parts of the data may have different formats because they are taken from different data sources. If the selected data are from flat files, voice message, images, and web pages, they need to be converted to a consistent and unified format. In general, data cleaning means to filter, aggregate, and fill in missing values (also known as imputation). By filtering the data, the selected variables are examined for outliers and redundancies. Outliers differ greatly from the majority of data, or data that are clearly out of range of the selected data groups. For example, if the age of a customer included in the data is 190, it should be a data entry error and should be identified and fixed (perhaps, taken out from the data mining project that examines the various aspects of the customers and age is perceived to be a critical component of the customer characteristics). Outliers may be caused by many reasons, such as human errors or technical errors, or may naturally occur in a dataset due to extreme events. Suppose the age of a credit card holder is recorded as "12." This is likely a data entry error, most likely by a human. However, there might actually be an independently wealthy pre-teen with important purchasing habits. Arbitrarily deleting this outlier could dismiss valuable information.

Redundant data are the same information recorded in several different ways. Daily sales of a particular product are redundant to seasonal sales of the same product, because we can derive the sales from either daily data or seasonal data. By aggregating data, data dimensions are reduced to obtain aggregated information. Note that although an aggregated dataset has a small volume, the information will remain. If a marketing promotion for furniture sales is considered in the next three or four years, then the available daily sales data can be aggregated as annual sales data. The size of sales data is dramatically reduced. By smoothing data, missing values of the selected data are found and new or reasonable values then added. These added values could be the average number of the variable (mean) or the mode. A missing value often causes no solution when a data-mining algorithm is applied to discover the knowledge patterns.

Step 4—Model Building

In this step, various modeling methods and algorithms are designed and applied to an already prepared dataset to address the specific business question, need, or purpose. The modeling step also encompasses the assessment and comparative analysis of the various types of models that can address the same type of data mining tasks (e.g., clustering, classification, etc.). Because there is not a universally known best method or algorithm for a specific data mining task, one should use a variety of viable model types along with a well-defined experimentation and assessment strategy to identify the "best" method for a given data mining problem. Even for a single method or algorithm, a number of parameters must be calibrated to obtain optimal results. Some methods may have specific requirements on the way that the data are to be formatted; thus stepping back to the data preparation step is often necessary.

Depending on the business need, the analytics task can be of a prediction (either classification or regression), an association, or a clustering or segmentation type. Each of these tasks can use a variety of analytics and data mining methods and algorithms. For instance, classification type data mining tasks can be accomplished by developed neural networks, decision trees, support vector machines (SVMs), or logistic regression, among others.

The standard procedure for modeling in data mining is to take a large preprocessed dataset and divide it into two or more subsets for training and validation or testing. Then, use a portion of the data (the training set) for development of the models (no matter what modeling technique or algorithms is used), and use the other portion of the data (the test set) for testing the model that is just built. The principle is that if you build a model on a particular set of data, it will of course test quite well on the data that is was built on. By dividing the data and using part of it for model development and a separate part of it for testing creates objective results for the accuracy and reliability of the model. The idea of splitting the data into components is often carried to additional levels with multiple splits in the practice of data mining. Further details about data splitting and other evaluation methods can be found in Delen (2015).

Step 5—Testing and Evaluation

In this step, the developed models are assessed and evaluated for their accuracy and generality. This step assesses the degree to which the selected model (or models) meets the business objectives and, if so, to what extent (i.e., do more models need to be developed and assessed). Another option is to test the developed models in a real-world scenario if time and budget constraints permit. Even though the outcome of the developed models is expected to relate to the original business objectives, other findings that are not necessarily related to the original business objectives, but that might also unveil additional information or hints for future directions, often are discovered.

The testing and evaluation step is a critical and challenging task. No value is attained by an analytics project unless the expected business outcomes are achieved through the newly discovered knowledge patterns. Determining the business value from discovered knowledge patterns is somewhat similar to playing with puzzles. The extracted knowledge patterns are pieces of the puzzle that need to be put together in the context of the specific business purpose. The success of this identification depends on the interaction among data scientists, business analysts, and decision makers (such as business managers). Because data scientists and analysts may not have the holistic understanding of the business objectives and what they mean to the business operations, and the business analysts and decision makers may not have the technical knowledge to interpret the results often presented in seemingly complex mathematical or numerical format, interaction among them is crucial. To properly interpret knowledge patterns, it is often necessary to use a variety of tabulation and visualization techniques (e.g., pivot tables, cross tabulation of findings, pie charts, histograms, box plots, scatter plots, etc.).

Step 6—Deployment

Development and assessment of the models is not the end of the analytics project. Even if the purpose of the model is to have a simple exploration of the data, the knowledge gained from such exploration will need to be organized and presented in a way that the end user can understand and benefit from it. Depending on the requirements, the deployment phase can be as simple as generating a report or as complex as implementing a repeatable computer-based decision support system across the enterprise (Delen, Sharda, and Kumar, 2007). In many cases, it is the customer, not the data analyst, who carries out the deployment steps. However, even if the analyst will not carry out the deployment effort, it is important for the customer to understand up front what actions need to be carried out to actually make use of the created models.

The deployment step may also include maintenance activities for the deployed models. Because everything about the business is constantly changing, the data that reflect the business activities also is changing. Over time, the models (and the patterns embedded within them) built on the old data becomes obsolete, irrelevant, or misleading. Therefore, monitoring and maintenance of the models are important, if the analytics results are to become a part of the day-to-day business decision-making environment. A careful preparation of a maintenance strategy helps to avoid unnecessarily long periods of incorrect usage of analytics results. To monitor the deployment of the analytics results, the project needs a detailed plan on the monitoring process, which may not be a trivial task for complex analytics models.

The CRISP-DM process is the most complete and most popular data mining methodology being practiced in industry as well as in academia. As opposed to using it as is, practitioners add their own insight to make it specific to their organization's style of practice.

Sample, Explore, Modify, Model, Assess Process

In order to be applied successfully, the data analytics projects must be viewed as a process rather than a set of tools or techniques. In addition to the CRISP-DM process, there is yet another well-known methodology developed by the SAS Institute, called sample, explore, modify, model, assess (SEMMA) process. Beginning with a statistically representative sample of your data, SEMMA intends to make it easy to apply exploratory statistical and visualization techniques, select and transform the most signify cant predictive variables, model the variables to predict outcomes, and finally confirm a model's accuracy. A pictorial representation of SEMMA is given in Figure 2.3.

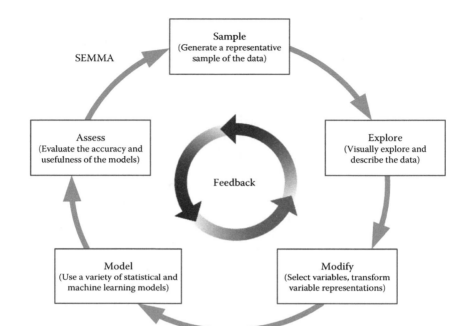

Figure 2.3 A graphical depiction of the SEMMA process.

By assessing the outcome of each stage in the SEMMA process, one can determine how to model new questions raised by the previous results, and thus proceed back to the exploration phase for additional refinement of the data. That is, as is the case in CRISP-DM, SEMMA also driven by a highly iterative experimentation cycle. Here are short descriptions for the five steps that constitute SEMMA.

Step 1—Sample

This is where a portion of a large dataset (big enough to contain the significant information yet small enough to manipulate quickly) is extracted. For optimal cost and computational performance, some (including the SAS Institute) advocate a sampling strategy, which applies a reliable, statistically representative sample of the full detail data. In the case of very large datasets, mining a representative sample instead of the whole volume may drastically reduce the processing time required to get crucial business information. If general patterns appear in the data as a whole, these will be traceable in a representative sample. If a niche (rare pattern) is so tiny that it is not represented in a sample and yet so important that it influences the big picture, it should be discovered using exploratory data description methods. It is also advised to create partitioned datasets for better accuracy assessment.

- Training, for model fitting
- Validation, for assessment and to prevent over fitting
- Test, for obtaining an honest assessment of how well a model generalizes

A more detailed discussion and relevant techniques for assessment and validation of data mining models can be found in Sharda et al. (2017).

Step 2—Explore

This is where a user searches for unanticipated trends and anomalies to gain a better understanding of the dataset. After sampling your data, the next step is to explore it visually or numerically for inherent trends or groupings. Exploration helps refine and redirect the discovery process. If visual exploration does not reveal clear trends, one can explore the data through statistical techniques including factor analysis, correspondence analysis, and clustering. For example, in data mining for a direct mail campaign, clustering might reveal groups of customers with distinct ordering patterns. Limiting the discovery process to each of these distinct groups individually may increase the likelihood of exploring richer patterns that may not be strong enough to be detected if the whole dataset is to be processed together.

Step 3—Modify

This is where the user creates, selects, and transforms the variables upon which to focus the model construction process. Based on the discoveries in the exploration phase, one may need to manipulate data to include information such as the grouping of customers and significant subgroups, or to introduce new variables. It may also be necessary to look for outliers and reduce the number of variables, to narrow them down to the most significant ones. One may also need to modify data when the "mined" data changes. Because data mining is a dynamic, iterative process, you can update data mining methods or models when new information is available.

Step 4—Model

This is where the user searches for a variable combination that reliably predicts a desired outcome. Once you prepare your data, you are ready to construct models that explain patterns in the data. Modeling techniques in data mining include artificial neural networks (ANN), decision trees, rough set analysis, SVMs, logistic models, and other statistical models—such as time series analysis, memory-based reasoning, and principal component analysis. Each type of model has particular strengths, and is appropriate within specific data mining situations depending on the data. For example, ANN are very good at fitting highly complex nonlinear relationships while rough sets analysis is known to produce reliable results with uncertain and imprecise problem situations.

Step 5—Assess

This is where the user evaluates the usefulness and the reliability of findings from the data mining process. In this final step of the data mining process, the user assesses the model to estimate how well it performs. A common means of assessing a model is to apply it to a portion of a dataset put aside (and not used during the model building) during the sampling stage. If the model is valid, it should work for this reserved sample as well as for the sample used to construct the model. Similarly, you can test the model against known data. For example, if you know which customers in a file had high retention rates and your model predicts retention, you can check to see whether the model selects these customers accurately. In addition, practical applications of the model, such as partial mailings in a direct mail campaign, help prove its validity.

The SEMMA process is quite compatible with the CRISP-DM process. Both aim to streamline the knowledge discovery process. Both were created as broad frameworks, which need to be adapted to specific circumstances. In both, once models are obtained and tested, they can then be deployed to gain value with respect to business or research application. Even though they have the same goal and are similar, the SEMMA and CRISP-DM processes have a few differences. Table 2.1 presents these differences.

Six Sigma for Data Analytics

Six Sigma is a popular business management philosophy that focuses on reducing the deviations (i.e., sigma) from the perfection by employing rigorous and systematic execution of proven quality control principles and techniques. This vastly popular management philosophy was first introduced by Motorola in the 1980s in the context of manufacturing management and has since then been adopted by many companies and organizations in a wide variety of business contexts beyond manufacturing. Ideally, it promotes zero defects and zero tolerance, which can roughly be translated in the business context as error free, perfect business execution. Six Sigma methodology manifested itself in the business world with the define, measure, analyze, improve, and control (DMAIC) process. Because of its success in many business problems and settings, DMAIC methodology is also applied to analytics projects. Figure 2.4 shows a DMAIC methodology as a simplified flow diagram.

Table 2.1   Comparison of CRISP-DM and SEMMA

Task	CRISP-DM	SEMMA	Comments
Project initiation	Business understanding	N/A	In this phase, CRISP-DM includes activities like project initiation, problem definition, and goal setting. SEMMA does not have a step for this phase.
Data access	Data understanding	Sample Explore	In this phase, both CRISP-DM and SEMMA have the steps to access, sample, and explore the data.
Data transformation	Data preparation	Modify	In this phase, both CRISP-DM and SEMMA process the data to make it amenable to machine processing.
Model building	Modeling	Model	In this phase, both CRISP-DM and SEMMA suggest building and testing various models.
Project evaluation	Evaluation	Assess	In this phase, both CRISP-DM and SEMMA suggest assessing the findings against the project goals.
Project finalization	Deployment	N/A	In this phase, CRISP-DM prescribes deployment of the results while SEMMA does not explicitly state it.

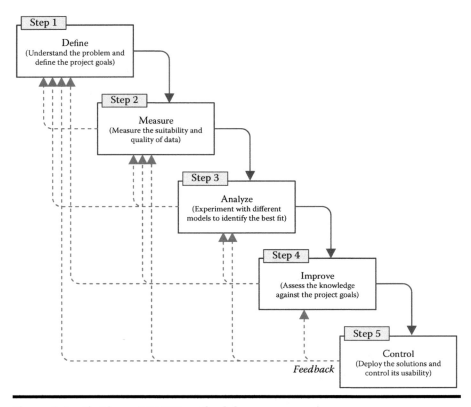

Figure 2.4 Six Sigma DMAIC methodology.

Step 1—Define

This is the first step in DMAIC where several tasks are to be accomplished to get the project set up and started. These tasks include (1) a thorough understanding of the business needs; (2) identifying the most pressing problem, (3) defining the goals and objectives, (4) identifying and defining the data and other resources needed to investigate the business problem, and (5) developing a detailed project plan. As you may have noticed, there is a significant overlap between this step and the "Business Understanding" which was the first step in the CRISP-DM process.

Step 2—Measure

In this step, the mapping between organizational data repositories and the business problem is assessed. Since data mining requires problem-relevant, clean, and usable data, identification and creation of such a resource is of critical importance to the success of the project. In this step, the identified data sources are to be consolidated and transformed into a format that is amenable to machine processing.

Step 3—Analyze

Now that the data are prepared for processing, in this step, a series of data mining techniques is used to develop models. Since there is not a single best technique for a specific data mining task (because there are many and most of them are machine learning techniques with many parameters to optimize), several probable techniques need to be applied and experimented with to identify and develop the most appropriate model.

Step 4—Improve

Once the analysis results are obtained, in this step, the improvement possibilities are investigated. Improvements can be at the technique level or they can be at the business problem level. For instance, if the model results are not satisfactory, other more sophisticated techniques (e.g., ensemble systems) can be used to boost the performance of the models. Also, if the modeling results are not clearly addressing the business problem, via a feedback loop to previous steps, the very structure of the analysis can be re-examined and improved, or the business problem can be further investigated and restated.

Step 5—Control

In this step a final examination of the project outcomes is assessed and if found satisfactory, the models and result are disseminated to decision makers and/or integrated into the existing business intelligence systems for automation.

The Six Sigma-based DMAIC methodology has a lot of resemblance to the CRISP-DM process. We do not have any evidence that suggests one is inspired from the other. That said, since what these two processes portray are rather logical and straightforward steps in any business system analysis effort, they may not have inspired each other. The users of DMAIC are rare compared to CRISP-DM and SEMMA.

Which Process Is the Best?

Even though some are more elaborate than others, there is not a sure way to compare these data analytics processes. They all have pros and cons in their own respect. Some are more problem focused, while others are more analysis driven. Businesses that does data mining adopts one of these methodologies, and often slightly modifies it to make it their own. To shed some more light to the question of "Which one is the best?" KDNuggets.com, a well-known and well-respected web portal for data mining, conducted a survey where it asked the very same question. Figure 2.5 shows the polling results of that survey (Piatetsky, 2014).

As the survey results indicate, CRISP-DM dominates the others as the most popular data mining process. Compared to the others, CRISP-DM is the most

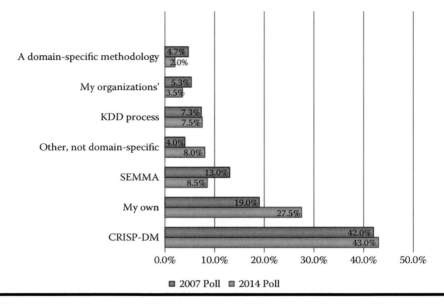

Figure 2.5 Preference poll for standard analytics processes. (From Piatetsky, G., CRISP-DM, still the top methodology for analytics, data mining, or data science projects, KDnuggets. Retrieved from http://www.kdnuggets.com/2014/10/crisp-dm-top-methodology-analytics-data-mining-data-science-projects.html, 2014.)

complete and most matured methodology for data analytics projects. Many of the ones that fall under "My Own" are also known to be small deviations (specializations) of CRISP-DM.

Application Case: Predicting Student Attrition with Data Analytics

Predicting attrition has always been an intriguing and challenging problem for data scientists and business managers. Attrition (a tool to better predict and manage retention) is an important subject in many application domains, and based on the domain, it may be called differently. For instance, in marketing, it is often called "customer churn", referring to the timely identification of at-risk customers (i.e., one about to leave your products or services). The marketing saying that goes something like "attracting a new customer cost ten times more than retaining the one you have" is a testament to the importance of predicting and properly managing churn. Attrition is also an important concept in employee retention, where accurate prediction and management of at-risk employees would save time and money for the organization in maintaining a productive and capable workforce.

In this application case, the focus will be on predicting student attrition for better management of student retention in higher education institutions.

Student retention is a critical part of many enrollment management systems. Affecting university rankings, school reputation, and financial well-being, student retention has become one of the most important priorities for decision makers in higher education institutions. Improving student retention starts with a thorough understanding of the reasons behind the attrition. Such an understanding is the basis for accurately predicting at-risk students and appropriately intervening to retain them. In this study, using five years of institutional data along with several data mining techniques (both individuals as well as ensembles), we developed analytical models to predict and to explain the reasons behind freshmen student attrition. The comparative analysis results showed that the ensembles performed better than the individual models, while a balanced dataset produced better prediction results than an unbalanced dataset. The sensitivity analysis of the models revealed that the educational and financial variables are among the most important predictors of the phenomenon.

Introduction and Motivation

Student attrition has become one of the most challenging problems for decision makers in academic institutions. In spite of all of the programs and services to help retain students, according to the U.S. Department of Education, Center for Educational Statistics (nces.ed.gov), only about half of those who enter higher education actually earn a bachelor's degree. Enrollment management and the retention of students has become a top priority for administrators of colleges and universities in the United States and other developed countries around the world. A high student drop-out rate usually results in overall financial loss, lower graduation rates, and inferior school reputation in the eyes of all stakeholders (Gansemer-Topf and Schuh, 2006). The legislators and policymakers who oversee higher education and allocate funds, the parents who pay for their children's education to prepare them for a better future, and the students who make college choices look for evidence of institutional quality and reputation to guide their decision-making processes (Delen, 2010).

The principal motivation for improving student retention is the economic and social benefits of attaining a higher education degree (Thomas and Galambos, 2004), both for individuals and for the public. Generally speaking, the economic and social attributes that motivate individuals to enter higher education are (1) public economic benefits, such as increased tax revenues, greater productivity, increased consumption, increased workforce flexibility, and decreased reliance on government financial support; (2) individual economic benefits such as higher salaries and benefits, employment, higher savings levels, improved working conditions, and personal and professional mobility; (3) public social benefits, such as reduced crime rates, increased charitable giving and community service, increased quality of civic life, social cohesion and the appreciation of diversity, and the improved ability to adapt to and use technology; and (4) individual social benefits, such as improved health and life expectancy, improved quality of life for offspring, better consumer decision-making, increased personal status, and more hobbies and leisure activities (Hermaniwicz, 2003)

Traditionally, student attrition at a university has been defined as the number of students who do not complete a degree in that institution. Studies have shown that the vast majority of students withdraw during their first year of college than during the rest of their higher education (Deberard, Julka, and Deana, 2004; Hermaniwicz, 2003). Since most of the student dropouts occur at the end of the first year (the freshman year), many of the student retention and attrition studies (including this study) have focused on first-year dropouts (or the number of students not returning for the second year). This definition of attrition does not differentiate between the students who may have transferred to other universities and obtained their degrees there. It only considers the students dropping out at the end of the first year voluntarily and not by academic dismissal.

Research on student retention has traditionally been survey driven (e.g., surveying a student cohort and following them for a specified period of time to determine whether they continue their education) (Caison, 2007). Using such a design, researchers worked on developing and validating theoretical models including the famous student integration model developed by Tinto (1993). Elaborating on Tinto's theory, others have also developed student attrition models using survey-based research studies (Berger and Braxton, 1998; Berger and Milem, 1999). Even though they have laid the foundation for the field, these survey-based research studies have been criticized for their lack of generalized applicability to other institutions and the difficulty and costliness of administering such large-scale survey instruments (Cabrera, Nora, and Castaneda, 1993). An alternative (and/or a complementary) approach to the traditional survey-based retention research is an analytic approach where the data commonly found in institutional databases is used. Educational institutions routinely collect a broad range of information about their students, including demographics, educational background, social involvement, socioeconomic status, and academic progress. A comparison between the data-driven and survey-based retention research showed that they are comparable at best, and to develop a parsimonious logistic regression model, data-driven research was found to be superior to its survey-based counterpart (Caison, 2007). But in reality, these two research techniques (one driven by surveys and theories and the other driven by institutional data and analytic methods) complement and help each other (Miller and Tyree, 2009). That is, the theoretical research may help identify important predictor variables to be use in analytical studies while analytical studies may reveal novel relationships among the variables that may lead to development of new and betterment of the existing theories.

To improve student retention, one should try to understand the non-trivial reasons behind the attrition. To be successful, one should also be able to accurately identify those students that are at risk of dropping out. So far, the vast majority of student attrition research has been devoted to understanding this complex, yet crucial, social phenomenon. Even though these qualitative, behavioral, and survey-based studies revealed invaluable insight by developing and testing a wide range of theories, they do not provide the much-needed instrument to accurately predict (and potentially improve) student attrition (Delen, 2011; Miller and Herreid, 2010; Veenstra, 2009). In this project, we propose a quantitative research approach where the historical institutional data from student databases is used to develop models that are capable of predicting, as well as explaining, the institution-specific nature of the attrition problem. Though the concept is relatively new to higher education, for almost a decade now, similar problems in the field of marketing have been studied using predictive data mining techniques under the name of "churn analysis," where the purpose is to identify among the current customers who are most likely to leave the company so that some kind of intervention process can be executed for the ones who are worthwhile to retain. Retaining existing customers is crucial because the related research shows that acquiring a new customer costs roughly ten times more than keeping the one that you already have (Lemmens and Croux, 2006).

Analytics Methodology

In this research, we followed a popular data mining methodology called CRISP-DM (Shearer, 2000), that, as explained in the previous section, is a six step process: (1) understanding the domain and developing the goals for the study, (2) identifying, accessing, and understanding the relevant data sources, (3) preprocessing, cleaning, and transforming the relevant data, (4) developing models using comparable analytical techniques, (5) evaluating and assessing the validity and the utility of the models against each other and against the goals of the study, and (6) deploying the models for use in decision-making processes. This popular methodology provides a systematic and structured way of conducting data mining studies, and hence increasing the likelihood of obtaining accurate and reliable results. The attention paid to the earlier steps in CRISP-DM (i.e., understanding the domain of study, understanding the data, and preparing the data) sets the stage for a successful data mining study. Roughly 80% of the total project time is usually spent on these first three steps.

The method evaluation step in CRISP-DM requires comparing the data mining models for their predictive accuracy. Traditionally, in this comparison process the complete dataset is split into two subsets, two-thirds for training and one-third for testing. The models are trained on the training subset and then evaluated on the testing subset. The prediction accuracy on the testing subset is used to report the actual prediction accuracies of all evaluated models. Since the dataset is split into two exclusive subsets randomly, there always is a possibility of those two datasets not being "equal." To minimize this bias associated with the random sampling of the training and testing data samples, we used an experimental design called *k*-fold cross-validation. In *k*-fold cross-validation, also called rotation estimation, the complete dataset is randomly split into *k* mutually exclusive subsets of approximately equal size. The classification model is trained and tested *k* times. Each time, it is trained on all but one fold and tested on the remaining single fold. The cross-validation estimate of the overall accuracy is calculated as simply the average of the *k* individual accuracy measures as in the following equation:

$$CV = \frac{1}{k} \sum_{i=1}^{k} PM_i$$

where, *CV* stands for the cross-validation result for a method, *k* is the number of folds used, and *PM* is the performance measure for each fold.

In this case study, to estimate the performance of the prediction models, a tenfold cross-validation approach was used. Empirical studies showed that 10 seems to be an optimal number of folds (that optimizes the time it takes to complete the test while minimizing the bias and variance associated with the validation process) (Kohavi, 1995). In tenfold cross-validation the entire dataset is divided into 10 mutually exclusive subsets (or folds). Each fold is used once to test the performance of the prediction model that is generated from the combined data of the remaining ninefolds, leading to 10 independent performance estimates.

A pictorial depiction of this evaluation process is shown in Figure 2.6. With this experimental design, if *k* is set to 10 (which is the case in this study and a common practice in most predictive data mining applications), for each of the seven model types (four individual and three ensembles), 10 different models are developed and tested. Combined with the tenfold experimentation conducted on the original (i.e., unbalanced) datasets using the four individual model types, the total number of models developed and tested for this study was 110.

Data Description

The data for this study came from a single institution (a comprehensive public university located in the Midwest region of the United States) with an average enrollment of 23,000 students, of which roughly 80% are residents of the same state and roughly 19% of the students are listed under some minority classification. There is no significant difference between the two genders in the enrollment numbers. The average freshman student retention rate for the institution is about 80%, and the average 6-year graduation rate is about 60%.

In this study we used five years of institutional data, which entailed 16,066 students enrolled as freshmen between (and including) the years of 2004 and 2008. The data was collected and consolidated from various university student databases. A brief summary of the number of records (i.e., freshman students) by year is given in Table 2.2.

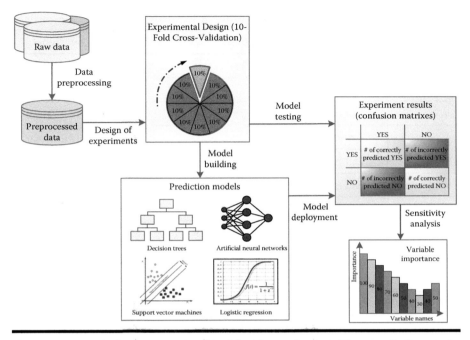

Figure 2.6 **Analytics process employed for the student attrition prediction study.**

Table 2.2 **Five-year Freshmen Student Data Used in this Study**

Year	Total Number of Freshmen Students	Returned for the 2nd Fall	Freshman Attrition (%)
2004	3249	2541	21.79%
2005	3306	2604	21.23%
2006	3234	2576	20.35%
2007	3207	2445	23.76%
2008	3070	2391	22.12%
	Total: 16066	Total: 12557	Average: 21.84%

The data contained variables related to the student's academic, financial, and demographic characteristics. A complete list of variables obtained from the student databases is given in Table 2.3. After converting the multidimensional student data into a flat file (a single file with columns representing the variables and rows representing the student records), the file was assessed and preprocessed to identify and remove anomalies and unusable records. For instance, we removed all

Table 2.3   Variables Obtained from Student Records

No	Variables	Data Type
1	College	Multi Nominal
2	Degree	Multi Nominal
3	Major	Multi Nominal
4	Concentration	Multi Nominal
5	Fall hours registered	Number
6	Fall earned hours	Number
7	Fall GPA	Number
8	Fall cumulative GPA	Number
9	Spring hours registered	Number
10	Spring earned hours	Number
11	Spring GPA	Number
12	Spring cumulative GPA	Number
13	Second fall registered (Y/N)	Nominal
14	Ethnicity	Nominal
15	Sex	Binary Nominal
16	Residential code	Binary Nominal
17	Marital status	Binary Nominal
18	SAT high score comprehensive	Number
19	SAT high score English	Number
20	SAT high score reading	Number
21	SAT high score math	Number
22	SAT high score science	Number
23	Age	Number
24	High school GPA	Number
25	High school graduation year and month	Date
26	Starting term as new freshman	Multi Nominal

(Continued)

Table 2.3 (*Continued*) Variables Obtained from Student Records

No	Variables	Data Type
27	TOEFL score	Number
28	Transfer hours	Number
29	CLEP earned hours	Number
30	Admission type	Multi Nominal
31	Permanent address state	Multi Nominal
32	Received fall financial aid	Binary Nominal
33	Received spring financial aid	Binary Nominal
34	Fall student loan	Binary Nominal
35	Fall grant/tuition waiver/scholarship	Binary Nominal
36	Fall federal work study	Binary Nominal
37	Spring student loan	Binary Nominal
38	Spring grant/tuition waiver/scholarship	Binary Nominal
39	Spring federal work study	Binary Nominal

international student records from the dataset because they did not contain some of the presumed important predictors (e.g., high school GPA and SAT scores). In the data transformation phase, some of the variables were aggregated (e.g., "Major" and "Concentration" variables aggregated to binary variables *MajorDeclared* and *ConcentrationSpecified*) for better interpretation for the predictive modeling. Additionally, some of the variables were used to derive new variables (e.g., *Earned/Registered* and *YearsAfterHighSchool*).

$$\text{EarnedByRegistered} = \frac{\text{EarnedHours}}{\text{RegisteredHours}}$$

$$\text{YearsAfterHighSchool} = \text{FreshmenEnrollmentYear} - \text{HighSchoolGraduationYear}$$

The *Earned/Registered* hours variable was created to have a better representation of the students' resiliency and determination in their first semester of the freshman year. Intuitively, one would expect greater values for this variable to have a positive impact on retention. The *YearsAfterHighSchool* variable was created to measure the impact of the time taken between high school graduation and initial college enrollment. Intuitively, one would expect this variable to be a contributor to the prediction of attrition. These aggregations and derived variables are determined based on a number of experiments conducted for a number of logical hypotheses. The ones that made more common sense and the ones that led to better prediction accuracy were kept in the final variable set.

Reflecting the population, the dependent variable (i.e., "Second Fall Registered") contained many more *yes* records (approximately 80%) than *no* records (approximately 20%). We experimented with the options of using and comparing the results of the models built with the original data (biased for the *yes* records) versus the well-balanced data.

Predictive Analytics Models

In this study, four popular classification methods (i.e., ANN, decision trees, SVMs, and logistic regression) along with three ensemble techniques (i.e., bagging, busting, and information fusion) are built and compared to each other using their predictive accuracy on the holdout samples. A large number of studies compare data mining methods in different settings. Most of these previous studies found machine-learning methods (e.g., ANN, SVMs, and decision trees) to be superior to their statistical counterparts (e.g., logistic regression and discriminant analysis) in terms of both being less constrained by assumptions and producing better prediction results. Our findings in this study confirm these results. What follows are brief descriptions of the individual and ensemble prediction models used in this study.

Artificial neural networks (ANN) are biologically inspired, analytical techniques, capable of modeling extremely complex nonlinear functions. Formally defined, neural networks are analytic techniques modeled after the processes of learning in the cognitive system and the neurological functions of the brain and capable of predicting new observations (on specific variables) from other observations (on the same or other variables) after executing a process of so-called learning from existing data. In this study we used a popular neural network architecture called multilayer perceptron (MLP) with a back-propagation, supervised learning algorithm. MLP, a strong function approximator for prediction and classification problems, is arguably the most commonly used and well-studied ANN architecture. Hornik, Stinchcombe, and White (1990) empirically show that given the right size and structure, MLP is capable of learning arbitrarily complex nonlinear functions to an arbitrary accuracy level. MLP is essentially the collection of nonlinear neurons (perceptrons) organized and connected to each other in a feedforward multilayered structure. A pictorial representation of the ANN architecture used in this study is shown in Figure 2.7.

Decision trees are powerful classification algorithms that are becoming increasingly more popular due to their intuitive explainability characteristics. Popular decision tree algorithms include Quinlan (1986, 1993)'s ID3,

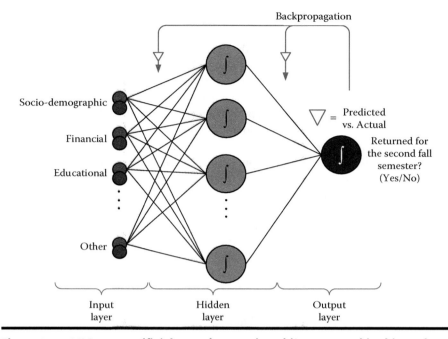

Figure 2.7   **MLP-type artificial neural network architecture used in this study.**

C4.5, C5, and Breiman et al. (1984)'s classification and regression trees (CART) and chi-squared automatic interaction detector (CHAID). In this study, we used the C5 algorithm, which is an improved version of the C4.5 and ID3 algorithms.

Logistic regression is a generalization of linear regression. It is used primarily for predicting binary or multiclass dependent variables. Because the response variable is discrete, it cannot be modeled directly by linear regression. Therefore, rather than predicting a point estimate of the event itself, it builds the model to predict the odds of its occurrence. While logistic regression has been a common statistical tool for classification problems, its restrictive assumptions on normality and independence led to an increased use and popularity of machine learning techniques for real-world prediction problems.

Support vector machines (SVMs) belong to a family of generalized linear models that achieves a classification or regression decision based on the value of the linear combination of features. The mapping function in SVMs can be either a classification function (used to categorize the data, as is the case in this study) or a regression function (used to estimate the numerical value of the desired output). For classification, nonlinear kernel functions are often used to transform the input data (inherently representing highly complex nonlinear relationships) to a high dimensional feature space in which the input data becomes more separable (i.e., linearly separable) compared to the original input space. Then, the maximum-margin hyperplanes are constructed to optimally separate the classes in the training data. Two parallel hyperplanes are constructed on each side of the hyperplane that separates the data by maximizing the distance between the two parallel hyperplanes. An assumption is made that the larger the margin or distance between these parallel hyperplanes, the better the generalization error of the classifier will be (Cristianini and Shawe-Taylor, 2000).

Ensembles and bagging (random forest): A random forest is a classifier that consists of many decision trees and outputs the class that is the mode of the classes output by individual trees. A random forest consists of a collection (ensemble) of deceivingly simple decision trees, each capable of producing a response when presented with a set of predictor values. A random forest has shown to run very efficiently on large datasets with a large number of variables. The algorithm for inducing a random forest was first developed by Breiman (2001).

Ensembles and boosted trees: The general idea of boosted trees is to compute a sequence of very simple trees, where each successive tree is built for the prediction residuals of the preceding tree. It learns from the previous tree, in order to construct the succeeding one so that the misclassification of cases is minimized. Detailed technical descriptions of this methods can be found in Hastie et al (2001).

Ensembles and information fusion: Information fusion is the process of "intelligently" combining the information (predictions in this case) provided by two or more information sources (i.e., prediction models). While there is an ongoing debate about the sophistication level of fusion methods, there is a general consensus that fusion (combining predictions) produces more accurate and more robust prediction results (Delen, 2015).

Sensitivity Analysis

In machine-learning algorithms, sensitivity analysis is a method for identifying the "cause and effect" relationship between the inputs and outputs of a prediction model (Delen et al, 2017). The fundamental idea of sensitivity analysis is that it measures the importance of predictor variables based on the change in modeling performance that occurs if a predictor variable is not included in the model. Hence, the measure of sensitivity of a specific predictor variable is the ratio of the error of the trained model without the predictor variable to the error of the model that includes this predictor variable. The more sensitive the network is to a particular variable, the greater the performance decrease would be in the absence of that variable, and therefore the greater the ratio of importance. This method is often followed in machine learning techniques to rank the variables in terms of their importance according to the sensitivity measure defined in the following equation (Saltelli, 2002):

$$S_i = \frac{V_i}{V(F_t)} = \frac{V(E(F_t|X_i))}{V(F_t)}$$

where $V(F_t)$ is the unconditional output variance. In the numerator, the expectation operator E calls for an integral over $X_{\sim i}$; that is, over all input variables but X_i, then the variance operator V implies a further integral over X_i. Variable importance is then computed as the normalized sensitivity. Saltelli et al. (2004) showed that the equation above is the proper measure of sensitivity to rank the predictors in order of importance for any combination of interaction and nonorthogonality among predictors. As for the decision trees, variable importance measures were used to judge the relative importance of each predictor variable. Variable importance ranking uses surrogate splitting to produce a scale which is a relative importance measure for each predictor variable included in the analysis. Further details on this procedure can be found in Breiman et al. (1984).

Results

In the first set of experiments, we used the original dataset which was composed of 16,066 records. Based on the tenfold cross-validation, the SVMs produced the best results with an overall prediction rate of 87.23%; the decision tree came out as the runner up with an overall prediction rate of 87.16%; followed by ANN and logistic regression with overall prediction rates of 86.45% and 86.12% respectively (see Table 2.4). A careful examination of these results reveals that the prediction accuracy for the "yes" class is significantly higher than the prediction accuracy of the *no* class. In fact, all four model types predicted the students who are likely to return for the second year with better than 90% accuracy while they did poorly on predicting the students who are likely to drop out after the freshman year with less

Table 2.4 Prediction Results for tenfold Cross-Validation with Unbalanced Dataset

	ANN(MLP)		DT(C5)		SVM		LR	
	No	Yes	No	Yes	No	Yes	No	Yes
No	1494	384	1518	304	1478	255	1438	376
Yes	1596	11142	1572	11222	1612	11271	1652	11150
SUM	3090	11526	3090	11526	3090	11526	3090	11526
Per-class Accuracy	48.35%	96.67%	49.13%	97.36%	47.83%	97.79%	46.54%	96.74%
Overall Accuracy	86.45%		87.16%		87.23%		86.12%	

than 50% accuracy. Since the prediction of the "no" class is the main purpose of this study, less than 50% accuracy for this class was deemed unacceptable. Such a difference in prediction accuracy of the two classes can be attributed to the skewness of the original dataset (i.e., approximately 80% "yes" and approximately 20% "no" samples). Previous studies also commented on the importance of having a balanced dataset for building accurate prediction models for binary classification problems (Wilson and Sharda, 1994).

In the next round of experiments, we used a well-balanced dataset where the two classes were represented equally. In realizing this approach, we took all of the samples from the minority class (i.e., the "no" class herein) and randomly selected an equal number of samples from the majority class (i.e., the "yes" class herein), and repeated this process ten times to reduce the bias of random sampling. Each of these sampling processes resulted in a dataset of 7,018 records, of which 3,509 were labeled as "no" and 3,509 were labeled as "yes." Using a tenfold cross-validation methodology, we developed and tested prediction models for all four model types. The results of these experiments are shown in Table 2.5. Based on the hold-out sample results, SVMs generated the best overall prediction accuracy with 81.18%, followed by decision trees, ANN, and logistic regression with overall prediction accuracy of 80.65%, 79.85%, and 74.26% respectively. As can be seen in the per-class accuracy figures, the prediction models did significantly better on predicting the "no" class with the well-balanced data than they did with the unbalanced data. Overall, the three machine learning techniques performed significantly better than their statistical counterpart, logistic regression.

Next, another set of experiments was conducted to assess the predictive ability of the three ensemble models. Based on the tenfold cross-validation methodology, the information fusion type ensemble model produced the best results with an overall prediction rate of 82.10%, followed by the bagging type ensembles and busting type ensembles with overall prediction rates of 81.80% and 80.21% respectively.

Table 2.5 Prediction Results for tenfold Cross-Validation with Balanced Dataset

		ANN(MLP)		DT(C5)		SVM		LR	
		No	Yes	No	Yes	No	Yes	No	Yes
Confusion	No	2309	464	2311	417	2313	386	2125	626
Matrix	Yes	781	2626	779	2673	777	2704	965	2464
	SUM	3090	3090	3090	3090	3090	3090	3090	3090
Per-class Accuracy		74.72%	84.98%	74.79%	86.50%	74.85%	87.51%	68.77%	79.74%
Overall Accuracy		79.85%		80.65%		81.18%		74.26%	

Table 2.6   Prediction Results for the Three Ensemble Models

	Boosting (Boosted Trees)		Bagging (Random Forest)		Information Fusion (Weighted Average)	
	No	Yes	No	Yes	No	Yes
No	2242	375	2327	362	2335	351
Yes	848	2715	763	2728	755	2739
SUM	3090	3090	3090	3090	3090	3090
Per-class Accuracy	72.56%	87.86%	75.31%	88.28%	75.57%	88.64%
Overall Accuracy	80.21%		81.80%		82.10%	

(See Table 2.6 for a complete list of results for the ensembles.) Even though the prediction results are slightly better than the individual models, ensembles are known to produce more robust prediction systems compared to a single-best prediction model (Delen, 2015).

In addition to assessing the prediction accuracy for each model type, a sensitivity analysis was conducted on the developed models to identify the relative importance of the independent variables (i.e., the predictors). In realizing the overall sensitivity analysis results, each of the four individual model types generated its own sensitivity measures ranking all of the independent variables in a prioritized list. Each model type generated slightly different sensitivity rankings of the independent variables. After collecting all four sets of sensitivity numbers, the sensitivity numbers were normalized and aggregated into a single table (see Table 2.7).

Using the numerical figures from Table 2.7, a horizontal bar chart is created to pictorially illustrate the relative sensitivity or importance of the independent variables (see Figure 2.8). In Figure 2.8, the y-axis lists the independent variables in the order of sensitivity or importance from top (most important) to bottom (the least important) while the x-axis shows the aggregated relative importance of each variable.

The x-axis denotes the normalized relative importance measure for independent variables.

Discussion and Conclusions

The results show that, given sufficient data with the proper variables, predictive analytics methods can predict freshmen student attrition with approximately 80% accuracy. Results also showed that, regardless of the prediction model employed,

Table 2.7 Aggregated Sensitivity Analysis Results

Variable Name	Ann	DT	SVM	LR	Sum
YearsAfterHighSchool	0.0020	0.0360	0.0030	0.0040	0.0450
Age	0.0085	0.0360	0.0000	0.0010	0.0455
HighSchoolGpa	0.0050	0.0360	0.0060	0.0000	0.0470
HighSchoolGraduationMonth	0.0070	0.0360	0.0030	0.0010	0.0470
StartingTerm	0.0075	0.0360	0.0000	0.0040	0.0475
Sex	0.0110	0.0360	0.0000	0.0010	0.0480
ConcentrationSpecified	0.0130	0.0360	0.0000	0.0000	0.0490
MajorDeclared	0.0065	0.0360	0.0000	0.0080	0.0505
ReceivedFallAid	0.0125	0.0360	0.0030	0.0010	0.0525
TransferredHours	0.0080	0.0360	0.0030	0.0080	0.0550
SatHighReading	0.0080	0.0360	0.0000	0.0120	0.0560
SatHighComprehensive	0.0216	0.0360	0.0000	0.0000	0.0576
SpringFederalWorkStudy	0.0220	0.0360	0.0000	0.0010	0.0590
ClepHours	0.0210	0.0360	0.0030	0.0010	0.0610
SatHighScience	0.0230	0.0360	0.0000	0.0040	0.0630
PermanentAddressState	0.0270	0.0360	0.0000	0.0010	0.0640
FallGrantTuitionWaiverScholarship	0.0280	0.0360	0.0000	0.0000	0.0640
FallFederalWorkStudy	0.0240	0.0360	0.0030	0.0080	0.0710
SatHighEnglish	0.0216	0.0360	0.0030	0.0180	0.0786
SatHighMath	0.0224	0.0360	0.0060	0.0200	0.0844
Ethnicity	0.0460	0.0385	0.0060	0.0010	0.0915
AdmissionType	0.0610	0.0360	0.0000	0.0010	0.0980
MaritalStatus	0.0800	0.0360	0.0060	0.0010	0.1230
FallStudentLoan	0.0700	0.0360	0.0200	0.0000	0.1260
FallRegisteredHours	0.0490	0.0360	0.0180	0.0300	0.1330

(Continued)

Table 2.7 (*Continued*) Aggregated Sensitivity Analysis Results

Variable Name	Ann	DT	SVM	LR	Sum
SpringGrantTuitionWaiverScholarship	0.0605	0.0360	0.1100	0.1800	0.3865
FallGpa	0.1800	0.0550	0.1750	0.0000	0.4100
SpringStudentLoan	0.0750	0.0360	0.1500	0.1900	0.4510
EarnedByRegistered	0.1100	0.0430	0.5100	0.5100	1.1730

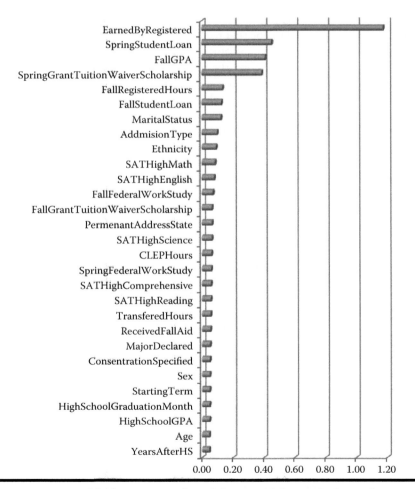

Figure 2.8   A graphical representation of the sensitivity analysis results.

the balanced dataset (compared to the unbalanced or original dataset) produced better prediction models for identifying the students who are likely to drop out of college prior to their sophomore year. Among the four individual prediction models used in this study, SVMs performed the best, followed by decision trees, neural networks, and logistic regression. From the usability standpoint, despite the fact that SVMs showed better prediction results, one might choose to use decision trees because compared to SVMs and neural networks, they portray a more transparent model structure. Decision trees explicitly show the reasoning process of different prediction outcomes, providing a justification for a specific outcome, whereas SVMs and ANN are mathematical models that do not provide such a transparent view of how they do what they do.

Recent trends in forecasting are leaning toward using a combination of forecasting techniques (as opposed to one that performed the best based on the test dataset) for a more accurate and more robust outcome. That is, it is a good idea to use these three models together for predicting the freshmen students who are about to drop out, as they confirm and complement each other. These types of models are often called ensembles. In this study, we developed prediction models using three main types of ensembles: bagging, busting, and information fusion. It is shown that the prediction results of ensembles are better that those of the individual ones. As mentioned before, the advantage of using an ensemble is to have not only slightly better prediction results but also a prediction system that is robust in its predictions.

Successful student retention practices at the institutional level follow a multi-step process, which starts with determining, storing (in a database), and using student characteristics to identify the at-risk students who are more likely to drop out, and ends with developing effective and efficient intervention methods to retain them. In such a process, data mining can play the critical role of accurately predicting attrition as well as explaining the factors underlying the phenomenon. Because machine learning methods (such as SVMs, ANN, and decision trees) are capable of modeling highly nonlinear relationships, they are more appropriate techniques to predict the complex nature of student attrition with a high level of accuracy.

The success of an analytics project relies heavily on the richness (quantity and quality) of the data representing the phenomenon under consideration. Even though this study used a large sample of data (covering 5 years of freshman student records) with a rather rich set of features, more data and more variables can potentially help improve the predictive modeling results. These variables, which are mentioned in recent literature as important, include the student's social interaction (being a member of a fraternity or other social groups), the student's prior expectation from his educational endeavors, and the student's parents' educational and financial background. Once the initial value of this quantitative analysis is realized by the institution, new and improved data collection mechanisms can be put in place to collect and potentially improve the analysis results.

As the sensitivity analysis of the trained prediction models indicate, the most important predictors for student attrition are those related to the past and present educational success of the student and whether they are getting financial help. To improve the retention rates, institutions may choose to enroll more academically successful students, and provide them with financial assistance. Also, it might be of interest to monitor the academic experience of freshmen students in their first semester through looking at a combination of grade point average and the ratio of completed hours over enrolled hours.

The focus (and perhaps the limitation) of this study is the fact that it aims to predict attrition using institutional data. Even though it leverages the findings of the previous theoretical studies, this study is not meant to develop a new theory; rather, it meant to show the viability of predictive analytics methods as a means to provide an alternative to understanding and predicting student attrition at higher education institutions. From the practicality standpoint, an information system encompassing these prediction models can be used as a decision aid to student success and management departments who are serious about improving retention.

Potential future directions of this study include (1) extending the predictive modeling methods and ensembles with more recent techniques such as rough set analysis and meta-modeling, (2) enhancing the information sources by including the data from survey-based institutional studies (which are intentionally crafted and carefully administered for retention purposes) in addition to the variables in the institutional databases, and (3) deployment of the system as a decision aid for administrators to assess its suitability and usability in the real world.

References

Berger, J.B. and Braxton, J.M. (1998). Revising Tinto's interactionalist theory of student departure through theory elaboration: Examining the role of organizational attributes in the persistence process, *Research in Higher Education* 39(2): 103–119.

Berger, J.B. and Milem, J.F. (1999). The role of student involvement and perceptions of integration in a causal model of student persistence, *Research in Higher Education* 40(6): 641–664.

Breiman, L. (2001). Random forests, *Machine Learning* 45(1): 5–32.

Breiman, L., Friedman, J.H., Olshen, R.A., and Stone, C.J. (1984). *Classification and Regression Trees*. Monterey, CA: Wadsworth and Brooks/Cole Advanced Books and Software.

Cabrera, A.F., Nora, A., and Castaneda, M.A. (1993). College persistence: Structural equations modeling test of an integrated model of student retention, *Journal of Higher Education* 64(2): 123–139.

Caison, A.L. (2007). Analysis of institutionally specific retention research: A comparison between survey and institutional database methods, *Research in Higher Education* 48(4): 435–449.

Chapman, P., Clinton, J., Kerber, R., Khabaza, T., Reinartz, T., Shearer, C., and Wirth, R. (2000). CRISP-DM 1.0 step-by-step data mining guide. https://www.the-modeling-agency.com/crisp-dm.pdf (accessed January 2018).

Cristianini, N. and Shawe-Taylor, J. (2000). *An Introduction to Support Vector Machines and Other Kernel-based Learning Methods*. London, UK: Cambridge University Press.

Deberard, S.M., Julka, G.I., and Deana, L. (2004). Predictors of academic achievement and retention among college freshmen: A longitudinal study, *College Student Journal* 38(1): 66–81.

Delen, D. (2010). A comparative analysis of machine learning techniques for student retention management, *Decision Support Systems* 49(4): 498–506.

Delen, D. (2011). Predicting student attrition with data mining methods, *Journal of College Student Retention: Research, Theory & Practice* 13(1): 17–35.

Delen, D. (2015). *Real-World Data Mining: Applied Business Analytics and Decision Making.* Upper Saddle River, NJ: FT Press.

Delen, D., Sharda, R., and Kumar, P. (2007). Movie forecast guru: A web-based DSS for Hollywood managers, *Decision Support Systems* 43(4): 1151–1170.

Delen, D., Tomak, L., Topuz, K., and Eryarsoy, E. (2017). Investigating injury severity risk factors in automobile crashes with predictive analytics and sensitivity analysis methods. *Journal of Transport & Health* 4: 118–131.

Fayyad, U., Piatetsky-Shapiro, G., and Smyth, P. (1996). From data mining to knowledge discovery in databases. *AI Magazine* 17(3): 37–54.

Gansemer-Topf, A.M. and Schuh, J.H. (2006). Institutional selectivity and institutional expenditures: Examining organizational factors that contribute to retention and graduation, *Research in Higher Education* 47(6): 613–642.

Hastie, T., Tibshirani, R., and Friedman, J. (2001). *The Elements of Statistical Learning: Data Mining, Inference, and Prediction.* New York: Springer.

Hermaniwicz, J.C. (2003). *College Attrition at American Research Universities: Comparative Case Studies.* New York: Agathon Press.

Hornik, K., Stinchcombe, M., and White, H. (1990). Universal approximation of an unknown mapping and its derivatives using multilayer feed-forward network, *Neural Networks* 3: 359–366.

Kohavi, R. (1995). A study of cross-validation and bootstrap for accuracy estimation and model selection, in *the Proceedings of the 14th International Conference on AI (IJCAI)*, San Mateo, CA: Morgan Kaufmann, pp. 1137–1145.

Lemmens, A. and Croux, C. (2006). Bagging and boosting classification trees to predict churn, *Journal of Marketing Research* 43(2): 276–286.

Miller, T.E. and Herreid, C.H. (2010). Analysis of variables: Predicting sophomore persistence using logistic regression analysis at the University of South Florida, *College and University* 85(1): 2–11.

Miller, T.E. and Tyree, T.M. (2009). Using a model that predicts individual student attrition to intervene with those who are most at risk, *College and University* 84(3): 12–21.

Piatetsky, G. (2014). CRISP-DM, still the top methodology for analytics, data mining, or data science projects, KDnuggets. Retrieved from http://www.kdnuggets.com/2014/10/crisp-dm-top-methodology-analytics-data-mining-data-science-projects.html

Quinlan, J. (1986). Induction of decision trees, *Machine Learning* 1: 81–106.

Quinlan, J. (1993). *C4.5: Programs for Machine Learning.* San Mateo, CA: Morgan Kaufmann.

Saltelli, A. (2002). Making best use of model evaluations to compute sensitivity indices, *Computer Physics Communications* 145: 280–297.

Saltelli, A., Tarantola, S., Campolongo, F., and Ratto, M. (2004). *Sensitivity Analysis in Practice—A Guide to Assessing Scientific Models.* Hoboken, NJ: John Wiley & Sons.

Sharda, R., Delen, D., and Turban, E. (2017). *Business Intelligence, Analytics, and Data Science: A Managerial Perspective* (4th ed.). London, UK: Pearson Education.

Shearer, C. (2000). The CRISP-DM model: The new blueprint for data mining, *Journal of Data Warehousing* 5: 13–22.

Thammasiri, D., Delen, D., Meesad, P., and Kasap, N. (2014). A critical assessment of imbalanced class distribution problem: The case of predicting freshmen student attrition, *Expert Systems with Applications* 41(2): 321–330.

Thomas, E.H. and Galambos, N. (2004). What satisfies students? Mining student opinion data with regression and decision tree analysis, *Research in Higher Education* 45(3): 251–269.

Tinto, V. (1993). *Leaving College: Rethinking the Causes and Cures of Student Attrition.* (2nd ed.). Chicago, IL: The University of Chicago Press.

Veenstra, C.P. (2009). A strategy for improving freshman college retention, *Journal for Quality and Participation* 31(4): 19–23.

Wilson, R.L. and Sharda, R. (1994). Bankruptcy prediction using neural networks, *Decision Support Systems* 11: 545–557.

Chapter 3

Transforming Knowledge Sharing in Twitter-Based Communities Using Social Media Analytics

Nicholas Evangelopoulos, Shadi Shakeri, and Andrea R. Bennett

Contents

> "I suggest that collective intelligence be taken seriously as a scientific and social goal" (Gruber, 2008, p. 5).

Introduction

In the early years of knowledge-based systems development, two influential approaches emerged within the artificial intelligence community. The first approach was that of expert systems, where knowledge was acquired by engineers through copious interviews with subject-matter experts. Typically, the engineer would spend as many as 100–200 hours performing the interviews, to collect the raw material needed to build a knowledge base that represented the declarative and procedural knowledge of the human experts in a format that facilitated its retrieval (e.g., Diederich, Ruhmann, & May, 1987; Gonzalez & Dankel, 1993; Rich, Knight, & Nair, 2009, p. 427; Sharda, Delen, & Turban, 2018, p. 13). A second approach was rule-induction systems, where algorithms processed training sets of examples and searched automatically for patterns that could be coded or quantified (e.g., Gonzalez & Dankel, 1993, p. 62; Rich et al., 2009, p. 355). This approach had an inherent advantage: interaction between knowledge engineers and human experts was no longer necessary. All that was necessary was a good algorithm, a fast machine, and a good amount of training data. Eventually, this approach led to the explosion of machine learning, data mining, and stock-trading bots. However, as the effort to acquire knowledge from individual experts was side-tracked, a new opportunity became practically possible: the acquisition of socially constructed, tacit knowledge from human communities and organizations.

In the era of social media, Big Data, and the internet of things, social media communities fully operate in cyberspace, their primary function is contributing to a social media document collection (corpus), and the knowledge of their members is inscribed in the corpus in ways that do not make it readily available without some processing. Social media analytics (SMAs) offer the opportunity to probe a social media corpus, extract its inscribed knowledge, and transform its tacit elements to explicit. In this chapter, we present a systematic approach to this probing and we demonstrate its application using vignettes that involve various Twitter-based communities (TBCs).

In many organizational environments, especially in the corporate world, the knowledge holder tends to hoard knowledge for the purposes of increasing his or her position in the organizational power structure (De Long & Fahey, 2000; Riege, 2005). However, as organizations understand the benefits of knowledge retention, they feel the need to provide incentives for knowledge transfer. While in the corporate domain, such a transfer may be a challenge, in the domain of social media communities, where social media platforms provide an environment for people to engage in knowledge sharing as a community practice, knowledge transfer is easier to address. In this chapter, we propose a dimensional model for the design of a data warehouse for the knowledge base, arguing that the underlying individual tacit knowledge and the corresponding collective tacit knowledge hidden in a user-generated social media corpus can be converted into explicit knowledge through the use of advanced analytics techniques. While a traditional data warehouse stores business facts that reflect what is important to a business, such as sales dollar amounts, inventory levels, or account balances, our data warehouse for community-based knowledge stores community knowledge facts, such as topics of interest and opinions that reflect what is important to a community. And, while traditional data warehouse facts are compiled by running queries on a transactional database, our knowledge base facts are compiled by running analytics on a community corpus. After such a knowledge base is built, it can be used by community members for a variety of purposes, such as raising awareness of a public health issue, promoting a movement, or participating in civic engagement. It can also be used by organizations that wish to include social media sites in their strategy for building a knowledge organization or simply for building an attractive workplace that is mindful of organizational communities and their needs. The vignettes we present in the next sections provide examples of the analytics that can help a community or organization move closer to these goals.

Collective Knowledge within Communities of Practice

The advancement of information technologies has revolutionized the way individuals communicate. The web has evolved from a static information-sharing environment to a platform where users can interact, collaborate, accomplish common goals, and collectively create knowledge. For instance, social networking sites, such as Twitter, allow users to form communities and create community-level knowledge through socialization, dialog, and conversation. Engaging in communities is inevitable, because they not only connect individuals of similar interests and build relationships that will result in feelings of belonging and mutual commitment (Wenger, 1998), but they also allow us to construct meaning and negotiate it with other members of our communities. Nonaka (1991) argues that "socialization is a rather limited form of knowledge creation" (p. 99), because interaction between members of a community allows the members to socially construct reality and constantly reshape it through discourse. This view subscribes to the constructionist perspective, which proposes that sense-making, or meaning-making, takes place when individuals constantly construct and reconstruct their understanding of the world through social interactions and relationships (Berger & Luckmann, 1966). Thus, community-level knowledge reflects the world-view of the whole community, as opposed to an individual member of that community.

Learning is integral to communities. Wenger, McDermott, and Snyder (2002) argue that "community creates the social fabric of learning" (p. 28), and the social theory of learning suggests that learning occurs through practice, community engagement, meaning creation, and identity development. In other words, after developing a sense of belonging to a specific community, individuals learn through engagement (i.e., direct experience of the world), imagination (i.e., their individual and collective images of the world), and alignment (i.e., coordinating their local activities with other processes) (Wenger, 1998). These three modes of belonging determine the degree to which an individual must engage in learning (Wenger, 2000) to remain competent within a particular context. Several learning theories emphasize the importance of context in learning. For example, the theory of situated learning posits that learning occurs in context (is situated) and is the process of participating in a community of practice (CoP) (Lave, 1988, 1991; Lave & Wenger, 1991).

Humans, as social creatures, belong to CoPs, which are among the basic building blocks of social learning systems (Davenport & Hall, 2002). Collaborative activities within a CoP result in the creation of a large body of language artifacts. Jubert (1999), in this regard, describes CoPs as "a flexible group of professionals, informally bound by common interests, who interact through interdependent tasks guided by a common purpose thereby embodying a store of common knowledge" (p. 166). CoPs present a context in which individuals can interact, build relationships, establish norms and mutual trust, and collectively accomplish tasks. As technologies have advanced, CoPs have evolved beyond their traditional boundaries (e.g., a physical classroom or office space). Recently, online CoP platforms have become more commonly used for collaborative work and social engagement. As a result, concepts such as computer-supported cooperative work (CSSW) or computer-supported cooperative learning (CSCL) have emerged in the literature (e.g., Grudin, 1994; Palmer & Fields, 1994). Research on these concepts aims to demonstrate the way online platforms should be designed to support CSSW and learning. The design of such platforms may include developing user-friendly interfaces, addressing the social dynamics of community activities, standardizing various processes, and coordinating the complex interactions between multiple tasks performed by multiple user communities.

As discussed above, communicating and collaborating at a community level, whether via physical or online CoPs, lead to the creation of collective knowledge (collective intelligence). Collective knowledge is created through an aggregation of many individual user contributions through participation and collaboration (Larsson, Bengtsson, Henriksson, & Sparks, 1998; Gruber, 2008). Collective knowledge systems (CKSs) are social ecosystems in which human (i.e., the intelligent user) and computer components interact, and such interaction leads to the accumulation of collective knowledge. Gruber (2008) defines CKSs as machines that collect and harvest a large amount of human-generated knowledge via the queries built by the human user. Such systems offer capabilities, including content sharing, cocreating, coediting, and construction of knowledge that reflect the collective intelligence of the users. Intelligent users are critical to CKSs, as they not only create content, which is later aggregated for generating collective intelligence, but also provide intelligent feedback to the system (e.g., which queries where the most effective at addressing the problem at hand).

Web 2.0 tools, referred to as collective intelligence tools, such as social networking applications, facilitate the creation of learning communities, which have specific socio-cultural contexts, and the production of collaborative knowledge. This knowledge is embedded in collections of artifacts generated by the community members as they interact and collaborate. The collections can be analyzed to glean insights that point to underlying elements of community knowledge (Gunawardena et al., 2009). The process of analyzing these collections can generate various intelligent solutions. The macro-micro information framework proposes that the knowledge obtained from the analysis of social media contents can be used for informing unintended users at both macro- (e.g., society or community) and micro- (e.g., individual) levels. For example, opinions expressed in Twitter postings are used to predict stock market value. Complex content analysis techniques (e.g., document summarization, topic extraction, and pattern recognition, etc.) are necessary for the analysis of the aggregated content (Evangelopoulos, Magro, & Sidorova, 2012).

Evolution of Analytics in Knowledge Management

Traditionally, knowledge management (KM) experts have concerned themselves with finding new methods of creating and maintaining corporate intellectual capital to meet the needs of the society (Wiig, 1997a). Effective KM processes require knowledge infrastructure (KI). KI may consist of "technology, structure, and culture along with a process architecture of acquisition, conversion, application, and protection" of knowledge (Gold, Malhotra, & Segars, 2001, p. 185). KI permits organizations to create, capture, and accumulate tacit knowledge in a repository or knowledge base. The knowledge base assists individuals within an organization in solving or addressing emerging and future problems, making evidence-based decisions and predictions, and creating innovative ideas later used for developing new products and services.

Data analytics tools and techniques, as part of KI, have improved access to rich sources of data and have improved knowledge discovery. They have also contributed to the development of knowledge bases extracted from vast amounts of unstructured, messy data found on social media and the web. The insights obtained from Big Data through the application of data or text analytics have enhanced the production of innovation and helped in sustaining the organization in the constantly changing business environment. Therefore, the emergence of analytics offers new avenues for knowledge managers for creating, disseminating, adopting, and applying the knowledge that was otherwise unusable. Enhanced with analytic tools, KM can support the decision-making process within an organization.

Figure 3.1 depicts the evolution of KM. It is around the early 2000s that modern analytics, such as data and text mining (Tsui, 2003), enter the fold. These tools, such as those employed in the studies featured in this chapter, enable analysts to discern meaningful patterns and associations within data (including words and phrases). Data and text mining tools are important for businesses seeking to engage in direct marketing, implement customer-relationship management applications, and generate business intelligence, because their outcomes can be utilized in processes such as decision-making, content management, and matching customer segments to products and services. Though Figure 3.1 begins in the 1970s, the evolution of KM is much older, emerging in early economies based on natural resources (Wiig, 1997b), as indicated by the emphasis on and appreciation of the knowledge held and employed by master craftsmen (e.g., blacksmiths,

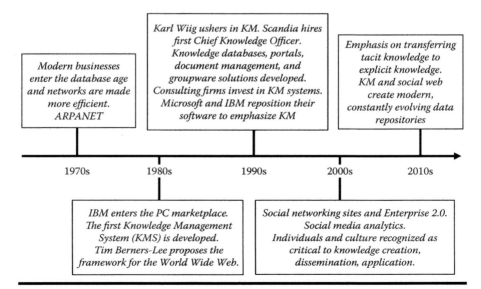

Figure 3.1 Evolution of knowledge management.

masons, tailors, etc.) and members of trade guilds. This specialized emphasis on knowledge remained largely consistent until the late twentieth century (where the time line of Figure 3.1 begins), when the mass availability of information technology (IT) systems meant that business leaders had more control over the efficiency of their firms' manufacturing, marketing, and logistics networks. Extensive information gathering on customers and other firms, and the ability to store this information in databases, led to business practices such as point-of-sale analysis and total quality management. The 1980s and the 1990s saw an overall shift in business emphasis away from products, manufacturing, and skills and toward the use of knowledge and other intellectual capital for increasing market share (Wiig, 1997b). This emphasis encouraged many organizations to pursue KM strategies, including working collaboratively with customers to understand their wants and needs.

The emergence of social networking sites (SNSs) in the early 2000s enabled individuals to build and maintain larger social networks, including with others whom they knew online and in person; self publish their ideas and opinions; and collaborate with others globally to generate and manage knowledge (as evidenced by the coproduction of Wikipedia entries) (Hemsley & Mason, 2013). These SNS-associated abilities mean that SNSs enable users to engage in collective sense making, construction of meaning, and maintenance of collective knowledge. The current and ongoing tools of social media analysis (SMA) allow for a contemporary understanding of knowledge as an object, a process, and as access to information. SMA allows for the conversion of tacit knowledge to an explicit knowledge object that can be stored, updated, and referenced. SMA allows for the process of tacit knowledge creation, sharing, distribution, and usage to be explicitly traced across social networks. And, because SMA enables the conversion of tacit knowledge to explicit, knowledge holders can oversee who has access to this codified knowledge.

Figure 3.2 depicts the evolution of business intelligence and analytics (BI&A) terminology since the 1970s, when the primary focus of such applications was the provision of structured reports that business leaders could reference during

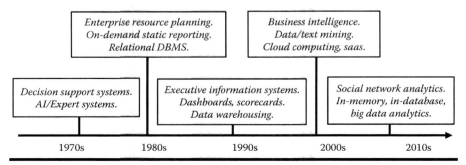

Figure 3.2 Evolution of decision support, business intelligence, and analytics.

decision-making (Sharda et al., 2018, p. 13). Though such reports—which might be generated daily, weekly, monthly, quarterly, or annually—were useful, managers of increasingly complex organizations needed more data. This need led to the advent of decision support systems (DSSs), which aimed to combine data and models to enable decision makers to solve unstructured problems. The early 1980s saw these models evolve into "rule-based expert systems" (Sharda et al., 2018, p. 13), which utilized heuristics to codify experts' knowledge for computer processing and warehousing. Businesses also began to capture organizational data differently in the 1980s, when enterprise resource planning (ERP) systems first entered the fray. Also during this time, relational database management (RDBM) replaced traditional sequential, nonstandardized data representation schemas. "These systems made it possible to improve the capture and storage of data, as well as the relationships between organizational data fields, while significantly reducing the replication of information" (Sharda et al., 2018, p. 14).

With the 1990s came the need for more versatile reporting and the development of executive information systems (EISs), graphics-based dashboards that provided executives with visually appealing snapshots of the most important key performance indicators (KPIs) for decision-making (Sharda et al., 2018, p. 14). Data warehouse (DW) repositories were also developed to maintain the integrity of Business Intelligence (BI) systems, while also enabling highly visual EIS reporting. Over a relatively short period, most medium-to-large firms had adopted EIS and DW.

The use of "BI systems" to refer to DW-driven DSSs emerged in the 2000s because the capabilities of these online repositories had evolved to handle the ever-changing storage needs of modern businesses (Sharda et al., 2018, p. 14). Specific DW evolutions were given names such as "real-time data warehousing" and "right-time data warehousing," which reflect vendors' and organizations' emphasis on the need for fresh data for decision-making. Data and text mining became popular in the new millennium, largely because of the advancements in DW. Specifically, because DWs could store massive amounts of data, companies began to recognize the benefits of mining that data for bits of institutional knowledge that could be used to improve business practices and processes. The mining of this data, in turn, created a need for even more data storage and processing power, problems that large businesses had the resources to solve but left smaller firms by the wayside. The needs of these small-to-medium firms led to the development of "service-oriented architecture and software and infrastructure-as-a-service analytics business models" (Sharda et al., 2018, p. 14), which provided organizations with as-needed access to analytics capabilities, whereby they were billed only for the services they used. Among the most challenging and interesting tools available to contemporary organizations are social media and social networking (Sharda et al., 2018, p. 15). Terms such as Big Data convey the complexity of the rich, unstructured information provided by these sites, in addition to the vast challenges inherent in the analysis of this data.

This chapter views a modern-day overlap in the evolutions of KM and BI&A, a convergence that has been steadily increasing since the turn of the twenty-first century. As depicted in Figures 3.1 and 3.2, the early 2000s brought to KM Enterprise 2.0 and its associated social networking sites and SMAs. Also during this time, individuals became recognized as critical to knowledge creation, dissemination, and application, while BI analysts began to mine DWs for the knowledge created by organizational members. Today, KM's emphasis on the conversion of tacit to explicit knowledge means that businesses aim to store such knowledge in constantly evolving DWs. All of these phenomena coincide with SMA, as undertaken by the vignettes presented here. SMA can infiltrate Enterprise 2.0 documents, uncover their underlying tacit knowledge, and expose overarching KPIs, including authors, times, locations, opinions, and themes, that might be useful to decision makers.

Social Media Analytics

Today's business climate is complex and in constant transformation. This mobility demands access to new insights for predicting the future and for making well-informed decisions. To support such decisions, the application of complex tools and methodologies (i.e., data analytic techniques) for deriving the collective knowledge of the corpus is paramount. BI refers to the application of analytic techniques to turn a business into an intelligent entity, which is capable of thinking, predicting, and making evidence and data-driven decisions. BI is a framework of various elements (e.g., architecture, tools, databases, analytical tools, etc.) for maintaining the sustainability and stability of the firm. SMA is part of the traditional BI framework; though the two might be different in scope and nature. For instance, the data used in traditional analytics is mostly structured and historical, whereas social media data is messy, vast, diverse, and real-time. (Sharda et al., 2018).

SMA is "the art and science of extracting valuable hidden insights from vast amounts of semi-structured and unstructured social media data to enable informed and insightful decision making" (Khan, 2015, p. 1). According to Sharda et al. (2018), SMA is defined as the systematic and scientific methods used for drawing "valuable insights hidden in all the user-generated content on social media sites" (p. 311). The knowledge gained from SMA techniques reflect both the individual and collective intelligence of a community. Sharda et al. (2018) explain that social media data is a "collective good" (p. 4), meaning that data on social media platforms and social technologies is user-generated, such that social interactions among people support the creation, sharing, and exchange of information, ideas, and opinions in virtual communities that are equivalent to CoPs.

Khan (2015) proposed a multi-dimensional SMA model described as seven layers of SMAs. The model implies that text, network, actions (e.g., like, dislike, share, etc.), hyperlinks, mobile, location, and search engines comprise SMA. Moreover, the model indicates that, while some of these aspects (e.g., text and actions) are easily identifiable in data, others are hidden (e.g., location and networks) and can only be uncovered by in-depth analytics. Similarly, Sharda et al. (2018) express that, by categorizing social media data into distinct groupings (i.e., keywords, content grouping, geography, time of day, and landing page), data analysts can benefit from the full potential of analytics techniques. For instance, dividing data based on the geographic locations of users enables analysts to identify the sources of other data points (e.g., topics, users, actions, etc.). Another example is that of the temporal segmentation of data, which assists in the understating of complex phenomena (e.g., the peak hours of the day for customer service requests) and in making data-driven decisions (e.g., the hours that customer service should be offered).

These discussions point to the need for a data warehouse that can support KM and SMA functions. We introduce a model for such a data warehouse later in this chapter, arguing that TBCs are developed around certain database modeling dimensions, and proposing author, time, topic, opinion (i.e., sentiment), and location as a starting set. These dimensions represent attributes of a corpus, and, at the same time, characteristics of the communities that generated the corpus. Furthermore, we recommend the process of compiling these dimensions in their physical form, as database tables, as a roadmap to the extraction of a community's collective intelligence through processing its corpus. In KM terms, the product of this process constitutes the community's collective tacit knowledge. Therefore, before we introduce our data warehouse model we provide some additional discussion on how TBCs build their collective tacit knowledge as they function as communities of practice and as organizations.

Twitter-Based Communities as Communities of Practice

We argue that TBCs are CoPs because they connect groups of individuals through shared expertise and passion for accomplishing mutual goals (Wenger & Syder, 1999). Like a CoP, a TBC is built for various reasons and can center around the creation of knowledge across diverse dimensions (Davenport, 2001). Twitter users form or join a TBC based on the common goals and interests of the community, and they engage in the practice of expanding knowledge as a collaborative activity (Gunawardena et al., 2009). Community knowledge is gradually accumulated through social interaction, experience sharing, observation, informal relationships, and mutual trust (Panahi, Watson, & Partridge, 2012), and this collective knowledge—the collective intelligence or wisdom of the crowd—can be employed for informing users beyond the community boundaries (i.e., at the broader society level). As Gruber (2008) argues, "true collective intelligence can emerge if the data collected from all those people is aggregated and recombined" to develop new insights and new learning that would otherwise be hard to achieve at the individual level (p. 5). The success in the achievement of this goal depends on the use of appropriate data analytic techniques. We present some of these techniques later in this chapter.

Twitter-Based Communities as Organizations

Even though TBCs are CoPs, they are also organizations of individuals in which shared meanings and actions play critical roles in shaping members' experiences. Being CoPs, TBCs involve communities of people with similar interests engaging in the practices of meaning construction and negotiation, learning, and producing shared repositories of collective resources (such as the TBC's corpus). In doing so, the community members form intimate ties and establish norms to develop a sense of belonging, trust, and joint enterprise. The stability of such organizations depends upon their members participating in events, exercising opinion leadership, and contributing to the formation of membership and other relationships. Central to all these functions is the creation of repositories of artifacts (i.e., the corpus) that constitute the fabric of social learning (Wenger, 2000). The construction of meaning is integral to the practice aspect of the communities and the knowledge that is collaboratively produced by their members. Wenger (1998) asserts that "practice is about meaning as an experience of everyday life" (p. 52). Similarly, Walsh & Ungson (1991) suggest that sense making and social functioning are essential aspects of one another. They view organizations as networks of "intersubjectively shared meanings that are sustained through the development and use of a common language and everyday social interaction" (p. 60).

Sense making is necessary for creating common knowledge and coordinating community actions based on the development of shared meanings (Weick, 1995, p. 17). Czarniawska-Jerges (1992) claims that "shared meaning is not what is crucial for collective action, but rather it is the experience of collective actions that is shared" (p. 188). Furthermore, Burns and Stalker (1961) emphasize the importance of "social cognition" by arguing that, in working organizations, decisions are made either as a collaborative social activity or by the knowledge provided by others' sense making and approval. Therefore, a TBC is a sense-making organization as it facilitates social interactions and the formation of social cognition (i.e., collective intelligence), shared meaning (e.g., commonly understood language artifacts), and shared action (e.g., collaborative decision-making). These functions are performed through members' participation in communal practice, as well as the accumulation of large corpora that represent the collective knowledge of the community.

Transforming Tacit Knowledge in Twitter-Based Communities

Tacit knowledge resides in individuals' minds. "Tacit knowledge is highly personal," hard to express (Nonaka, 1991, p. 98), and generated through the internalization of explicit knowledge (i.e., captured and codified knowledge such as the contents of journal papers, documents, etc.). Conversely, explicit knowledge is articulated tacit knowledge. The process of converting tacit knowledge to explicit and vice versa is explained through the knowledge spiral model, developed by Nonaka and Takeuchi (1995), which suggests that knowledge can be placed on a spectrum from completely tacit to completely explicit.

As individuals in TBCs collaborate, they develop implicit ways of experiencing and learning together (Lave & Wenger, 1991). In the process of achieving shared goals, TBC members share and merge their tacit knowledge. As a result, a collective tacit knowledge emerges (Leonard & Sensiper, 1998; Kabir & Carayiannis, 2013). In other words, the knowledge each person carries is necessarily complementary. Orchestra and sports players are good examples of how the combination of individuals' tacit knowledge could lead to innovation beyond personal knowledge (Leonard & Sensiper, 1998). Knowledge managers believe that organizations that are more successful in discovering and incorporating the unexpressed tacit knowledge are more capable of designing innovative products and services (Zappavigna-Lee, 2006). Extending these ideas (e.g., Iedema, 2003; Zappavigna-Lee & Patrick, 2004; Zappavigna-Lee et al., 2003), we propose that TBCs encompass a kind of collective tacit knowledge that, from the point of view of the knowledge engineer (or the analyst), cannot be acquired at the level of isolated, individual members. In this chapter, we discuss how one can begin to uncover collective tacit knowledge by performing analytics at the corpus level.

Representing Twitter-Based Community Knowledge in a Dimensional Model

Document management systems (DMSs) have a long history in organizations where a significant part of operations is based on collections of documents. Such organizations include not only publishers or digital libraries, but also educational institutions, insurance firms, governmental agencies, and so on. A DMS typically includes scanning, storage, and workflow routing functions (Sullivan, 2001, p. 130). As DMSs move into the realm of KM, it becomes necessary for them to adopt a document warehouse model with a well-designed architecture. This is where metadata repositories become essential elements for the organization of a document collection. In general, metadata can be broken down into four categories (Sullivan, 2001, pp. 144–150):

■ Document content metadata, such as author, title, subject/keywords, description/abstract, publication source, publisher, document type, format, language, publication date, edition, etc.

- Search and retrieval metadata, such as search keys and indices, document source, versioning, URL, source type, waiting time, username, password, directory, search engine, etc.
- Text mining metadata, such as keywords, topics, clusters, other features, summaries, opinion categories, etc.
- Storage metadata, such as store document index, summary index, URL index, pathname, translation information, etc.

SNS such as Facebook, Twitter, LinkedIn, Snapchat, and so on, as computer-mediated environments that facilitate the creation and sharing of information, base a significant part of their operations around documents and other forms of media. However, because SNS are typically open and community-based, their knowledge base is not managed as formally as that of a corporate organization. Since this type of knowledge is built around a virtual CoP, where members interact, share experiences, and learn through transactions with a corpus, we refer to the community as a corpus-based community, and its knowledge as corpus-based knowledge. The rest of this chapter focuses on Twitter, where a TBC is treated as a special type of a corpus-based community.

As mentioned earlier, within a TBC, as within any knowledge organization, knowledge consists of explicit and tacit elements. How does one begin to represent the architecture of a data warehouse that supports KM in a TBC? Figure 3.3 presents a dimensional model that includes some explicit elements. This overall layout, also known as the star schema (e.g., Adamson & Venerable, 1998, pp. 9–17; Kimball & Ross, 2002, p. 16), has a fact table in the middle, surrounded by

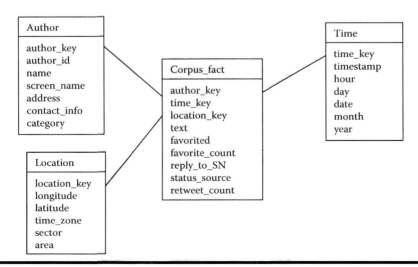

Figure 3.3   Dimensional model for a Twitter corpus.

dimension tables. The fact table records each Tweet with references to who is the author, when it was posted, and from which location. Each reference is expanded as an entry in a dedicated table, or dimension, where details for the referenced authors, locations, and time periods are provided. This dimensional model provides a platform that facilitates the efficient generation of reports and visualizations that constitute a well-established mechanism for providing organizational intelligence.

Occasionally, custom fact tables and custom dimensions are included in the dimensional model (Adamson & Venerable, 1998, p, 20). As we will show later in this chapter, our additional facts and dimensions are produced with the use of social analytics. Such knowledge might be derived days or months after the recording of the original core facts, (or, in the case of real-time analytics, ex post, that is, a fraction of a second after the core facts are captured). To present the model, we employ database management terminology to refer to new facts as derived facts and their corresponding dimensions as derived dimensions. Figure 3.4 shows the process of producing these derived facts and dimensions.

The process shown in Figure 3.4 starts with the extraction of topics or opinions expressed in Tweets, using text analytics. In a preliminary phase, the results of these analytics are stored as Tweet metadata, or, in database modeling terms, as derived attributes of the fact table. Subsequently, data warehouse extract, transform, and load (ETL) operations are used to organize topics and opinions in separate tables, so that they can be managed independently and serve as a standardized, common reference across the entire data warehouse. As shown in Figure 3.5, the data warehouse design now looks like an expanded star schema that includes derived (i.e., computed using analytics or database queries) facts, as well as derived dimensions. We will refer to this configuration as a knowledge base data warehouse (KBDW) schema.

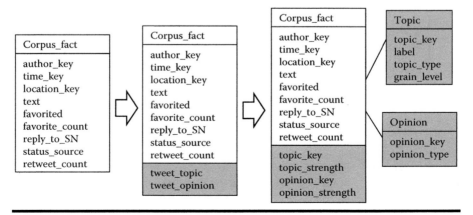

Figure 3.4   Process of producing derived facts and dimensions.

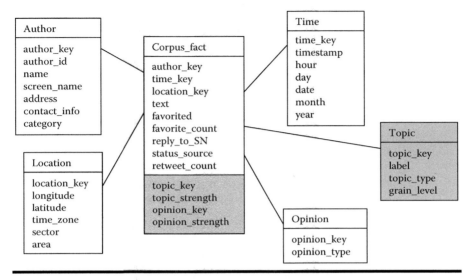

Figure 3.5 Expanded dimensional model with derived dimensions and their interactions.

Our proposed dimensional model for a KBDW offers the opportunity to create a knowledge base that can measure the community member contributions, identify time trends, locate the distribution of events and information across geographical or other types of places, and track the diffusion of sentiments and ideas across time, space, and communities. On a conceptual level, our model bears a lot of similarity with the analytic variant of a customer relationship management data warehouse (Kimball & Ross, 2002, p. 144). However, the focus of this chapter is on a corpus that reflects the activity, the accomplishments, and the knowledge of a CoP, the analytics that can uncover this knowledge, and the database models that can organize this knowledge for easy access, retention, and transfer outside the minds of the most entrenched community members. In the sections that follow, we examine various dimensions of the KBDW and their interactions. We begin with the user dimension, presented in the next section.

User Dimension

The author of a document is arguably its most important attribute. On a collective level, multiple authors produce the user dimension of a corpus. Understanding the authors includes an investigation of who they are as community members or organizational stakeholders. Authors can vary significantly by level of involvement and document authoring activity. Vignette 3.1 presents an example of how the user dimension contributes to the knowledge base of a specific TBC.

VIGNETTE 3.1   Top Users Posting Zika Tweets in April 2016

Zika is a Flavivirus-borne disease primarily communicated to humans through infected bites of the mosquito species *Aedes aegypti*. The disease also spreads through blood transfusion, sex, and from a pregnant woman to her fetus (CDC, 2017). The virus was initially isolated in the Zika forest of Uganda in a rhesus macaque monkey in 1947. While scientists have known the Zika virus for seven decades, it only recently became a public health crisis, first in the Americas and then internationally. Of the reported cases of Zika virus, about 80% are asymptomatic (Duffy et al., 2009), but in a few people, mild symptoms such as headaches, rash, joint pain, conjunctivitis, fever, and muscle pain are reported. As Zika reached epidemic levels, with a prevalence rate of 99.7 cases per 100,000 people in different parts of the world, including the United States, 1246 new cases were reported in 2015 alone.

The Zika virus turned into a public health crisis when it became associated with birth defects. Medical researchers claim that Zika infections during pregnancy can lead to various conditions for fetuses and infants, including a severe birth defect called microcephaly (CDC, 2017). The state of emergency caused by the Zika virus lasted for a year. However, after Zika cases reached a 95% reduction in April 2017, compared to the same month in 2016, the Brazilian government announced an end to the Zika crisis (The Atlantic, 2017). The World Health Organization (WHO), however, emphasizes that like all mosquito-borne diseases, Zika is a seasonal disease and may return to countries where *Aedes aegypti* mosquitoes commonly live.

Tweets about Zika were downloaded from the Twitter user interface on the web using search terms that specified Zika as a keyword and April 1–30, 2016 as the time interval. A set of 17,422 Tweets was obtained. These were produced by 7503 unique users. Table 3.1 lists the top 10 users who posted the highest number of Tweets in the data set.

Figure 3.6 shows the cumulative Tweet percentages by cumulative user rank. For example, the top 20% of Twitter users who posted messages about Zika in April 2016 contributed 60.98% of all Tweets about Zika that month, while the top 80% of users contributed 91.38%. The relationship between user rank and Tweeting frequency appears to follow a power law. Figure 3.7 shows a graph of this relationship for the top 1000 rank-ordered users.

Table 3.1 Top 10 Users Posting Zika Tweets in April 2016

Rank	User Handle (Name)	User Type and Description	Frequency	Cumulative %
1	Hniman (Henry L. Niman)	Individual; tracks and analyzes infectious diseases (e.g., H1N1, Zika, etc.).	340	1.95
2	Zika_News (Zika News)	Organization; detects and shares News about Zika virus.	292	3.63
3	TheFlaviviruses (Flavivirus Virus)	Organization; detects and share news about diseases caused by the Flaviviridae family.	227	4.93
4	Greg_folkers (Greg Folkers)	Individual; reports on infamous diseases.	205	6.11
5	AdamNeira (WorldPeace2050)	Individual; Tweets on a variety of topics, including disease pandemics, terrorism, etc.	181	7.15
6	MackayIM (Ian M. Mackay)	Individual; virologist who provides medical advice.	151	8.01
7	ironorehopper (ironorehopper)	Individual; (no information about the author was available).	136	8.79
8	Crof (Crawford Kilian)	Individual; Retired college teacher & writer.	127	9.52
9	Halt_Zika (#Halt_Zika)	Organization; affiliated with a company working on anti-Zika drugs.	92	10.05
10	CDCDirector (Dr. Anne Schuchat)	Individual; The CDC director.	90	10.57

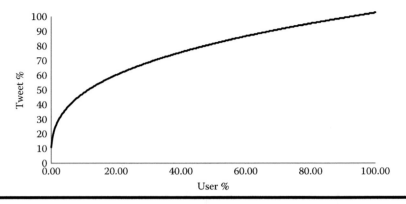

Figure 3.6   Cumulative Tweet percentages by cumulative user rank.

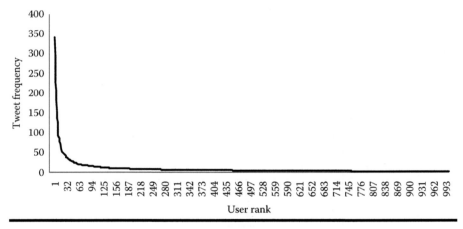

Figure 3.7   Power law in user contributions.

A careful examination of the list of the top contributors presented in Table 3.1. Based on their profiles, they are all interested in the Zika virus disease. Some of the accounts are personal and others are non-personal (group accounts). For example, Henry L. Niman (screen name: *hniman*) who posted 340 Tweets during April 2016 and ranks at the top for that month, is a medical doctor who tracks and analyzes infectious diseases such as H1N1 and Zika. Similarly, the Zika News (screen name: *Zika_News*, 292 Tweets posted in April 2016) and Flavivirus Virus (screen name: *TheFlaviviruses*, 227 Tweets posted in April 2016), ranked No. 2 and 3, respectively. Both are non-personal Twitter accounts, dedicated to the detection and distribution of Zika-related news or

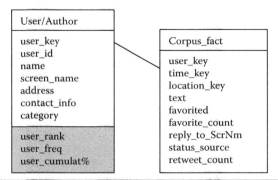

Figure 3.8   The user dimension in the data warehouse model, with added derived attributes.

information. A third category of users are non-human (bots, etc.). We assume that at the current phase of AI evolution all non-human users act as representatives of corresponding humans or human organizations. In the future, it may become possible to have completely autonomous AI agents who act on their own interests.

The analysis and the output presented here helps uncover some of the collective knowledge about who the users are, typically held by the core members of this community (the Zika TBC). The data on the most active users described in this vignette allows researchers to determine the top contributors to the TBC. This is a form of tacit knowledge that we are able to uncover through an analysis of the entire corpus. Though we can obtain the knowledge represented by Table 3.1 using spreadsheet manipulations on the entire corpus, a more efficient approach is to run structured query language (SQL) queries on a data warehouse. Figure 3.8 shows the logical design of a small segment of such a KBDW as applicable to Vignette 3.1: The derived attributes *user_rank*, *user_frequency*, and *user_cumulative%* in the User/Author dimension. Table 3.1 can now be generated by running a straightforward SQL query.

Interaction among Users

Social networks refer to a social structure composed of groups of individuals or organizations, linked to one another with some connections or relationships (Sharda et al., 2018, p. 304). Based on this definition, social network analysis (SNA) provides techniques and tools for investigating social structures by measuring network attributes, including centrality measures, density, clustering coefficient, distance, and so on. Although SNA concentrates on methodology more than theory, it enables a comprehensive understanding of users' social activities and relationships both at individual and collective levels. For instance, using SNA techniques, we can learn which users manifest the properties of opinion leaders (i.e., authorities and influential users) by studying their communication patterns. Research suggests that such individuals with high connectivity and issue involvement are successful in affecting the perceptions and attitudes of the rest of the network (e.g., Ibarra & Andrews, 1993; Xu et al., 2014).

One common characteristic of social networks is clustering or network transitivity (Mishra et al., 2007). Clustering refers to the property of two users (nodes) that are neighbors with a third user, to have an increased probability of being related (friends). In statistical terms, the clustering coefficient measures "the probability that two of one's friends are friends themselves," which is based on triples of nodes (i.e., three connected nodes). The global clustering coefficient is the number of closed triples over the total number of triples (both closed and open) in the graph. In a graph where all nodes are entirely connected to one another, the coefficient equals 1. However, in most real-world networks (e.g., social networks), the coefficient would range between 0.1 and 0.5. A local clustering coefficient, on the other hand, determines how close a particular node and its neighbors are to forming a "closed triple" (Girvan & Newman, 2002, p. 7821). Clusters or communities emerge as a subset of highly (fully) connected individuals that are loosely related to other users in a network. For a cluster to form, each user or node should have more connections within the community than with the rest of the graph (Radicchi et al., 2004). The communities typically appear as dark, dense spots in a graph representation.

Based on the above discussion, we argue that clusters may form closed social circles (i.e. cliques) where each user is directly linked to every other user in the community. Because cliques are a locked network, their members may be exposed to redundant information. Therefore, cliqued users need to connect to the rest of the network through weak ties to access nonredundant sources of information. A weak tie is a user (node or vertex) that links users from across various communities. The theory of the strength of weak ties suggests that the probability of weak ties to act as bridges between communities is greater than that of strong ties (Granovetter, 1983). Because trusted weak ties play a critical role in providing access to non-redundant information, they contribute even more positively to the transfer and receipt of useful knowledge than strong ties (Levin & Cross, 2004).

To detect communities in a social network structure, one needs to apply techniques and tools developed for performing such tasks. As a result, researchers across various domains (e.g., applied mathematics and computer science) have advanced algorithms and standards to undertake such duties. Traditionally, community detection algorithms (CDAs) compute the edge weight for every pair of users in the network graph and create a weight index. The edge weights are the calculated edge betweenness measures (i.e., the number of shortest paths between pairs of vertices), adopted from Freeman (1978)'s vertex betweenness centrality. Then, the edges are added to a network of n vertices with no edges between them in order of their weights starting with the strongest weights to the weakest (weighted index). As a result, communities begin to appear, as users with the most powerful network ties are linked to one another. Unlike the traditional community detection model, Girvan and Newman (2002) propose a method that can also determine the structure of the communities using a different clustering approach. The method focuses on identifying of the community peripheries (using the edge betweenness), rather than the ones formed between highly-connected individuals (the individuals with high betweenness centralities).

As discussed earlier, CDAs compute important network features and detect communities. As the algorithms have different capabilities in handling networks of different sizes and complexities, and are thus utilized for different purposes, their performances have to be tested and evaluated by through the use of benchmarks or benchmark graphs. The GN (for Girvan and Newman) benchmark, for example, identifies communities within very small-sized graphs by computing only specific network features (e.g., distribution, community sizes, etc.) Although most algorithms perform well on the GN benchmark graphs, they might not produce good results for extremely large, heterogamous networks with overlapping communities (Yang, Algesheimer, & Tessone, 2016).

Alternatively, the LFR benchmark, which stands for Lancichinetti, Fortunato, & Radicchi, produces artificial networks to enhance the performance of CDAs in detecting communities in large, complex networks (Lancichinetti, Fortunato, & Radicchi, 2008; Lancichinetti & Fortunato, 2009). For instance, Yang, Algesheimer, and Tessone (2016) employ LFR benchmark graphs to examine and compare the accuracy and computing time of eight CDAs (i.e., edge betweenness, fastgreedy, infomap, label propagation, leading eigenvector, multilevel, spinglass, and walktrap) available in the R package igraph. The study results indicate that the multilevel algorithm was superior to all the other algorithms on the LFR benchmark used for the analyses. When the algorithms were strictly tested for accuracy, (computing time is irrelevant for computing communities in small networks), infomap, label propagation, multilevel, walktrap, spinglass, and edge betweenness algorithms outperformed the others. However, for large networks, where computing time is a critical criterion in selecting an algorithm, infomap, label propagation, multilevel, and walktrap were revealed to be superior options.

In Vignette 3.2, we present an example of identified communities within the network of users featured in Vignette 3.1 using the leading eigenvector algorithm.

VIGNETTE 3.2   Identifying User Communities Within the Zika Network

We continue the scenario in Vignette 3.1, addressing Tweets about Zika, a Flavivirus-borne disease primarily communicated to humans through infected bites of the mosquito species *Aedes aegypti*. The analysis of the network graph formed by users' communications about the Zika virus in April 2016 (see Figures 3.9 and 3.10) sheds light on their communication patterns. Figure 3.9 depicts the users (i.e., who interacts) and the intensity of their interactions (i.e., how much information is transferred between the pair) in April 2016. For instance, the CDC is a recognized health organization that has developed a large number of connections (mostly in-degree) as it distributes news and information on the status of the Zika virus. The width of the links (edges) in the network represents the number of times the two users exchanged information between them through replying, mentioning, and Tweeting interactions. To better explain why each user is placed in a particular spot in the network, we calculated the betweenness and degree centralities of each node. Degree centrality refers to the number of connections each node has, whereas betweenness centrality denotes the number times each node lies in geodesic paths between other pairs of users. Our interpretation of the centrality measures is as follows:

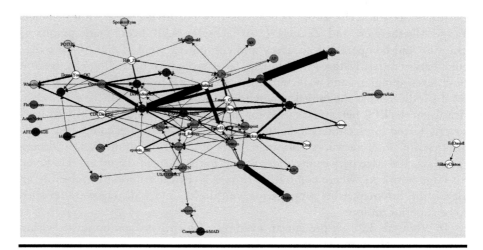

Figure 3.9   Network of users within the Zika social network in April 2016.

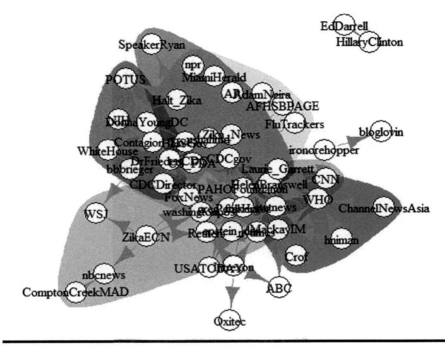

Figure 3.10   Communities of users within the Zika social network in April 2016.

The investigation of the degree centrality measures of users suggests that we can categorize network users into two distinct groupings: information authorities and information disseminators. The information authorities are the user accounts (e.g., @CDCgov, @WHO, @CDCDirector), which are highly cited as credible sources of information by other network users when disseminating information. These users, as a result, have developed many in-degree connections over time. The information disseminators (e.g., @ironore-hoper, @bloglovin), on the other hand, are the nodes with many out-degree links, because they distribute information they receive from authoritative outlets. We observe that the connections of users (i.e., the degree centrality) impact their positioning in the network. For example, the @CDCgov's many in-degree ties have positioned it almost at the heart of the network. Whereas, @bloglovin with only one out-degree connection is pulled toward the margins. Similarly, calculating the vertex betweenness centrality (as proposed by Freeman, 1978) reveals that the network has an overall low betweenness centrality. We hypothesize that this is occurring because most nodes have a betweenness of zero (i.e., these users are not connecting any user and community across the graph), with only a few having betweenness of more than 10 (e.g., @Zika_News, @PeterHotez, @HelenBranswell). Remarkably, while both degree and betweenness centrality influence the positioning of vertices in networks, nodes with high betweenness centrality might not necessarily have high degree centrality. This phenomenon agrees with the theory of the strength of weak ties, as it suggests that weak ties (i.e., information bridges) typically have a high betweenness with very few links (Granovetter, 1973). A good example of this type of user in this following network is the @CDCgov, which has a zero betweenness centrality with a relatively high degree centrality (i.e., 17).

Figure 3.10 presents the clusters (communities) of users in the graph. As defined earlier, network clusters are groups of individuals with dense internal interactions and sparse external friendships (Mishra et al., 2007). To identify the clusters, we employed various community cluster detection algorithms suggested by Yang, Algesheimer, and Tessone (2016), including edge betweenness, fastgreedy, infomap, label propagation, leading eigenvector, multilevel, spinglass, and walktrap, all implemented in the R package igraph. After obtaining the clustering results, we observed that the leading eigenvector algorithm outperformed the others. Consequently, upon using the leading eigenvector algorithm, we identified six network clusters (see Table 3.2). After analyzing the composition of each cluster, we observed that each community consists of a mixture of various stakeholder groups, (e.g., healthcare, press, politics, etc.). This observation is consistent with the stakeholder view of organizations, which posits that multiple groups of individuals are needed to influence how a CoP achieves its goals. For instance, if the community has the goal of raising awareness within the society at large by providing the most accurate, detailed information about a disease epidemic, the contribution of different types of members (stakeholders) might be necessary to increase the quality of the provided information, thus facilitating the achievement of the goal. The contribution of different stakeholders might increase the credibility of, and trust in, the shared information.

Figure 3.11 shows the logical design of a small segment of a KBDW that is applicable to Vignette 3.2. After obtaining the

Table 3.2 Network Clusters for Zika Tweets in April 2016

Rank	Member Count	Member User Handles
1	12	AdamNeira, ushahmd, AFHSBPAGE, US_FDA, npr, CDCgov, Contagion_Live, PeterHotez, bbbrieger, greg_folkers, FluTrackers
2	11	CNN, nbcnews, epstein_dan, Reuters, PAHOFoundation, statnews, ComptonCreekMAD, nytimes, FoxNews, ZikaECN, WSJ, washingtonpost
3	7	Zika_News, AP, Laurie_Garrett, MiamiHerald, Halt_Zika, HHSGov, SpeakerRyan
4	6	CDCDirector, WhiteHouse, NIH, DonnaYoungDC, DrFriedenCDC, POTUS
5	6	Crof, MackayIM, hniman, ChannelNewsAsia, WHO, HelenBranswell
6	4	Intrexon, USATODAY, Oxitec, ABC

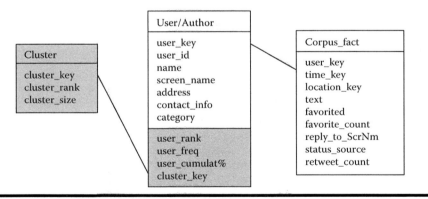

Figure 3.11   The derived cluster dimension in the data warehouse model.

size and the rank of each cluster of users through SNA, clusters are now standing on their own, represented as separate derived dimensions. Using a data warehouse with this configuration, we could generate Table 3.2 by running a straightforward SQL query that draws information from all user clusters and their associated users.

Time Dimension

The time at which a document is created not only communicates the author's thoughts at that specific date, hour, and minute, but also provides clues to the environmental context on which he or she is commenting. The time period in which documents in a collection were generated comprises the time dimension of a corpus. Examples of time-stamped documents include news stories (which are "news" only with reference to a specific point in time), emails and other communication documents, server logs, customer service calls, and social media postings.

Once the time dimension is established, various corpus statistics can be tracked over time, forming a data structure known as a time series. These include counts, averages, ranges, and so on, that can be easily obtained by executing database queries. Time series models include regression with linear and polynomial time trends, auto-regressive models, exponential smoothing models, time series decomposition, and autoregressive-integrated-moving average (ARIMA) models. For more details on these time series analytics, refer to a standard time series forecasting text, such as Bowerman, O'Connell, & Koehler (2005). Vignette 3.3 presents an example of how the time dimension contributes to the knowledge base of a TBC.

VIGNETTE 3.3   Tweets on International Women's Day 2017

International Women's Day (IWD) is held globally on March 8 of each year as a celebration of the political and social achievements of women, while also aiming to bring attention to instances of gender-based discrimination and call for their rectification (University of Chicago, 2014; Sanghani, 2017; United Nations, 2017). IWD is not affiliated with a single group; participants include members of governments, women's organizations, corporations, and charities, in addition to individual citizens (Sanghani, 2017).

After the success of the Women's March on Washington during the January 2017 inauguration of U.S. President Donald Trump—an event that *The Washington Post* called "likely the largest single-day demonstration in recorded U.S. history" (Chenoweth & Pressman, 2017) —event organizers coordinated an event called A Day Without Women to coincide with IWD 2017 (Sanghani, 2017). Utilizing social media and other outlets, the organizers encouraged women to participate in a worldwide labor strike (Sanghani, 2017; Zarya, 2017; Pruitt,

2017), to refrain from shopping, and to wear red in solidarity with the movement (Sanghani, 2017; Pruitt, 2017). These specific calls for participation, in addition to the tumultuous global environment—as many citizens were still shocked at the United Kingdom's "Brexit" from the European Union and Donald Trump's victory in the U.S. presidential election—set the scene for massive participation and demonstrations during IWD 2017 (Sanghani, 2017).

Tweets for IWD 2017 were collected via the hashtags #IWD, #IWD2017, #InternationalWomensDay, and #DayWithoutAWoman using the R package twitteR and Twitter's API interface. An hourly trend analysis revealed that at the height of the day—between noon and 5 p.m.—traffic would likely reach about 70,000 Tweets per hour; thus, a formula was generated to determine the number of Tweets that should be collected each hour to generate a representative sample. The Tweets collected for March 8, 2017, totaled 98,900, representing a 10% sample, stratified over time. Subsequently, the Tweets underwent additional cleaning and formatting processes, with the final sample of complete Tweets totaling 82,857.

The time dimension provides a snapshot of knowledge generated within specified parameters. Researchers seeking to conduct longitudinal studies in KM could trace patterns, congruences, and incongruences among information disseminated over time using such data. The time dimension also offers researchers the ability to potentially pinpoint when a specific piece of knowledge enters the corpus, in addition to the ability to trace the rate of its adoption into the knowledge base, as indicated by its being shared by various users over time. Practitioners, such as the organizers of the Day Without Women strike, can utilize the time dimension to determine the ideal timeframe for introducing an event or idea to the corpus, such that the optimal level of digital word-of-mouth marketing is generated.

Figure 3.12 shows the distribution of Tweets about IWD 2017 over the 24-hour course of March 8, 2017. As would be expected, Twitter activity climbs from the early morning to midday and is highest between 1 p.m. and 5 p.m., which is when most of the rallies and events were taking place in the United States.

Figure 3.13 shows the logical design of a small segment of a KBDW that applies to Vignette 3.3. We assume that the time grain is set at the hour level, which means that there is unique timekey for each hour of the day and all hours are listed in the timetable where the hourly frequency is added as a derived

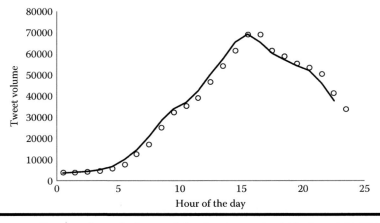

Figure 3.12 Tweet volume by hour of the day (1 = 1:00 AM, 13 = 1:00 PM, etc.).

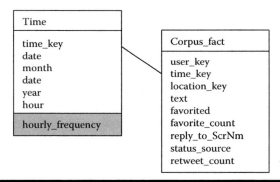

Figure 3.13 The time dimension in the data warehouse model.

attribute. Using a data warehouse with this configuration, we could generate the information depicted in Figure 3.12 by running a straightforward SQL query that draws information from the timetable.

Location Dimension

The proliferation of geographic information systems (GIS) has brought attention to the spatial dimension of various processes in the social, political, financial, and scientific domains. Efficient use of spatial data, including users' geographic location, place of origin, place of destination, or place of interest in general, help uncover and add to the knowledge base the location dimension, which can often be hidden in a corpus of collective tacit knowledge. The location at which a document is created provides not only situational and contextual clues but also allows for understanding the patterns of occurrence of certain events. Through tracking of the geographic locations of events, the stakeholder can proactively address current, emerging, and potential problems, and explain certain behaviors associated with a social event, such as a social movement or a disease outbreak. The combination of the location dimension with other documents' dimensions (e.g., authors, opinion, topics, or time) enable further decoding and extracting other hidden knowledge layers. Vignette 3.4 presents an example of how the location dimension contributes to the knowledge base of a TBC.

VIGNETTE 3.4   Zika Tweets in May and June 2017

We continue the scenario from Vignettes 3.1 and 3.2, addressing Tweets about Zika, a Flavivirus-borne disease primarily communicated to humans through infected bites of the mosquito species *Aedes aegypti*. The Zika virus grew into a public health crisis in 2015 and 2016, which demanded the development of preventive and predictive measures to control the spread of the disease. In May 2017, due to a 95% decrease in the number of newly infected individuals, the Brazilian government declared an end to the Zika virus emergency (BBC News, 2017). Even though the Zika crisis is over, health authorities are concerned about the potential of another Zika emergency; thus, asking for remaining vigilant about the disease status and dynamics. This might entail new insights provided by the analysis of various sources of data on the web and social media. Several researchers (e.g., Bodnar & Salathé, 2013; Carneiro & Mylonakis, 2009; Culotta, 2010; Salathé & Khandelwal, 2011) have employed Twitter data to improve their understanding of predicting disease occurrences and dynamics. Twitter data is a rich source for extracting the collective knowledge of TBCs, and, as such, this vignette presents mining of Twitter spatial data (i.e., Tweets' latitudes and longitudes), a form of collective tacit knowledge within a TBC.

In August 2009, Twitter began supporting the per-Tweet geotagging functionality. When this feature is activated by the user, location coordinates (i.e., latitude and longitude) are associated with the Tweet. This enables the tracking of the Tweet author's location. The spatial data is findable through the Twitter API, unless the authors opt to delete past location data from displaying in their Tweets. Since this feature is off by default, only a small portion of Tweets have the geo-location tag.

To access the collective knowledge hidden in the Twitter geographical data produced by the Zika community, we collected the Zika Tweets posted in the months of May and June 2017 using the Zika hashtag as a query keyword and Twitter API. In the data cleaning process, we removed Tweets lacking the latitude and longitude tags. As a result, we obtained a sample of 200 data points, which was equal to roughly 0.1% of all the collected Tweets.

Visualization of geo-location data and spatial analytics can be done via various packages, such as the commercial GIS package ArcGIS, that can match each data point (latitude and longitude) on a map. Figure 3.14 depicts the distribution of the location data of the 200 geo-located Zika Tweets across a world map. A similar visualization can be produced using Tableau®, a commercial analytics and visualization package. As observed in the visualization, most Tweets were created by individuals in the United States and South America (especially Brazil, the center of the epidemic), while some originate in South Africa, various European locations, and India. We postulate that the Tweets' geotags represent the countries that have been most affected by the infection, thus actively involved in the management of the crisis (e.g., the United States, Brazil, and India). India is a good example to support our hypothesis, as we believe that the Tweets originating in India contain information about the three laboratory-confirmed cases of Zika virus disease reported in the Bapunagar area, Ahmedabad district, and Gujarat state (The New York Times, 2017; WHO, 2017). These Tweets might carry important information about

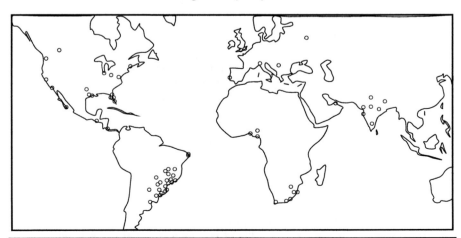

Figure 3.14   Dispersion of Tweets about Zika on a world map.

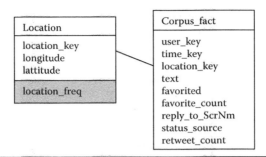

Figure 3.15   The location dimension in the KBDW model.

the preventive and protective actions that should be taken into serious consideration, as the country's climate is suitable for the thriving of the mosquitos that carry the virus.

The analysis explained above is a case-in-point of how the analysis of one dimension of a corpus allows for converting the individual tacit knowledge into the hidden collective tacit knowledge using data analytics tools. The extraction of the collective tacit knowledge permits us to obtain a deeper insight of the data (Tweets collected in May and June 2017), as we are able to explain the authors' reasoning for content creation and to better understand the dynamics of the Zika virus.

Figure 3.15 shows the logical design of a small segment of a KBDW that applies to Vignette 3.4. For KM, it is more useful to aggregate locations from the unique point level to a larger geographical area. Areas can be city blocks, neighborhoods, postal codes, counties, states, and so on. There is then a unique location_key for each area, and all areas are listed in the location table. Using a data warehouse with this configuration, we could generate tables listing various statistics for each location area unit and create, for instance, input data for diffusion analysis, by running a straightforward SQL query that draws information from the location table. An example of such a location statistic is location_freq, or location frequency, shown in Figure 3.15 as a derived attribute.

Topic Dimension

As community members communicate their ideas by exchanging oral or written documents, word usage patterns tend to exhibit characteristics of a spontaneous order: From long and seemingly flat lists of words, a structure of organized, socially constructed, corpus-level topics emerges. The modeling of such topics has taken two main approaches: the generative, or probabilistic, approach, and the variational, or statistical, approach.

The generative approach assumes that document authors are already aware of a number of topics that express issues relevant to their community, and are aware of the probability distribution of each topic across all terms in the dictionary. They then generate words for the documents they author by selecting a topic from a mix of topics that characterizes the particular document, and selecting a term from the mix of terms that characterizes the particular topic. Latent Dirichlet allocation (LDA), the most widely-cited algorithm that follows the probabilistic approach to topic modeling, uses a Bayesian approach with Markov chain Monte Carlo simulations to estimate the parameters of the multinomial distributions of topics across the documents and terms across the topics (Blei, Ng, & Jordan, 2003; Blei, 2012). The variational approach assumes that humans acquire meaning by being repeatedly exposed to multiple examples of documents containing such meaning (Kintsch & Mangalath, 2011). Because of a need to minimize cognitive effort, they create a mental space that quantifies the patterns of co-occurrence of terms within similar contexts as linear combinations, or vectors, of terms and documents. Terms and documents are then projected into that space in a lower dimensionality that drops thousands of dimensions that describe the specific document content and keeps only a few abstract concepts that describe terms and documents in a broad sense. Latent Semantic Analysis (LSA), the most widely-cited algorithm that follows the variational approach to topic modeling, uses the matrix operation of singular value decomposition to project the original term frequency matrix to a space of principal components that explain maximum variation using a minimum number of dimensions (Deerwester et al., 1990; Dumais, 2004). The two approaches seem to be somewhat complementary, since LDA explains how documents are produced from topics without explaining how topics were acquired and LSA explains how topics are acquired without explaining how they were produced. And, in practice, the two algorithms often extract very similar topics after processing the same corpus. We now continue with Vignette 3.5, where we follow the LSA approach and perform text analytics to extract topics from a corpus of Tweets.

VIGNETTE 3.5   Topics in Tweets on International Women's Day 2017

We continue the scenario in Vignette 3.3, addressing Tweets about IWD, which is held globally on March 8 of each year as a celebration of the political and social achievements of women. After compiling the set of 82,857 IWD Tweets as described in Vignette 3.3, we followed established text analytic preparation and term filtering steps (see, e.g., Coussement and Van Den Poel, 2008). Since our goal was to understand the latent semantic structure of our corpus, we implemented the factor analysis variant of LSA (Sidorova, Evangelopoulos, Valacich, & Ramakrishnan, 2008; Evangelopoulos, Zhang, & Prybutok, 2012). In our analysis, we followed the main LSA steps outlined in Kulkarni, Apte, and Evangelopoulos (2014), as well as Evangelopoulos, Ashton, Winson-Geideman, & Roulac (2015) as listed below.

- *LSA step 1, term frequency matrix:* The first step of the analysis is to compile a term frequency matrix, also known as the vector space model, by using software that parses each passage into individual words and removes trivial English terms such as "the," "of," "and," and so on, as well as low-frequency terms. For the implementation of this step, we used the commercial software package SAS® Text Miner, available free of charge to academics (SAS, 2017). Following standard text mining practice, we weighted the raw term frequencies using a transformation called inverse document frequency, which promotes the frequency of rare terms and discounts the frequency of common terms and stemmed the terms to combine terms such as organize, organized, organizes, organization, and so on. We also performed *n*-gram analysis to identify groups of words that tend to appear together as a phrase and use them as a single, composite term.
- *LSA step 2, decomposition:* The term frequency matrix was subjected to singular value decomposition (SVD). The output of this step included a set of term loadings and a set of document loadings on the extracted latent semantic dimensions.
- *LSA step 3, dimensionality selection and rotation of the dimensions:* Following Kulkarni et al. (2014) and Evangelopoulos et al. (2015), we produced a scree plot of the eigenvalues. This is shown in Figure 3.16. An "elbow" point is identified at $k = 4$, which means four topics are retained.
- *LSA step 4, factor labelling:* To understand the extracted topics, we examined the two sets of factor loadings, that is, the loadings for terms and the loadings for Tweets (documents). Each topic was then labeled based on the related high-loading terms and Tweets. Overall, the factors were easy to understand, as the high-loading Tweets for each topic were very similar to each other, and we were able to complete the labeling process without controversy.

Topic analysis for the IWD 2017 Tweets produced four overarching topics. The first topic—Here is to Strong Women—included

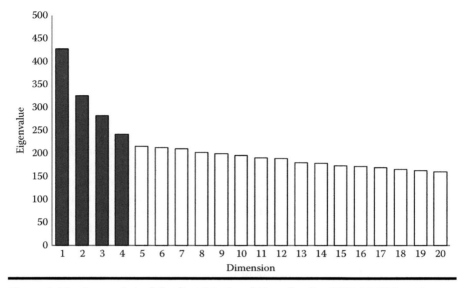

Figure 3.16   Scree plot of the first 20 eigenvalues for the IWD 2017 Tweets.

terms such as "strong," "strong woman," "know," "beautiful," and "raise." Many of the individual Tweets associated with this topic were postings of the phrase "Here's to strong women. May we know them. May we be them. May we raise them." The second topic—Celebrate Women—comprised individual Tweets that mostly related to the sentiment of IWD in general and included terms such as "celebrate," "world," "today," "good," and "amazing." The third topic—Love—was generated by various messages of support for women, which included terms such as "love," "world," "amazing," "life," and "inspire." The final topic—Wear Red in Support—directly reflects the popularity of the Day Without Women strike, as indicated by terms such as "today," "red," "wear," "support," and "work." Table 3.3 is a list of the topics, terms, and the number of documents associated with each topic.

Figure 3.17 shows the logical design of a small segment of a KBDW that applies to Vignette 3.5. Since the topics were extracted from the corpus through text analytics, table topic is modeled as a derived dimension. Topic attributes include its ID field topic_key, the label, a list of top-loading terms, and the total count of all documents in the corpus that are associated with the topic. If a data warehouse with this configuration is built, we can generate Table 3.3 by running a straightforward SQL query that simply lists the contents of the topic table. Regarding

Table 3.3   Topics and Terms for IWD 2017 Tweets and Their Associated Document Counts

Topic	Label	Top Loading Terms	Doc. Count
T1	Here is to strong women	+strong,+strong woman,+know, beautiful,+raise	4157
T2	Celebrate women	+celebrate,+world, today,+good, amazing	5641
T3	Love	+love,+world, amazing,+life,+inspire	3650
T4	Wear red in support	today,+red,+wear,+support,+work	5065

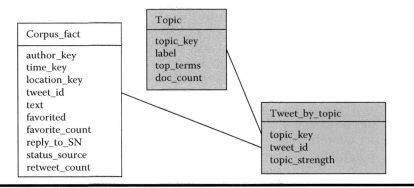

Figure 3.17   Topic as a derived dimension in the KBDW model.

the association between Tweets and topics, an initial modeling approach might record topic_key and a corresponding topic_ strength as derived attributes in the fact table. This, however, assumes that each Tweet can only be associated with one topic. In reality, Tweets may cross-load on multiple topics, which highlights one of the strengths of topic modeling compared to, say, clustering, where each document is typically assigned to a unique cluster. Therefore, Figure 3.17 shows a Tweet_By_Topic table that models the many-to-many relationship between the Corpus_Fact and Topic tables. Topic strength can be based on a factor loading scale, which shows the degree to which a Tweet is correlated with a topic, or derived in the form of a binary indicator that tags the Tweet as "relevant" to the specific topic.

Topic-Time Interaction

Topics do not have a static preference among TBC members. In a constantly changing environment, the community members adapt discourse and let the topics in the corpus wax and wane. In KBDW terms, the dynamic behavior of topics across time can be studied by considering the interaction between the topic and time dimensions. Vignette 3.6 provides an example of this interaction.

VIGNETTE 3.6   International Women's Day 2017: Tweet Topics by Hour

We continue the scenario in Vignettes 3.3 and 3.5, addressing Tweets about IWD 2017. After compiling the set of 82,857 IWD Tweets as described in Vignette 3.3, we performed topic extraction as described in Vignette 3.5. At this point, each Tweet was associated with one of the 24 hours of the day and one or more of the four extracted topics. The next step was to perform spreadsheet manipulations of the table of 82,857 Tweets that included an hour column and the 82,857-by-4 matrix that provided the Tweet-to-topic association and produced a 24-by-4 cross-tabulation of Tweet counts per hour, by topic, shown in Table 3.4. Tweets for hours 2, 4, 6, 8, and 10 were not collected, therefore count data for these hours was interpolated and then rounded to the nearest integer. Such a cross-tabulation (also known as contingency table) can help us identify certain elements of the topics' dynamic behavior by performing additional analytics. Since the columns of the table are essentially distributions of each topic across the 24 hours of the day, one analytic option is to perform chi-square tests that compare these distributions. The test is highly significant, producing a Pearson chi-square statistic equal to 259.832, and a p-value less than 0.001. This indicates that the distributions of the four topics across the 24 hours, shown in Figure 3.18 as trend lines, are significantly different.

The four trend lines shown in Figure 3.18 help us tell the story of IWD 2017: In the late morning hours, calls

Table 3.4   Cross-Tabulation of the Four Topics by Hour of the Day

Hour	T1	T2	T3	T4
0	12	25	7	17
1	10	12	10	14
2	19	48	17	32
3	27	84	23	49
4	40	95	37	57
5	53	105	50	65
6	84	153	81	96
7	114	201	112	127
8	170	331	159	210
9	225	461	205	293
10	285	501	245	329
11	345	541	285	365
12	300	416	276	399
13	283	328	234	392
14	554	756	461	762
15	278	379	257	378
16	470	600	387	534
17	222	296	220	299
18	55	61	46	58
19	218	291	191	271
20	238	281	228	244
21	255	243	195	245
22	233	277	225	263
23	265	284	238	290

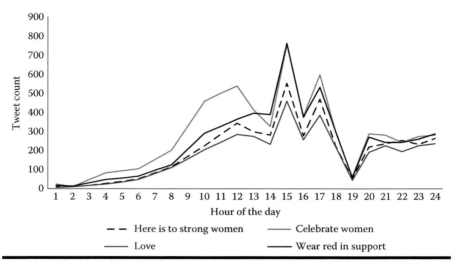

Figure 3.18 Time trends for the four IWD Tweet topics over the 24-hour day on March 8, 2017.

to celebrate women dominate the discourse, but, as we move into the early afternoon and various demonstration and celebration events unfold, calls to wear red in support take over. We expect, of course, these patterns to be common knowledge among the most organized community members who participated—and Tweeted—in IWD 2017, but that is community tacit knowledge. With the help of SMAs, we can extract this knowledge from the community corpus and make it explicit in the form of Table 3.4 and Figure 3.18.

Going back to the design of our KBDW, we note that the cross-tabulation shown in Table 3.4 can be generated in a more straightforward way if, instead of spreadsheet manipulations, we used database operations. The hour attribute of the time dimension shown in Figure 3.13 associates each Tweet with one of the 24 hours of the day, and the Tweet_By_Topic table shown in Figure 3.17 associates each Tweet with one or more of the four extracted topics. Figure 3.19 combines these segments to model the indirect association between the time and topic dimensions through the fact and tweet-by-topic tables. Table 3.4 can now be generated by running a SQL query on these tables.

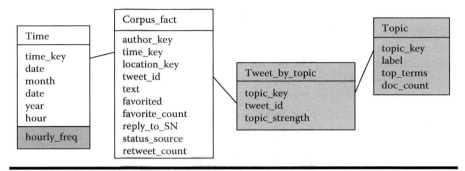

Figure 3.19 The dimensions of topic and time, and their relationship in the KBDW model.

Opinion Dimension

Textual data can generally be divided into the categories of fact or opinion, with facts defined as "objective expressions about entities, events, and their properties," and opinions defined as "subjective expressions that describe people's sentiments, appraisals, or feelings toward entities, events, and their properties" (Liu, 2010, p. 627). Sentiments, therefore, represent a specific type of textual opinion. The study of opinions is important, because individuals and organizations seek opinions (or word-of-mouth recommendations) when engaged in decision-making, and because an individual's opinion reflects the knowledge that he or she has gleaned from various experiences. The process of uncovering the sentiments conveyed by documents is called sentiment analysis or opinion mining, which is defined as "the computational study of opinions, sentiments, and emotions expressed in text" (Liu, 2010, p. 629).

Opinionated texts contain an expression on their left and either an explicit (overt) or implicit (implied) opinion on their left (Liu, 2010, p. 649). Opinions can be expressed on a gamut of things, including people, places, organizations, events, and topics. In sentiment analysis, the entity targeted by an opinion is termed its object (Liu, 2010, p. 629). Objects are comprised of components (parts) and attributes (properties), which can be further deconstructed into continuous layers of subcomponents and subattributes, such that the objects of opinionated texts can be depicted using "part of" the relationships represented by taxonomies, hierarchies, and trees. In addition to the object of the opinion, sentiment analysis seeks to uncover its orientation (i.e., whether it is positive, negative, or neutral) and its source (i.e., the opinion holder, or the person or organization that originally expresses the opinion). Analysts seeking to conduct sentiment analysis can either find public opinions about a specific object or find opinions from specific sources about an object.

The advent of the internet has been paramount in enabling the study of opinions, and the popularity of user-generated content that began with Web 2.0 has made the mass opinions conveyed via product reviews, forums, discussion groups, and blogs readily available for analysis (Liu, 2010, p. 627). Though an opinion holder's sentiment might be explicit in a product review, it is likely more implicit in longer texts and more conversational documents, such as Twitter posts. Therefore, specialized techniques and applications are required to uncover the underlying sentiments of these documents.

The feature-based sentiment analysis model (Hu & Liu, 2004; Liu, 2006; Liu, Hu, and Cheng, 2005) instructs researchers to mine direct opinions for discovering all quintuple groupings of opinion objects, object features, polarity, holders, and time and identifying synonyms for words and phrases that convey opinions that occur within the documents (Liu, 2010, p. 632). Results can then be conveyed using opinion summaries and visualized with bar graphs and pie charts (Pang & Lee, 2004). However, to successfully implement the model, researchers must also engage in sentiment classification and the generation of an opinion lexicon.

Sentiment classification seeks to classify opinionated documents as expressing positive, negative, or neutral opinions. The sentiment classification process is five-fold and consists of identifying terms and their frequencies and then tagging those terms according to their parts of speech. From there, opinion words and phrases are isolated and evaluated for both their syntactic dependency (the degree to which the interpretation of a word or phrase is dependent on its surrounding words or phrases) and the effects of any negation in the phrase (the degree to which the inclusion of a negative word, such as "not" or "but," affects whether the expressed opinion is positive or negative) (Liu, 2010, p. 638). For example, application rules dictate the following:

- Negative adjective = negative opinion
- Positive adjective = positive opinion
- Negation word + negative adjective = positive opinion
- Negation word + positive adjective = negative opinion

An opinion lexicon is the equivalent to a dictionary to which the text of the opinionated documents will be compared, and its generation depends on the identification and compilation of opinion phrases and idioms, in addition to base-type and comparative-type positive and negative opinion words (Liu, 2010, pp. 641–642). Positive-opinion words express desired states, while negative-opinion words express undesired states. Base-type opinion words are basic adjectives, such as "good" or "bad," while comparative-type opinion words express opinions of comparison or superlatives using terms such as "better," "worse," "best," and "worst," which are derived from their associated base-type words. Comparative-type opinion words are unique because they state an opinion that two objects differ from each other on a certain feature, rather than stating a simple opinion on a single object. Documents can contain four types of comparative relationships, which can be subcategorized into gradable and nongradable comparisons. Gradable comparisons express opinions of non-equality (that the qualities of one object are greater or less than those of another), equality (that the qualities of two objects are the same or similar), or superlative (that one object is the best or worst of all comparable objects in respect to a quality). Nongradable comparisons convey difference, but without any positive or negative connotations.

To collect and compile the opinion words and phrases for the lexicon, analysts may employ either a dictionary-based or corpus-based approach (Liu, 2010, p. 642). When utilizing a dictionary-based approach, researchers manually collect small sets of opinion words and then search for their synonyms and antonyms via a platform such as WordNet (Hu & Liu, 2004; Kim & Hovy, 2004). The ongoing accumulation of synonyms and antonyms expands the initial set of words and phrases, and this bootstrapping-type process continues until no new words are found. Finally, the analysts review the complete set of opinion words and phrases and manually remove or correct any errors. The implementation of the corpus-based approach begins with a list of seed opinion adjectives, which are used, along with sets of linguistic constraints and conventions regarding connectives and sentence structures, to identify additional opinion words and their orientations. The reliance upon rules of syntax means that conjunctions are especially important for establishing sentiment consistency via the corpus-based approach. For example, Tweets containing "and" indicate complimentary opinions between the phrases that the conjunction connects. On the other hand, phrases connected by words such as "or" or "but" might not convey similar opinions. Once words and phrases are collected and compiled and sentiment consistency is established, the final set is clustered on a graph to produce positive and negative subsets for the lexicon.

Opinions are expressions, which means they are necessarily conceptual; therefore, they represent concepts that opinionated text sources can express in a myriad of ways (Liu, 2010, p. 649). Thus, regardless of the lexicon-generation method employed, sentiment analysts must be aware of various linguistic rules that govern the structuring of sentences, if they are to accurately interpret the underlying sentiment that the opinionated text is expressing. For example, instead of purely negative or positive opinions, text documents might refer to deviations from an expected norm. In such instances, documents referring to an outcome that is within the range of desired values can be interpreted as positive expressions, whereas documents referring to outcomes outside of the desired range of values (either too low or too high) can be interpreted as negative expressions. Opinionated texts might also refer to increases and decreases in quantities of opinionated items, such that:

- Decrease in a negative item = Positive opinion
- Decrease in a positive item = Negative opinion
- Increase in a negative item = Negative opinion
- Increase in a positive item = Positive opinion

Though this chapter deals specifically with expressions of positive, negative, and neutral sentiment, opinion analyses, in general, can be undertaken to assess more complex phenomena, including emotions (Aman & Szpakowicz, 2007) such as joy, sadness, anger, fear, surprise, and disgust (Strapparava & Mihalcea, 2008;

Chaffar & Inkpen, 2011), political leanings (Grimmer & Stewart, 2013; Ceron et al., 2014), and media slant (Gentzkow & Shapiro, 2010).

Analyses that extend beyond opinion polarity must employ the tactics mentioned here, in addition to those for identifying, classifying, and cataloging a lexicon of words that align with the various emotions the analysts aim to uncover (Strapparava & Mihalcea, 2008). This process requires that researchers uncover not only the words that refer directly to emotional states (e.g., "fearful" or "excited"), but also the ones that denote an indirect contextual reference (e.g., words like "laugh" or "cry," which indicate emotional responses to a stimulus). Then, the researchers primarily employ the same methods for analysis. Regarding assessing political leanings (e.g., Ceron et al., 2014), there is no doubt that parties, candidates, and news media worldwide are interested in the promise of utilizing the opinions expressed on SNS to predict the outcomes of elections by forecasting voting behavior. Advances in this area even have the potential to render obsolete more traditional means of forecasting, such as public opinion polling. Finally, the modern proliferation of perceived media biases (Gentzkow & Shapiro, 2010) offers applications of opinion-analysis techniques to discern whether text-based reports actually contain a media slant, or whether such leanings are mere matters of readers' and pundits' interpretations.

Vignette 3.7 provides an example that illustrates a simple approach to opinion mining, dictionary-based, sentiment analysis.

VIGNETTE 3.7 Seattle Tweets in February 2017: Tweet Sentiment

As the residents of a certain city undertake their daily lives, social media provides outlets to express their opinions and feelings. Examining individual Tweets, positive or negative sentiment is certainly influenced by the context of the discussion. However, when Tweet sentiment is considered at a high level of aggregation, individual reasons for positive or negative feelings disappear, and what is left is the mood of an entire community, affected by social events, local problems, or local reasons to celebrate. To investigate the sentiment of Seattle residents, we used the Twitter API and the R library twitteR (Gentry, 2016) to collect all Tweets that mentioned Seattle over a period of 20 days in February 2017. A total of 113.9 K Tweets posted between 6 p.m. and midnight were collected. Of these, 6,754 Tweets, or about 6%, contained geo-location data. We scored these Tweets for sentiment using the dictionary-based sentiment analysis approach (Hu and Liu, 2004; Liu 2010). Examples of positive dictionary entries include words such as accomplishment, amazing, approve, awesome, bliss,

charming, comfort, and so on. Examples of negative dictionary entries include words such as alienate, allergy, annoying,
angry, badly, boring, careless, and so on. The 2,006 positive
entries and 4,784 negative entries compiled by Hu and Liu
(2004) were used for this purpose, and the analysis was done
in R using the Breen sentiment scoring algorithm. The results
of sentiment scoring are shown in Table 3.5 and Figure 3.20.
The sentiment scores are primarily positive.

**Table 3.5   Frequencies of Sentiment
Scores in the Geolocated Sample**

Sentiment Score	Frequency
−4	0
−3	2
−2	25
−1	265
0	3156
1	2664
2	186
3	15
4	2

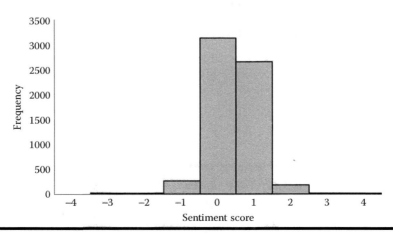

Figure 3.20   Histogram of sentiment scores in Seattle Tweets, February 2017.

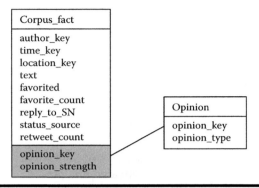

Figure 3.21 Opinion as a derived dimension in the KBDW model.

Figure 3.21 shows the logical design of a small segment of a KBDW that pertains to Vignette 3.7. Since the sentiments were extracted from the corpus through text analytics, table opinion is modeled as a derived dimension. Opinion attributes include an ID field, opinion_key, and opinion_type, which could take the value sentiment, as in Vignette 3.7, or other values, such as joy or fear, that may apply to other types of opinion analysis. The derived attribute opinion_strength, added to the fact table, reflects the degree of positive or negative sentiment, the degree of joy, and so forth. Assuming that a data warehouse with this configuration is built, we can generate the frequency distribution table that produced the histogram shown in Figure 3.20 by running a very straightforward SQL query that combines information from the contents of the fact and the opinion tables.

Opinion-Location Interaction

Since human communities tend to self-organize and cluster together in space with similar others, opinions can vary significantly across geographical locations. In KBDW terms, the spatial dynamics of opinion can be studied by considering the interaction between the opinion and location dimensions. Vignette 3.8 provides an example of such interaction.

VIGNETTE 3.8   Seattle Tweets in February 2017:
Tweet Sentiment by Geographical Sector

How does a city resident's mood vary by location? Following up on the data collection of 113.9 K Tweets that mentioned the term, Seattle in a 20-day period in February 2017, described in Vignette 3.7, we identified 6,754 Tweets (about 6% of all the collected Tweets) that contained geo-location data. Since each Tweet is already scored for sentiment as described in Vignette 3.7, a spreadsheet that listed the 6,754 Tweets together with their sentiment score and their geo-location codes was assembled. We then proceeded with spatial analytics. After some initial exploration of the longitude and latitude coordinates of the 6,754 Tweets, the boundaries of a rectangle that contains the city of Seattle were defined by the four points of NW (–122.406, 47.735), NE (–122.245, 47.735), SE (–122.245, 47.5345), and SW (–122.406, 47.5345), where NW stands for northwest and so forth. This rectangular area was subsequently divided into 80 sectors, based on a 10-by-8 grid. With the boundaries of each one of the 80 sectors defined, each geo-located Tweet was classified under one of the 80 sectors x1y1, x1y2, and so forth. The sector of this grid was used as the aggregation grain, and summary statistics were obtained for each sector. Table 3.6 lists the average sentiment score by sector, with the rows of the table being laid out from north to south and the columns from west to east so that the table places the sectors as they would appear

Table 3.6 Average Sentiment Score in the 80 Sectors in the 10-by-8 Grid of Seattle Locations

		West							East
		X1	X2	X3	X4	X5	X6	X7	X8
North	Y10			0.48	0.11	0.50	0.43		
	Y09		0.25	0.78	0.55	0.63	1.00		
	Y08	–0.25	0.38	0.21	0.00	0.25	0.38	0.00	0.00
	Y07	–0.10	0.17	0.00	0.35	0.22	0.38	0.43	
	Y06	0.06	0.44	0.30	0.50	0.50	0.82		0.25
	Y05		0.29	0.29	0.22	0.37	0.50	0.40	
	Y04			0.40	0.50	0.27	0.50	0.50	
	Y03	1.17	0.36		0.38	0.32	0.49		0.00
	Y02	0.23	0.07	0.00	–0.13	–0.08	0.22	0.43	2.00
South	Y01	0.00		0.00	0.44	0.05	0.71	0.60	0.00

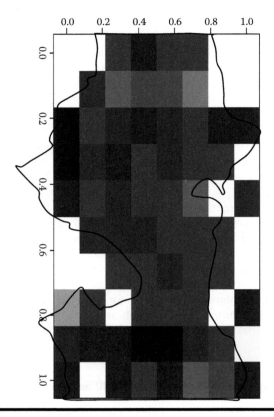

Figure 3.22   Distribution of average sentiment scores across the grid of 80 Seattle sectors.

on a map. Figure 3.22 was produced by overlaying an outline of the Seattle map on a heatmap produced in R by first reading the data in Table 3.6 as a matrix, and then executing the R function image (sentiment). Dark gray sectors indicate smaller sentiment scores and light gray sectors indicate larger (i.e., more positive) sentiment scores. White sectors indicate no sentiment score data.

Going back to the design of our KBDW, we note that Table 3.6 can be generated in a more straightforward way if, instead of spreadsheet manipulations, we used database operations. In Figure 3.23 we show a segment of a KMDW model that pertains to Vignette 3.8. The location grain is set to the sector level, and, for each sector, attributes longitude_bound1, longitude_bound2, latitude_bound1, and latitude_bound2 define the bounds of the sector. Considering the average sentiment of each location sector as a slowly changing attribute, we can accommodate it as the derived attribute location_ave_sentiment

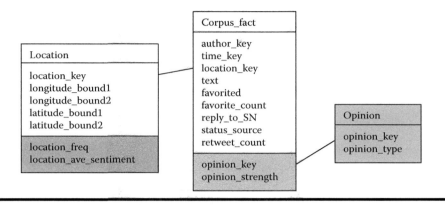

Figure 3.23   The dimensions of opinion and location, and their relationship in the KBDW model.

within the location dimension. Values for this attribute would have to be computed by executing a database query that combines information from the opinion, location, and fact tables. This is currently shown in Figure 3.23. However, if the local sentiment is to be assessed periodically, then an additional Location_By_Time table should be added, accommodating the many-to-many relationship between location and time, and recoding an opinion statistic (such the average sentiment of our Vignette 3.8) for each location-sector-time-period combination.

Conclusion and the Road Ahead

We have introduced a dimensional model approach to the design of a knowledge base for TBCs. From the theoretical point of view, we argue that a community of humans who actively communicate with each other through social media postings builds tacit knowledge that is embedded in a representative social media corpus. We have then illustrated how different kinds of SMAs, including social network analytics, time series analytics, text analytics, and spatial analytics, operate on the corpus and convert its embedded tacit knowledge into explicit, in the form of reports, tables, and visualizations. We have finally proposed a data warehouse model for a knowledge base, which starts with core data warehouse fact and dimension tables that correspond to explicit knowledge, and iteratively builds additional dimension and fact tables that uncover the community's tacit knowledge and make it accessible to the knowledge base user. With the proposed KBDW model, tacit knowledge is now made explicit, as it is stored, managed, and easily shared. Figure 3.24 presents a schematic representation of this proposed framework, where a CoP builds socially constructed knowledge and embeds it in a corpus. With the

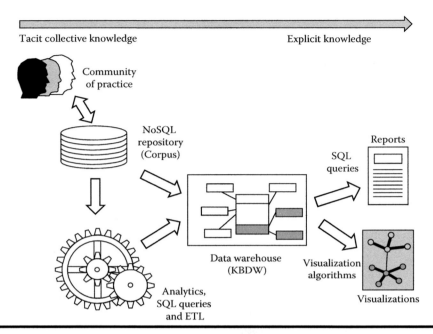

Figure 3.24 Framework for a corpus-based knowledge management system and transition from tacit collective knowledge to explicit knowledge.

help of analytics and a data warehouse architecture to accommodate the knowledge base, a corpus-based KM system can systematically transform tacit knowledge to explicit. This process is presented in Figure 3.24 as a transition from left to right.

Without the proposed KBDW dimensional model, all the analytics we have illustrated in our vignettes would produce separate reports that would include our Tables 3.1 through 3.6, stored as separate spreadsheets, and the corresponding visualizations. However, this approach would keep us in the early days of information and KM, when knowledge was stored in the form of disconnected spreadsheets and unstructured reports. A modern KM system needs a well-designed knowledge base that can be queried to produce reports on the fly. We believe that a dimensional model, such as the one we have presented in this chapter, can serve this purpose well.

The partial designs of a KBDW shown in Figures 3.8, 3.11, 3.13, 3.15, 3.17, 3.19, 3.21, and 3.23, can be extended to accommodate a variety of settings. The Twitter user can be replaced by the author of any document or the creator of any artifact. The case of Tweets examined in this chapter did not require any dedicated treatment, curation, or management of individual documents, and this is why we modeled the Tweets as transactional entries in the fact table. In different settings, such as rare document collections or art collections, an upgrade of the document entity to its own dimension might be appropriate. New dimensions can be added to account for document clusters, other types of document categories, or schools

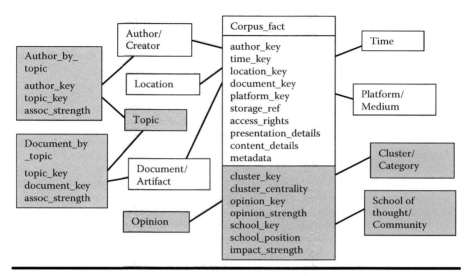

Figure 3.25 Extensions to the dimensional model for corpus-based knowledge management.

of creative thought associated with the document or artifact. Additional fact tables can account for interactions between the author and topic dimensions, or the document and topic dimensions. Some of these ideas are depicted in Figure 3.25.

We conclude this chapter by reiterating the importance of communities of practice, the collective knowledge they accumulate, and the proper management of their knowledge through well-designed KM systems. We believe that KM researchers and practitioners need to pay more attention to community knowledge. To put it in the words of Gruber, such collective knowledge should be "taken seriously as a scientific and social goal" (Gruber, 2008, p. 5). We hope our chapter has shed some light on analytic and modeling considerations as one begins to accomplish this goal.

Summary

In this chapter, we examine the knowledge that develops within a community of Twitter users. Focusing our discussion on a community of users that is built around a certain social or scientific interest, and a subset of actively involved contributors, we view such a Twitter-based community (TBC) as an online CoP. We view a corpus of Tweets produced by the TBC as the community's store of common knowledge. Using various kinds of SMAs that operate on such a corpus, we uncover the collective tacit knowledge that is embedded in it and discuss the process of its transfer to a data warehouse. We present modeling elements of this data warehouse based on the dimensions of user, time, location, topic, and opinion. We then discuss how physical database designs would include these dimensions and their interactions as database tables and how the execution of simple database queries would then transform the TBC's tacit collective knowledge into an explicit form. We include eight illustrative vignettes that examine various aspects of collective knowledge built within TBCs that discuss the Zika virus, International Women's Day, and the city of Seattle.

References

Adamson, C., & Venerable, M. (1998). *Data warehouse design solutions*. New York, NY: John Wiley & Sons.

Aman, S., & Szpakowicz, S. (2007). Identifying expressions of emotion in text. In V. Matoušek and P. Mautner (Eds.), *Text, speech and dialogue, lecture notes on artificial intelligence* (Vol. *4629*, pp. 196–205). Berlin, Germany: Springer-Verlag.

The Atlantic. (2017). Brazil declares an end to its Zika health emergency. Retrieved from https://www.theatlantic.com/news/archive/2017/05/brazil-ends-zika-emergency/526509/.

BBC News. (2017). Zika virus: Brazil says emergency is over. *BBC News*, May 12, 2017. Retrieved from http://www.bbc.com/news/world-latin-america-39892479.

Berger, P.L., & Luckmann, T. (1966). *The social construction of reality*. New York, NY: Random House.

Blei, D.M. (2012). Probabilistic topic models. *Communications of the ACM 55*(4), 77–84.

Blei, D.M., Ng, A., & Jordan, M. (2003). Latent dirichlet allocation. *Journal of Machine Learning Research, 3*, 993–1022.

Bodnar, T., & Salathé, M. (2013). Validating models for disease detection using twitter. *Proceedings of the 22nd International Conference on World Wide Web Rio de Janeiro, Brazil* (pp. 669–702). doi:10.1145/2487788.2488027.

Bowerman, B.L., O'Connell, R., & Koehler, A. (2005). *Forecasting, time series, and regression* (4th ed.). Stamford, CT: Thomson Learning.

Burns, T., & Stalker, G. M. (1961). *The management of innovation*. London, UK: Tavistock.

Carneiro, H.A., & Mylonakis, E. (2009). Google trends: A web-based tool for real-time surveillance of disease outbreaks. *Clinical Infectious Diseases, 49*(10), 1557–1564.

CDC (2017). [Zika] Symptoms. *Centers for disease control and prevention*. Retrieved from https://www.cdc.gov/zika/symptoms/symptoms.html.

Ceron, A., Curini, L., Iacus, S.M., & Porro, G. (2014). Every Tweet counts? How sentiment analysis of social media can improve our knowledge of citizens' political preferences with an application to Italy and France. *New Media & Society, 16*(2), 340–358.

Chaffar, S., & Inkpen, D. (2011). Using a heterogeneous dataset for emotion analysis in text. In C. Butz & P. Lingras (Eds.), *Advances in artificial intelligence, lecture notes in artificial intelligence* (Vol. *6657*, pp. 62–67). Berlin, Germany: Springer-Verlag.

Chenoweth, E., & Pressman, J. (2017, February 7). Analysis: This is what we learned by counting the women's marches. *Washington Post*. Retrieved from https://www.washingtonpost.com/news/monkey-cage/wp/2017/02/07/this-is-what-we-learned-by-counting-the-womens-marches/.

Coussement, K., & Van Den Poel, D. (2008). Improving customer complaint management by automatic email classification using linguistic style features as predictors. *Decision Support Systems, 44*(4), 870–882.

Culotta, A. (2010). Towards detecting influenza epidemics by analyzing Twitter messages. *Proceedings of the First Workshop on Social Media Analytics, USA.* doi:10.1145/1964858.1964874.

Czarniawska-Jerges, B. (1992). *Exploring complex organizations: A cultural perspective.* Newbury Park, CA: SAGE Publications.

Davenport, E. & Hall, H. (2002). Organizational knowledge and communities of practice. *Annual Review of Information Science and Technology, 36*(1), 170–227.

Davenport, E. (2001). Knowledge management issues for online organizations: Communities of practice as an exploratory framework. *Journal of Documentation, 57*(1), 61–57.

De Long, D.W., & Fahey, L. (2000). Diagnosing cultural barriers to knowledge management. *Academy of Management Executive, 14*(4), 113–127. doi:10.5465/AME.2000.3979820.

Deerwester, S., Dumais, S., Furnas, G., Landauer, T., & Harshmans, R. (1990). Indexing by latent semantic analysis. *Journal of the American Society for Information Science, 41*(6), 391–407.

Diederich, J., Ruhmann, I., & May, M. (1987). KRITON: A knowledge-acquisition tool for expert systems. *International Journal of Man-Machine Studies, 26*, 29–40.

Duffy, M.R., Chen, T.H., Hancock, W.T., & Hayes, E.B (2009). Zika virus outbreak on Yap Island, Federated States of Micronesia. *The New England Journal of Medicine, 360*, 2536–2543. doi:10.1056/NEJMoa0805715.

Dumais, S.T. (2004). Latent semantic analysis. *Annual Review of Information Science and Technology, 38*, 189–230.

Evangelopoulos, N., Ashton, T., Winson-Geideman, K., & Roulac, S. (2015). Latent semantic analysis and real estate research: Methods and applications. *Journal of Real Estate Literature, 23*(2), 355–380.

Evangelopoulos, N., Magro, M., & Sidorova, A. (2012). The dual micro/macro informing role of social network sites: Can twitter macro messages help predict stock prices? *Informing Science: The International Journal of an Emerging Transdiscipline, 15*, 247–268. Retrieved from http://www.inform.nu/Articles/Vol15/ISJv15p247-268Evangelopoulos0630.pdf.

Evangelopoulos, N., Zhang, X., & Prybutok, V.R. (2012). Latent semantic analysis: Five methodological recommendations. *European Journal of Information Systems, 21*(1), 70–86.

Freeman, L.C. (1978). Centrality in social networks conceptual clarification. *Social Networks, 1*, 215–239.

Gentry, J. (2016, August 29). *Package 'TwitteR'.* CRAN repository. Retrieved from https://cran.r-project.org/web/packages/twitteR/twitteR.pdf, accessed on July 3, 2017.

Gentzkow, M., & Shapiro, J.M. (2010). What derives media slant? Evidence from U.S. daily newspapers. *Econometrica, 78*(1), 35–71.

Girvan, M., & Newman, M. E. J. (2002). Community structure in social and biological networks. *Proceedings of the National Academy of Sciences of the United States of America 99*(12), 7821–7826. doi:10.1073/pnas.12265379.

Gold, A.H., Malhotra, A., & Segars, A.H. (2001). Knowledge management: An organizational capabilities perspective. *Journal of Management Information Systems, 18*(1), 185–214.

Gonzalez, A.J., & Dankel, D.D. (1993). *The engineering of knowledge-based systems: Theory and practice.* Englewood Cliffs, NJ: Prentice Hall.

Granovetter, M.S. (1973). The strength of weak ties. *American Journal of Sociology, 78*(6), 1360–1380.

Granovetter, M.S. (1983). The strength of weak ties: A network theory revisited. *Sociological Theory, 1*, 201–233.

Grimmer, J., & Stewart, B.M. (2013). Text as data: The promise and pitfalls of automatic content analysis methods for political texts. *Political Analysis, 21*, 267–297.

Gruber, T. (2008). Collective knowledge systems: Where the social web meets the Semantic Web. *Journal of Web Semantics, 6*(1), 4–13.

Grudin, J. (1994). Computer supported cooperative work: History and focus. *IEEE, 27*(5), 15–17.

Gunawardena, C.N., Hermans, M.B., Sanchez, D., Richmond, C., Bohley, M., & Tuttle, R. (2009). A theoretical framework for building online communities of practice with social networking tools. *Educational Media International, 46*(1), 3–16.

Hemsley, J., & Mason, R.M. (2013). Knowledge and knowledge management in the social media age. *Journal of Organizational Computing and Electronic Commerce, 23*(1–2), 138–167.

Hu, M., & Liu, B. (2004). Mining and summarizing customer reviews. *Proceedings of the ACM SIGKDD Conference on Knowledge Discovery and Data Mining (KDD)* (pp. 168–177). Seattle, WA.

Ibarra, H., & Andrews, S. (1993). Power, social influence, and sense making: Effects of network centrality and proximity on employee perceptions. *Administrative Science Quarterly 38*(2), 277–303. doi:10.2307/2393414.

Iedema, R. (2003). *Discourses of post-bureaucratic organization.* Amsterdam, the Netherlands: John Benjamins Publishing Company.

International Women's Day. (2017). About International Women's Day (March 8). Retrieved from https://www.internationalwomensday.com/About

Jubert, A. (1999). Communities of practice. *Knowledge Management, 3*(2), 1999.

Kabir, N., & Carayannis, E. (2013). Big data, tacit knowledge and organizational competitiveness. *Journal of Intelligence Studies in Business, 3*(3), 54–62.

Khan, G., F. (2015). *Seven layers of social media analytics: Mining business insights from social media text, actions, networks, hyperlinks, apps, search engine, and location data.* Lexington, KY: CreateSpace Independent Publishing Platform.

Kim, S.M., & Hovy, E. (2004). Determining the sentiment of opinions. *Proceedings of the International Conference on Computational Linguistics (COLING).* Geneva, Switzerland.

Kimball, R., & Ross, M. (2002). *The data warehouse toolkit,* (2nd ed.). New York, NY: John Wiley & Sons.

Kintsch, W., & Mangalath, P. (2011). The construction of meaning. *Topics in Cognitive Science, 3*(2), 346–370.

Kulkarni, S., Apte, U., & Evangelopoulos, N. (2014). The use of latent semantic analysis in operations management research. *Decision Sciences, 45*(5), 971–994.

Lancichinetti, A., & Fortunato, S. (2009). Benchmarks for testing community detection algorithms on directed and weighted graphs with overlapping communities. *Physical Review E, 80*, 1–9. doi:10.1103/PhysRevE.80.016118

Lancichinetti, A., Fortunato, S., & Radicchi, F. (2008). Benchmark graphs for testing community detection algorithms. *Physical Review E, 78*, 1–6. doi:10.1103/PhysRevE.78.046110

Larsson, R., Bengtsson, L., Henriksson, K., & Sparks, J. (1998). The interorganizational learning dilemma: Collective knowledge development in strategic alliances. *Organization Science, 9*(3), 285–305.

Lave, J. (1988). *Cognition in practice: Mind, mathematics, and culture in everyday life.* New York, NY: Cambridge University Press.

Lave, J. (1991). Situated learning in communities of practice. In L. Resnick, J. M. Levine, & S. D. Teasley (Eds.), *Perspectives on socially shared cognition* (pp. 63–82). Washington, DC: American Psychological Association.

Lave, J., & Wenger, E. (1991). *Situated learning: Legitimate peripheral participation.* Cambridge, MA: Cambridge University Press.

Leonard, D., & Sensiper, S. (1998). The role of tacit knowledge in group innovation. *California Management Review, 40*(3), 112–132. doi:10.2307/41165946.

Levin, D. Z., & Cross, R. (2004). The strength of weak ties you can trust: The mediating role of trust in effective knowledge transfer. *Management Science, 50*(11), 1477–1490.

Liu, B. (2006). *Web data mining: Exploring hyperlinks, contents, and usage Data.* Berlin, Germany: Springer-Verlag.

Liu, B. (2010). Sentiment analysis and subjectivity. In N. Indurkhya & F. J. Damereau (Eds.), *Handbook of natural language processing* (pp. 627–666). Boca Raton, FL: CRC Press.

Liu, B., Hu, M., & Cheng, J. (2005). Opinion observer: Analyzing and comparing opinions on the web. *Proceedings of WWW,* Chiba. Japan.

Mishra, N., Schreiber, R., Stanton, I., & Tarjan, R.E. (2007). Clustering social networks. In A. Bonato & F.R.K. Chung (Eds.) *Algorithms and models for the web-graph. WAW 2007. Lecture notes in computer science (vol. 4863).* Heidelberg, Germany: Springer.

The New York Times. (2017, June 3). India acknowledges three cases of Zika Virus. *The New York Times.* Retrieved from https://www.nytimes.com/2017/06/03/world/asia/india-zika-virus.html.

Nonaka, I. (1991). The knowledge-creating company. *Harvard Business Review 69*(6), 96–104.

Nonaka, I., & Takeuchi, H. (1995). *The knowledge-creating company: How Japanese companies create the dynamics of innovation.* New York, NY: Oxford University Press.

Palmer, J. D., & Fields, A. N. (1994). Computer supported cooperative work. *IEEE, 27*(5), 15–17.

Panahi, S., Watson, J., & Partridge, H. (2012). Social media and tacit knowledge sharing: Developing a conceptual model. *World Academy of Science, Engineering and Technology Index 64, International Journal of Social, Behavioral, Educational, Economic, Business and Industrial Engineering 6*(4), 648–655. http://scholar.waset.org/1307-6892/5672.

Pang, B., & Lee, L. (2004). A sentimental education: Sentiment analysis using subjectivity summarization based on minimum cuts. *Proceedings of the 42nd annual meeting on Association for Computational Linguistics* (p. 271). Barcelona, Spain: Association for Computational Linguistics.

Pruitt, S. (2017, March 6). The surprising history of International Women's Day. *The History Channel.* Retrieved from http://www.history.com/news/the-surprising-history-of-international-womens-day.

Radicchi, F., Castellano, C., Cecconi, F., Loreto, V., & Parisi, D. (2004). Defining and identifying communities in networks. *Proceedings of the National Academy of Sciences, 101,* 2658–2663.

Rich, E., Knight, K., & Nair, S.B. (2009). *Artificial intelligence,* (3rd ed.). New Delhi, India: Tata McGraw-Hill.

Riege, A. (2005). Three-dozen knowledge-sharing barriers managers must consider. *Journal of Knowledge Management, 9*(3), 18–35.

Salathé, M., & Khandelwal, S. (2011). Assessing vaccination sentiments with online social media: Implications for infectious disease dynamics and control. *PLOS Computational Biology, 7*(10), e1002199.

Sanghani, R. (2017, March 8). What is international women's day? *The Telegraph.* Retrieved from http://www.telegraph.co.uk/women/life/international-womens-day-2017-did-start-important/.

SAS (2017). SAS® OnDemand for Academics. Retrieved from http://support.sas.com/software/products/ondemand-academics/.

Sharda, R., Delen, D., & Turban, E. (2018). *Business intelligence, analytics, and data science: A managerial perspective,* (4th ed.). Upper Saddle River, NJ: Pearson Education.

Sidorova, A., Evangelopoulos, N., Valacich, J.S., & Ramakrishnan, T. (2008). Uncovering the intellectual core of the information systems discipline. *MIS Quarterly, 32*(3), 467–482 & A1–A20.

Strapparava, C., & Mihalcea, R. (2008). Learning to identify emotions in text. *Proceedings of the 2008 ACM symposium on Applied computing, Fortaleza, Ceara, Brazil* (pp. 1556–1560). New York, NY: ACM.

Sullivan, D. (2001). *Document warehousing and text mining.* New York, NY: John Wiley & Sons.

Tsui, E. (2003). Tracking the role and evolution of commercial knowledge management software. In C.W. Holsapple (Ed.), *Handbook of knowledge management.* Heidelberg, Germany: Springer-Verlag.

United Nations. (2017). International Women's Day March 8. Retrieved from http://www.un.org/en/events/womensday/.

University of Chicago. (2014). International Women's Day History. Retrieved from https://iwd.uchicago.edu/page/international-womens-day-history.

Walsh, J. P., & Ungson, G. R. (1991). Organizational memory. *Academy of Management Review, 16,* 57–91.

Weick K. E. (1995). *Sensemaking in organizations.* Thousand Oaks, CA: SAGE Publications.

Wenger, E. (1998). *Communities of practice: Learning, meaning, and identity.* Cambridge, MA: Cambridge University Press.

Wenger, E. (2000). Communities of practice and social learning systems. *Organization, 7*(2), 255–246.

Wenger, E., McDermott, R.A., & Snyder, W. (2002). *Cultivating communities of practice: A guide to managing knowledge.* Boston, MA: Harvard Business Press.

Wenger, E.C., & Snyder, W.M. (2000). Communities of practice: The organizational frontier. *Harvard Business Review, 78*(1), 139–145.

Wiig, K.M. (1997a). Knowledge management: An introduction and perspective. *Journal of Knowledge Management, 1*(1), 6–14. doi:10.1108/13673279710800682.

Wiig, K.M. (1997b). Knowledge management: Where did it come from and where will it go? *Journal of Expert Systems with Applications, 13*(1), 1–14.

World Health Organization. (2017). Zika virus infection—India. *World health organization,* May 26, 2017. Retrieved from http://www.who.int/csr/don/26-may-2017-zika-ind/en/.

Xu, W.W., Sang, Y., Blasiola, S., & Park, H.W. (2014). Predicting opinion leaders in Twitter activism networks: The case of the wisconsin recall election. *American Behavioral Scientist, 58*(10), 1278–1293.

Yang, Z., Algesheimer, A., & Tessone, C.J. (2016). A comparative analysis of community detection algorithms on artificial networks. *Scientific Reports, 6*(30750). doi:10.1038/srep30750.

Zappavigna-Lee, M.S. (2006). Tacit knowledge in communities of practice. In *Encyclopedia of communities of practice in information and knowledge management* (pp. 508–513). Hershey, PA: IGI Global.

Zappavigna-Lee, M.S., & Patrick, J. (2004). Literacy, tacit knowledge, and organizational learning. *Proceeding of the 16th Euro-International Systematic Functional Linguistic Workshop.* Madrid, Spain.

Zappavigna-Lee, M.S., Patrick, J., Davis, J., & Stern, A. (2003). Assessing knowledge management through discourse analysis. *Proceedings of the 7th Pacific Asia Conference on Information Systems.* Adelaide, South Australia.

Zarya, V. (2017). A brief but fascinating history of International Women's Day. *Fortune,* March 7, 2017. Retrieved from http://fortune.com/2017/03/07/international-womens-day-history.

Chapter 4

Data Analytics for Deriving Knowledge from User Feedback

Kuljit Kaur Chahal and Salil Vishnu Kapur

Contents

Introduction

User feedback plays a critical role in the evaluation and enhancement of the quality of any organizational setup. User feedback is the meaningful information given with the purpose of sharing experiences, suggesting improvements, and expressing opinions regarding a system. In modern times, in the electronic online ecosystems, gathering user feedback is not a tedious process. However, there is a need to identify suitable methods to discover and interpret meaningful patterns and knowledge from these huge datasets. Getting and then acting on user feedback is very important for running an organization successfully in the present day competitive market. In this chapter, we will identify the role of analytics in understanding user feedback of an organizational setup.

The Merriam-Webster dictionary defines feedback as the transmission of evaluative or corrective information about an action, event, or process to the original or controlling source. User feedback is the reaction of a user after using a service or product. The reaction can be positive, negative, or neutral depending upon the user's experience with the service or product. Given the proliferation of the Internet and social media, user feedback can be instantly captured. People now have unprecedented opportunities to raise their opinions on public platforms. Public opinion is useful and right for organizations as they can get to know about the success and failure of their policies and products from the user feedback easily available on social media.

However, most of the time, people express their opinions indirectly. Using natural languages, they mix facts with their opinions (Pawar et al. 2016). A fact is an objective expression regarding an event, experience, policy, or product, whereas an opinion is a subjective expression (of emotion, perceptions, or perspectives). Analysis of this subjective expression of the opinion holder is a key consideration while mining knowledge from publicly shared discourses. Opinion mining, or sentiment analysis, is the discipline that combines the techniques from different fields such as natural language processing, text mining, and information retrieval to segregate facts from opinions as they appear in unstructured text. Extracting user opinion from billions of user posts and then analyzing it to use it for decision-making is an insurmountable task, which needs automatic, rather than manual, ways to accomplish it successfully.

In the past, investigations reported in the research literature in this area have been largely restricted to reviews of business establishments such as hotels and restaurants, or electronic products such as cameras, mobile phones, and so on (Hridoy et al., 2015). Nowadays, there is tremendous interest in gauging public sentiment during elections to understand their response toward electoral issues and interest in different political parties and politicians (Mohammad et al., 2015). However, there is a need to analyze public response regarding government policies or other issues of public interest. Government policies sometimes draw a lot of flak from the general public. Therefore, it will be interesting to watch the Twitter stream to see the public opinion regarding a government policy. Data analytics of user feedback on social media may indicate the public's responsiveness as well as their interest in a government's policies. In this chapter, we analyze public sentiment as expressed on Twitter regarding the Indian government demonetizing high currency notes in November, 2016 to curb corruption. We stress the importance of extracting knowledge from user feedback, and then integrating and reusing this knowledge from the crowd in the decision-making process.

The study looks at what is known and not known about user feedback and seeks to identify the user feedback processes as a valuable resource for improved understanding of user needs for better decision-making. This chapter suggests an approach by which progress toward better integration of user feedback within the decision-making process through a knowledge management system in an organization can be realized. Such improvements will assist policy-makers because knowledge obtained from user feedback can be preserved as an asset for future use with the help of a knowledge management system.

The major goals of this chapter are as follows:

1. To propose a three-step approach for integrating user feedback into the knowledge management system of an organization to support policy-makers in better decision-making.
2. To demonstrate the proposed approach by applying it to the process of understanding public response to a change in a government policy. This example is regarding a sudden change in currency (demonetization) by the Indian government in November 2016.

The rest of the chapter is organized as follows: The next section discusses possible ways to collect user feedback; the third section gives details of opinion mining and data mining as techniques for user feedback analysis; the fourth section presents the existing work for user feedback analysis; our proposed approach is mentioned in the fifth section; and the last section concludes the chapter.

Collecting User Feedback

User feedback can be obtained through explicit or implicit means. In explicit feedback, a user can provide ratings, opinions, or comments through questionnaires or interviews or post them on social media and blogs. Implicit feedback involves observing user actions to understand the user's interaction with a system. We discuss the different ways of collecting user feedback in the following paragraphs.

■ *Observing user actions*: This is a kind of implicit feedback in which a user's feedback is inferred on the basis of his or her actions while interacting with a system. An online shopping system may track the purchase history of a user. Frequent buyers may be considered users with positive feedback. Another example can be when a user visits a web page time and again, and spends considerable time on the web page. The number of mouse clicks on a web page, or a usage pattern of various buttons on a page can also indicate to some extent the level of satisfaction a user may have regarding a system. However, negative feedback is not truly reflected through implicit means. The system provider may misinterpret negative feedback or may simply ignore it if it does not happen in the way he or she understands.

■ *Through questionnaires, surveys, interviews*: A user is asked to fill a survey form or answer a set of questions through an interview. Here the questions are pre-prepared. Organizations often use online or paper-based surveys to collect customer feedback. However, this fixed form approach is not an efficient approach as people find it boring and, therefore give only half-hearted responses. Moreover, only a very small number of customers can be approached. These traditional data collection methods are time consuming as well. Collecting data even of the order of thousands of responses is sometimes impractical due to the surmounting effort involved in the process. Open ended questions may receive weird answers. Moreover, results are often biased as either only positive or negative responses are possible.

■ *User response in social media networks, forums, comments sections*: Here social media acts as an interface between users and governments or organizations. People post their opinions about products or services on social networking sites such as Facebook or Twitter, and get a quick response from the concerned product or service providers. For example, recently a popular Indian television news personality Tweeted about her (bad) experience with a mobile company, which was instantly followed by posts from her followers with similar experiences. It was interesting to see an immediate response from the mobile company. It instantly reacted to the negative feedback and promptly promised to resolve the issues. Social media facilitates companies to reach out to a large customer base instantly. Opinion mining or sentiment analysis helps to reflect user response toward a product or service. Previous research shows that data obtained through social media is good indicator of public opinion, and reflects public sentiment accurately (Hridoy et al., 2015). However, with this we may miss the responses of certain population segments, such as people who are not using social media (like old people) people with limited literacy levels, or those who are less tech-savvy.

Analyzing User Feedback

Social media is a public platform where users express their opinions regarding products, services, or policies. Several online platforms such as social networks, blogs, review sites, wikis, and so on fall in the ambit of social media. Due to the arrival of Web 2.0 in 2000 and the widespread adoption of social media thereafter, users started generating huge volumes of content online. Businesses saw a lot of potential in this content for targeted marketing policies. Social media captures two types of data regarding its user base: user generated content and links or interactions between the communicating entities. A social network is modeled as a graph in which nodes represent the communicating entities (e.g., users), and edges represent the relationships or links (interactions) between the communicating entities. To understand user feedback, interested parties can analyze the content part to get the users' sentiments. Along with this, links help to gauge the spread of user sentiment through the social interactions of the user. Therefore, it is important to understand user sentiment through content analysis, augmented with link analysis to get meaningful insights. This section discusses opinion mining as an approach to content analysis followed by data mining to perform link analysis.

Opinion Mining

Opinion mining, or sentiment analysis, is an emerging tactic to analyze user feedback (Appel et al. 2015). Sentiment analysis pertains to extracting user expression from user feedback expressed in a natural language. It makes use of text analysis and computational linguistics to identify and extract attitudes expressed in user feedback. A user may give positive, negative, or neutral feedback. An opinion is expressed by a person (opinion holder) who expresses a viewpoint (positive, negative, or neutral) about an entity (target object, e.g., person, item, organization, event, policy, and service) (Khan et al., 2014). Therefore, an opinion broadly has the following parts: (1) opinion holders: people who hold the opinions; (2) sentiments: positive or negative; (3) opinion targets: entities and their features or aspects. As opinion may change with time, Liu (2011) adds a dimension of time, as well.

Therefore, an opinion must have five parts: Opinion holder, a target entity, aspects or features of the target, sentiment, and time. Appel et al. (2015) give a formal expression: an opinion is represented as a quintuple (e_j, a_{jk}, so_{ijkl}, h_i, t_l), where: e_j is a target entity, a_{jk} is an aspect or feature of the entity e_j; so_{ijkl} is the sentiment value of the opinion from the opinion holder h_i on feature a_{jk} of entity e_j at time t_l; so_{ijkl} is positive, negative, neutral, or can have more granular ratings; h_i is an opinion holder; and t_l is the time when the opinion is expressed.

Identifying all parts of an opinion quintuple from a given user discourse is a challenging problem. An opinion holder is a user who has expressed an opinion. Sometimes the opinion holder is mentioned explicitly. Otherwise, it has to be presumed to be of the author of the piece. A user may not express intent explicitly and rather use pronouns and context to relate different parts of the quintuple. Target entity can be a product or policy. Sentiment, also known as polarity of opinion, tells the orientation of an opinion, such as positive, negative, or neutral. Furthermore, an opinion can be expressed in the form of a document or a collection of sentences. It could also be that a single sentence contains a multitude of opinions on different features of an object. There is a need to develop aspect-based systems to identify sentiments in such situations (Mohammad, 2015).

Extracting an opinion from an unstructured text expressed in a natural language involves some ground work. The following steps are involved in the process of opinion mining (Appel et al., 2015):

Subjectivity classification: In the first step, a user's statements are categorized into opinionated and nonopinionated ones. An opinionated statement is the one that carries an opinion explicitly or implicitly. It is a subjective sentence. An objective sentence contains only factual information. Subjectivity classification deals with distinguishing between subjective sentences and objective sentences.

Sentiment classification: After applying the subjectivity classification function, objective sentences are dropped from further analysis. Sentiment classification focuses on subjective sentences to identify positive or negative sentiments expressed in them. In order to identify the polarity of a sentiment, several supervised, as well as unsupervised, learning methods can be employed. In supervised learning based sentiment classification, Naïve Bayesian classification algorithms, nearest neighbor algorithm, decision tree classifier, artificial neural networks, and support vector machines are the popular methods.

For identifying sentiments, creation of a sentiment resource beforehand is also necessary (Joshi et al. 2015). A sentiment resource is the knowledge base that a sentiment analysis tool can learn from. It can be in the form of a lexicon or a dataset. A lexicon is a collection of simple units such as words or phrases annotated with labels representing different sentiments. A sentiment dataset is a corpus of a higher order collection of words (e.g., sentences, documents, blogs, and Tweets) annotated with one or more sentiments. So a sentiment resource has two components: a textual unit and labels.

Annotation is the process of associating textual units with a set of predetermined labels. Sentiment lexicon annotation maps textual units with labels representing different sentiments. There are three schemes to accomplish sentiment lexicon annotation (Joshi et al. 2015): absolute, overlapping, and fuzzy. In absolute annotation, only one out of multiple labels is assigned to a textual unit. The overlapping scheme is used when labels are related to emotions. Multiple emotions may correspond to one positive (or negative) sentiment, for example, an unexpected guest or gift not only makes us happy but surprises us as well. Emotions are more complex to represent than sentiments. In the third scheme, a label is assigned on the basis of likelihood of a textual unit belonging to the label. Assigning a label using a distribution of positive: 0.8, and negative: 0.2, means that the textual unit occurs more in a positive sense but is not completely positive.

In sentiment annotated datasets, major sources are Tweets and blogs available on social media. It includes sentence-level datasets, and document (discourse)-level datasets. Existing labeling techniques include manual annotation and distant supervision. Twitter hashtags, as provided by Twitter users, can be used for the purpose of annotation in the distant supervision scheme.

An alternative to the manual approach, which is time and labor intensive, a machine learning-based system can be trained to classify an unseen corpus as per the observed sentiments in the training data. It predicts labels for the words, sentences, and documents in the unseen corpus of data. The lexicon-based methods use a predefined collection (lexicon) of words to begin with. These words are picked while keeping in mind their ability to represent different sentiments. Machine-learning classifiers can be trained to discover valuable insights in user reviews.

Mohammad (2015) refers to certain scenarios that are still challenging for the current sentiment annotation schemes: an example is assigning a single label to a sentence that portrays sarcasm. It depicts a positive emotional state of mind of the speaker who actually has a negative attitude. See Mohammad (2016) for some annotation schemes to address such issues. Furthermore, extracting sentiments from multilingual text is another challenge at this stage. Researchers have attempted to use the text after translating it to the English language (Salameh et al. 2015), and the results are encouraging, provided the translation quality is good.

Sentiment composition: Subjectivity classification and sentiment classification at the sentence level can be further aggregated to calculate document-level classification to determine if they are opinionated or nonopinionated (Van der Meer et al., 2011), and if they are opinionated, then to find a broad (document-level) positive or negative sentiment. The approach is to follow a recursive model to compose a higher order abstraction from smaller units (Dong et al., 2014). Words when put together in an order make a sentence, and multiple sentences constitute a document.

If a document contains a mixed response for an entity or object, then different features of the entity or object are identified. Aspect-based techniques are used to recognize all sentiments within a document. Positive or negative sentiments for these features are aggregated separately. But it is not a trivial task, and demands a computationally intensive solution.

Link Analysis

Data mining uses several techniques such as association rule mining, Bayesian classification algorithms, rule-based classifiers, and support vector machines to identify hidden patterns in a dataset (Han, 2006). Data mining techniques, when applied to link analysis on online social media, can provide insights into user behavior to help policy-makers understand user feedback in a more meaningful way (Barbier and Liu, 2011).

In social media, data mining helps to identify influential users. When a user with a large number of followers gives negative feedback for a policy, product, or service, the organization representatives have to be proactive to solve the problem, otherwise, it may leave a bad impression on a large user segment. The ability to identify influential users can also help in targeting marketing efforts on people with greater influence who are most likely to gain support for a policy, product, or service. Therefore, it is important to understand the factors that determine the influence of an individual in an online social network community. A simple factor can be to look at the structure of the social network that the user is a part of. For example, a user whose Twitter account has a large number of followers, or whose Tweets get a significant number of replies, or are retweeted widely may indicate popularity and influence of that individual in the community. Agarwal and Liu (2009) identify four measures to determine the influence of a blogger in a blog community. They are recognition, activity generation, novelty, and eloquence. Recognition is the number of inlinks to a blog post. Activity generation means the number of comments a blog receives. Novelty follows a blog's outlinks and the influence value of the blog post to which the outlink points to. If the outlinked blog post is an influential post, then the novelty of the current post is less. Lastly, the length of a blog post determines its eloquence.

Participation of influential users in a topic increases the longevity of the topic in the social domain. Yang and Leskovec (2011) studied temporal patterns of online content using time series clustering. They observed that the attention that a piece of content receives depends upon many factors including the participants who talk about it and the topic that it relates to.

Clustering also helps to identify a community on a social network. A social network representing a large user base can be partitioned into smaller subnetworks on the basis of similarity of user interactions thus creating small communities of users.

Closely related to data mining are two other multidisciplinary areas: digital ethnography and netnography. These approaches give a scientific description of a human society, for example, its culture or its demography. It can help to take more informed decisions if we know the culture (integrity and honesty of the contributing users to determine authenticity of the content) of the community to which the users belong, or their demography (young or old).

User Feedback Analysis: The Existing Work

In this section, our goal is to identify and report publications that contain a technique or approach to analyze user feedback.

Abookire et al. (2000) studied user feedback as a tool for improving the quality of a clinic software system that was a newly implemented computerized physician order entry system in 1998 at Massachusetts General Hospital, Massachusetts. The study analyzed end-user comments regarding the software system collected in the first year of its becoming operational. A total of 785 user comments were classified into 18 different categories to understand the software functions the users reported about. The study then analyzed the impact of this feedback on different aspects of the software system such as efficiency, cost, patient safety considerations, and so on. The authors opined that user feedback can play an important role in learning and improving the system in future. In a recent work in this domain, Vergne et al. (2013) proposed to analyze user feedback of a software project (e.g., open source projects receive feedback from the user community) for the purpose of understanding user requirements after the software project is released to end users. When the feedback is analyzed and combined with the most effective requirement suggestions for improving the software project, it helps to identify expert users that can contribute effectively to the requirements analysis.

Qiu et al. (2011) reported sentiment analysis dynamics of the online forum, American Cancer Society Cancer Survivors Network (CSN), in which the users are the cancer survivors (or their caregivers) who share their experiences of the disease and the treatments that they took. Unlike the previous research in this domain, this study applies computational techniques in collecting and analyzing user responses. The dataset comprises approximately half a million posts from the forum participants for a 10-year period of time from July, 2000 to October, 2010. The study uses machine learning classifiers to identify positive or negative sentiments reflected in the posts. It also analyzes the sentiment dynamics and contributing factors of this change over the period of time, that is, the change in sentiment of a user's posts from positive to negative or vice versa, and identifying contributing factors, such as the number of positive and negative replies to the post. The study reports that 75%–85% of users had a positive change in sentiment after learning about the experiences of others with the disease and its treatment options. However, in another study related to user feedback regarding the use of certain types of drugs, it is found that online user reviews of the drugs provided an exaggerated account of the impact of the drug treatment, and are much different from the results obtained from clinical trials (Bower, 2017). This may be due to the fact that such reviews are from people who have perhaps benefitted the most from the drugs.

Recommendation systems also use user-provided ratings of a product to recommend a product to other similar users. Collaborative filtering is the method that identifies the ratings that different user provide for a product. Products with good ratings are guessed to be eligible for recommendation to other users. Another scenario is when a user rates an item as good, other similar items (identified using content or context filtering) are also recommended to that user. Such automatic rating-based recommendation systems have applications in several areas including electronic commerce and entertainment. In electronic commerce, recommending popular items to potential buyers has been found to help in increasing sales. In the entertainment industry, movie and music ratings on the basis of user feedback help to recommend artists to a user on the basis of the user's taste. Many people look for reviews by movie critics before deciding to watch a movie in the theater. In addition to the number of reviews from movie critics for a movie, Sun (2016) takes into consideration 27 other attributes such as the number of faces in the poster, and applies data analytics to a dataset of more than 5000 movies to understand the goodness of the movies.

It is easier to find popular items from a large inventory on the basis of user feedback as the number of popular items is very small. However, it becomes difficult when the inventory is too big (e.g., online catalogues) and one has to filter to get items which obviously are not very popular (and lie in the long tail) but may be kind of "hidden jewels" that some users may like.

As the discussion in this section shows, most of the research studies apply analytics to analyze and acquire information from data. Currently, as far as we know, there is no link between user feedback and the knowledge management system in an organization. A knowledge management system may support the preservation of information acquired from user feedback as a knowledge asset, and policy-makers in the organization can use these knowledge assets in decision-making. This paper proposes a three-stage approach to create a link between user feedback and the knowledge management system of an organization.

Deriving Knowledge from User Feedback

Decision-making in an organization depends upon the ability of the decision maker to apply his or her knowledge of the past or of similar or related events to the current situation. This ability to apply knowledge is known as wisdom, and wisdom cannot be programmed. However, technology can be used to help the decision maker by providing his or her knowledge regarding the past or present events in the form of a dashboard. Knowledge is derived from information. Information is further derived from data. Data is a collection of raw discrete facts about certain events, which when processed (e.g., aggregated) create information. For example, several user ideas in a user feedback system can lead to a unified idea (see Figure 4.1).

Figure 4.1   Individual ideas aggregate to a unified idea.

Automated data processing leads to realizing knowledge-based systems. A knowledge-based system involves knowledge management. Nowadays, technology is available to process large volumes of data to obtain information, to process information to obtain knowledge, and then to put the knowledge instantly before the decision maker for better decision-making. The previous section mentioned automated support for user feedback analysis. However, in the context of using user feedback to feed a knowledge-based system of an organization, there is the need for a systematic approach to acquire user feedback and derive knowledge from it. We propose a three-stage process to extract meaning from the mounds of data and then preserve the knowledge for future use (see Figure 4.2). The first step, called data management from various sources, is to acquire, store, and (clean and) prepare the data for analysis (Labrinidis and Jagadish, 2012). Aggregating and filtering the user data for analytics is the second step in which the data is analyzed to extract and interpret the meaning. This mountain of data contains only 1% gold; the rest of the 99% is garbage. Aggregated and filtered data here actually refers to information.

Effective knowledge management is the third step in which we discuss which processes should be involved so that the 1% gold is preserved and maintained for future use. At the end, this knowledge-of-the-crowd may provide support to policy-makers in decision-making.

We demonstrate the use of data analytics for obtaining knowledge from user feedback with the help of a case of the Indian government in which the government all of a sudden introduced a major change in its currency. The Indian Prime Minister, Mr. Narendra Modi, on November 8, 2016 announced at 8:00 p.m. on national television the immediate culling of two currency notes in the denomination of Rs. 1000 and Rs. 500. This sudden move was targeted at curtailing money on the black market, and to putting an end to corruption in the country. But this "demonetization" caused the general public, dealing mostly in a cash economy, a lot of inconvenience. There was a lot of commotion on social media regarding this. People expressed their opinions on social media platforms. In the beginning, there were some voices in favor of the new government policy. But with time when people had to face several hardships like long ATM lines, and a cash crunch, they started criticizing the government. People felt that

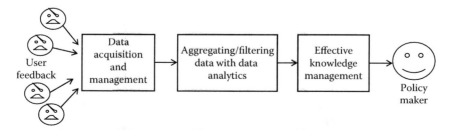

Figure 4.2 Deriving knowledge from user feedback.

implementing the new policy was not in the right spirit. However, for some of them, there was no change in opinion. They appreciated the government for this bold step.

We envisage a process model in which user feedback is first analyzed from two perspectives—from the sentiment point of view and from the user point of view. At a later stage, the knowledge pieces available in the information extracted from the previous step are preserved as knowledge assets in a knowledge management system that can further support policy-makers in decision-making. We can explain our proposed model for extracting knowledge from user feedback using data analytics with the help of a reference model (see Figure 4.3) in knowledge management as suggested by Botha et al. (2008).

As Figure 4.3 shows, the user is the main focus of the model by Botha, Kourie, and Snyman (2008). The model senses the knowledge contributed by users, organizes it, and then disseminates the knowledge thus created using the available technology stack. Similarly, in the model envisaged in this research, the user is the main focus. A government organization can sense the public mood by analyzing their response regarding a public policy on social media. Data analytics can be employed to extract knowledge from this feedback, and then use knowledge management tools to maintain the knowledge assets. Lastly, the policy-makers can use and share the knowledge repository for decision-making in the public interest.

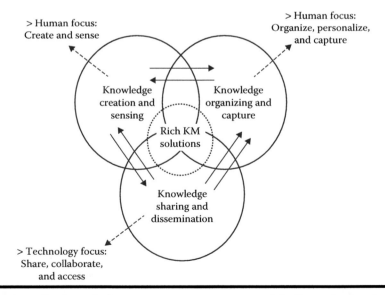

Figure 4.3   The knowledge management process model. (From Botha, A. et al., *Coping with Continuous Change in the Business Environment: Knowledge Management and Knowledge Management Technology*, Chandice Publishing, Oxford, UK, 2008.)

The major challenges involved in different stages of the proposed process model are identified as follows:

Data Management

Acquisition and recording user feedback: In this age of Web 2.0, users generate a lot of data every second; examples are Facebook, Twitter, and RSS feeds of news sources and blogs. Some social media sites such as Twitter and Facebook provide Application Programming Interfaces (APIs. An example is Twitter Auth, which helps to retrieve data from data sources. However, these sites normally put a limit on the number of API transactions per day. For example, for Twitter, a user cannot get history data beyond one to three weeks. And for Facebook, the developer application can make 200 calls per hour, per user in aggregate. This limit gives the data service companies (e.g., podargos or Gnip) a place in the data market, and they trade data as a business entity. Given the vast size of social media data available, an individual cannot afford to store volumes of data even if a crawler can collect it. Moreover, a business cannot wait for days to collect data when historical data can be purchased in minutes or hours. Data service companies lay down the infrastructure to collect data, and then offer data to researchers or businesses at the rate of thousands of dollars for some months of data. Some free data sources, such as Reddit and Kaggle, are also available.

Extracting, cleaning, and data annotation: After collecting data, some post-processing is required to validate and clean up the data. Huge amounts of subjective, opinionated, text documents are available in a dataset. To improve the quality of the input, it needs to be cleaned up by removing all the residual HTML tags; unwanted tags; stop words such as is; repeated letters such as the "I" in hiiiii; and punctuation marks. The data is often collected from different data sources, each of which may have the data in a different representation and format. So, all of this data will have to be cleaned and prepared for the process of data analysis. Annotation is the process of associating textual units with a set of predetermined labels. Tweets can be labeled positive, negative, or neutral on the basis of the words in the text. There are still challenges for annotating a Tweet that portrays sarcasm. A sample Tweet is shown below:

*"Great!! When can we expect next round of #**Demonetization**?"*

It depicts a positive emotional state of mind of a speaker who actually has a negative attitude. Furthermore, extracting sentiments from Tweets written in multilingual text is another challenge. Check this sample tweet:

"@narendramodi Shriman ji Jaali currency khatm hui kya... Aatankwad khatm hua kya... Kaladhan aaya kya... #Demonetization =D>:-? O:):-/:@"

A solution is to translate the text before using it. But the solution quality depends on quality of the translator.

Data Analytics

Such large volumes of data are worthless if they are not analyzed and used. Data analytics refers to the analysis of structured and unstructured data for the purpose of mining meaningful information (intelligence) out of millions of entries in a dataset. If we focus on data analytics on user feedback available on social media, the analytics can be put into two categories: content analytics and link analytics. In this context, content analytics corresponds to Tweet-level sentiment analysis, and in link analysis the focus is on user-level analysis.

Various metrics have emerged to extract information from user posts.

- Tweet-level Analysis
  - *Sentiment score*: Sentiment score is a numeric metric that presents the polarity of the sentiment expressed in a Tweet. A lower score indicates negative sentiment. The sentiment score can be further filtered on the basis of location and gender to understand the demographic structure of the users providing feedback.
  - *Understand the temporal change*: The time dimension considers how frequent Tweets are on a topic. Tweet frequency as a function of time may fade with time. Content on online social media is very volatile, and popularity increases in a matter of hours, and also fades away at the same speed.
  - *Change in sentiment score over a period of time*: User feedback can turn from positive to negative or from negative to positive over a period of time. Therefore, it is interesting to see the change in user response. It happened in the case of demonetization. People in favor of the move in the beginning criticized the government when they started facing the cash crunch, and had to stand in long ATM lines to get cash. They blamed the government for poorly implementing the policy. Though they consider that the move may benefit the economy in the long run, poor implementation puts them off. Sample these Tweets:

    "#Demonetization All most three months passed away 90% ATM's are still closed as no cash, PMO must intervene and solve the issue"

    "Modi is good at theory, bad in implementation. #Demonetization effect on the poor & unorganized economy is proof"

    "How much Black Money u recovered by #demonetization? who responsible for 200+ death, job loss, low GDP growth"

— *Response length*: User feedback in the form of longer posts may be more help-ful. A user who spends a long time writing a response may be more involved or have a detailed idea of the situation. But one has to look for the use of tags, or repeated words in a post. For microblogging sites like Twitter, with word limit of 140 characters, the response length cannot be long.

— *What are the commonly used words*: Use of words in a post can indicate not only the emotions, but it can also point toward the issues affecting the users. For example, the following Tweet highlights the status of the cottage industry in the wake of demonetization:

"#Demonetization turned a 700-year-old cottage industry comatose."

A word cloud can be created to indicate major issues highlighted in the user feedback. Figure 4.4a shows a word cloud created on November 16, 2016 (available at http://rpubs.com/rajeev3176). It shows many words such as "wedding" along with words expressing emotions such as anger, surprise, joy, and so on. Figure 4.4b shows a word cloud created a few months later. It shows the shift with time in the issues toward ATM and digital or cashless transactions.

■ *User analysis*: Analysis of the community structure shows links and inter-actions in users. Link analysis can help to identify community dynamics. The following metrics can be used to detect the community and the role of individuals in the community:

— *Popular users*: This metric tells how popular the users are who are Tweeting on the topic. Influence may depend on the number of followers of the user. It shows the reach of the individuals in the community.

— *Influential users*: This metric shows the influence of the users when they Tweet on a topic. Influence can be indicated by the number of replies and retweets their Tweets get.

(a)

(b)

Figure 4.4 (a) A word cloud on the day of demonetization; (b) a word cloud created after a few days.

- *Understand the spatial change*: Spatial change shows how the sentiment was in different states in a country and in the world. People living in small cities may have different opinions than people living in big cities.

However, the topic of the Tweet also plays a role in spreading the topic in the community that people start talking about it. On the other hand, participation of influential users also increases longevity of a topic in the community discussion.

Knowledge Management

The purpose of knowledge management is to ensure that the right information is delivered to the appropriate place or person for making informed decisions. User feedback is a valuable source of meaningful information for an organization or enterprise. There is a need to create a knowledge management system to capture knowledge from the information extracted from the user feedback in the previous step (explained in the previous section) of the process. The primary purpose of knowledge management is to support policy-makers in efficient decision-making. A knowledge base stores and maintains knowledge assets. A knowledge management system follows standard processes to contribute knowledge assets to the knowledge base, and to retrieve knowledge assets from the data base for reuse in unanticipated situations.

At present, many organizations have knowledge management systems in which employees share their experiences in a central repository. Other members of the organization can search the knowledge base, and reuse that knowledge in other situations. For example, Infosys employees use K-shop as the online knowledge portal available on the company intranet. Such a system is restricted to employees only. We propose to extend it to users to incorporate their feedback in the knowledge base. Every project should have an integrated model with a knowledge management system at place to learn from the current implementation.

For example, in our case study in this chapter, the government can learn a lot from the user feedback available on social media. We believe that the policy-makers had brainstormed before introducing the demonetization move to understand the impact of the policy on the general public. But if we look at the rule changes introduced by the government after the demonetization, it seems that the policy-makers learned a lot later when people actually started facing the problems. There were several changes in rules regarding cash deposits, withdrawals, and exchanges until the demonetization drive ended on March 31, 2017. Interestingly, the time period around November 8 is supposed to be auspicious for weddings in India. Several users mentioned the word in their Tweets on the very first day (see word cloud in Figure 4.4). But, an exception rule was issued on November 17 (a week later) to allow families with weddings scheduled to withdraw up to Rs 2,50,000 from their bank accounts. It shows that policy-makers were either in a hurry or could not foresee the hardships the genuine public was going to face.

This makes a case for the need of a knowledge management system as a component of the complete process that caters to the extraction of knowledge from user feedback, and then manages this knowledge for future use. Policy-makers can learn from and preserve the commonly used words in the user feedback (see Figure 4.4). In the future, when such a drive takes place again, they can consult the knowledge base and avoid repetition of the problems in future implementations.

In addition to this, expert users can also be identified on the basis of their involvement in topics and their understanding of the issues. When policy-makers work on suggestions provided in user feedback for current policies or for similar policies in the future, such users can be included in the discussion group. To select expert users as knowledge assets for a topic, there is a need to identify potential users who can contribute to that topic. For example, to discuss the impact of demonetization on "weddings," a set of suitable users can be identified and then ranked on the basis of their involvement in the topic. Most suitable users can be involved in the discussion later on.

As an example of expert users, a news editor named Narayan Ammachchi intuited almost a year ago the government's move of demonetization and Tweeted about it. His Tweet is reproduced below (source Twitter).

> *"Reserve Bank of India looks to be doing everything possible to persuade electronic payment systems such as PayTM. These are the signs that high-denomination notes, such as Rs. 500 and Rs. 1000, are on their way out. I have ling predicted that Modi government will use mobile payment system to root out corruption and curb black money once and for all."*

When the demonetization actually happened, that person was recognized as an expert to discuss the topic. Such experts can be added to the knowledge base to invite them for suggestions on similar policies.

> " ***@timesofindia*** *invited our News Editor,* ***@NAmmachchi*** *to @SpeakForIndia after he predicted* ***#India's #Demonitization*** *a year in advance"*

Conclusions and Future Work

In this chapter we discussed a process that makes it possible to derive knowledge from user feedback by using data analytics, and integrate that knowledge with the decision-making process. User feedback can be a valuable resource of knowledge for an organization. We found in the research literature that the studies in this domain just use data analytics to extract information from user feedback, which is not integrated with the decision-making process. Moreover, the majority of the existing work focuses on user reviews and ratings regarding electronic products, hotels, restaurants, movies, and so on. No study, to the best of our knowledge, focuses on user feedback regarding public policies to gauge the involvement of the general public in a government's policy decisions.

For our analysis, we chose the Indian government's demonetization move in November, 2016 as a policy decision that affected almost the entire population of the country. There was a lot of anger on the microblogging site Twitter as people complained about the poor implementation of the policy. However, the same government swept state elections post demonetization which indicates popularity of the government in a country that follows a democratic system of governance. Perhaps it points toward the digital divide in the country. Low levels of literacy combined with a low penetration of smartphones keeps a significant portion of the population away from the digital main stream. Furthermore, women and girls in rural areas are deprived from using digital technology due to the dominant ideology of patriarchy.

This study is limited to user feedback available only on the microblogging site Twitter. However, it shows a promising direction in which work can be further extended to integrate knowledge derived from user feedback with the decision-making process in a public organization.

References

Abookire, S., Martin, M., Teich, J., Kuperman, G., and Bates, D. (2000). Analysis of user-feedback as a tool for improving software quality. In *Proceedings of the AMIA Symposium*.

Agarwal, N., and Liu., H. (2009). Modeling and data mining in blogosphere. *Synthesis Lectures on Data Mining and Knowledge Discovery* (Vol. 1), pp. 1–109. https://doi.org/10.2200/S00213ED1V01Y200907DMK001.

Appel, O., Chiclana, F., and Carter, J. (2015). Main concepts, state of the art and future research questions in sentiment analysis, *Acta Polytechnica Hungarica*, 12(3), 87–108.

Barbier, G. and Liu, H. (2011). Data mining in social media. In C. Aggarwal (Ed.), *Social Network Data Analytics*, Springer, New york.

Botha, A., Kourie, D., and Snyman, R. (2008). *Coping with Continuous Change in the Business Environment: Knowledge Management and Knowledge Management Technology*, Chandice Publishing, Oxford, UK.

Bower, B. (2017). Online reviews can make over-the-counter drugs look way too effective "Evidence-based hearsay" and personal anecdotes overshadow clinical trial results, March 14, 2017, Science and the Public. Available at https://www.sciencenews.org/ accessed on April 14, 2017.

Dong, L., Wei, F., Zhou, M., and Xu, K. (2014). Adaptive multi-compositionality for recursive neural models with applications to sentiment analysis. In *Twenty-Eighth AAAI Conference on Artificial Intelligence (AAAI)*, Québec City, Québec.

Han, J. (2006). *Data Mining Concepts and Techniques*. Morgan Kaufmann, San Diego, CA.

Hridoy, S., Ekram, M. et al. (2015). Localized twitter opinion mining using sentiment analysis. *Decision Analytics* 2:8, 1–19.

Joshi, A., Ahir, S., and Bhattacharyya, P. (2015). Sentiment resources: Lexicons and datasets. In D. Das and E. Cambria (Eds.), *A Practical Guide to Sentiment Analysis*, Springer, Cham, Switzerland.

Khan, K., Baharudin, B., Khan, A., and Ullah, A. (2014). Mining opinion components from unstructured reviews: A review. *Journal of King Saud University—Computer and Information Sciences* 26, 258–275.

Labrinidis, A., and Jagadish, H. (2012). Challenges and opportunities with big data. *Proceedings of the VLDB Endowment*, 5(12), 2032–2033.

Liu, B. (2011). *Sentiment Analysis Tutorial*, AAAI-2011, University of Illinois, Chicago, IL.

Mohammad, S. M. (2015). Challenges in sentiment analysis. In E. Cambria, D. Das, S. Bandyopadhyay and A. Feraco (Eds.), *A Practical Guide to Sentiment Analysis.*, Springer International Publishing.

Mohammad, S. M. (2016). A practical guide to sentiment annotation: Challenges and solutions. In *Proceedings of the Workshop on Computational Approaches to Subjectivity, Sentiment and Social Media Analysis*. San Diego, CA.

Pawar, A. B., Jawale, M. A., and Kyatanavar, D. N. (2016) Fundamentals of sentiment analysis: Concepts and methodology. In W. Pedrycz and S. M. Chen (Eds.), *Sentiment Analysis and Ontology Engineering: Studies in Computational Intelligence*, Vol. 639. Springer, Cham.

Qiu, B., Zhaoy, K., and Mitra, P. (2011). *Get Online Support, Feel Better—Sentiment Analysis and Dynamics in an Online Cancer Survivor Community*. IEEE.

Salameh, M., Mohammad, S. M., and Kiritchenko, S. (2015). Sentiment after translation: A case-study on arabic social media posts. In *Proceedings of the North American Chapter of Association of Computational Linguistics*, Denver, CO.

Sun, C. (2016). Predict Movie Rating. Available at https://blog.nycdatascience.com/student-works/github-profiler-tool-repository-evaluation/ retrieved on April 10, 2017.

Van der Meer, J., Boon, F., Hogenboom, F., Frasincar, F., and Kaymak, U. (2011). A framework for automatic annotation of web pages using the google rich snippets vocabulary. In *26th Symposium On Applied Computing (SAC 2011), Web Technologies Track*, pp. 765–772. Association for Computing Machinery.

Vergne, M., Morales-Ramirez, I., Morandini, M. et al. (2013). Analysing user feedback and finding experts: Can goal-orientation help? In *Proceedings of the 6th International i* Workshop (iStar 2013)*, CEUR, Valencia, Spain, Vol. 978.

Yang, J., and Leskovec, J. (2011). *Patterns of Temporal Variation in Online Media*. WSDM'11, February 9–12, 2011, Hong Kong, China.

Chapter 5

Relating Big Data and Data Science to the Wider Concept of Knowledge Management

Hillary Stark and Suliman Hawamdeh

Contents

Introduction

The exponential growth in data and information generated on a daily basis has impacted both personal and organizational work environments. The acquisition, analysis, and subsequently the transformation of data into information and actionable knowledge is dependent to a large extent on advances in technology and the development of highly effective and efficient data-driven analytical tools. Analytical data-driven tools have become paramount in the sorting and deciphering of large amounts of digital information. The general consensus today is that we are living in the information age and are overwhelmed by Big Data and the amount of information generated on a daily basis, whether that information is at the personal or organizational level, also known as information overload. Information overload is a reality and it has potential negative implications both on the health of the individual and the health of the organization.

It is becoming increasingly clear that the acquisition, analysis, and transformation of Big Data into actionable knowledge will depend to a large extent on creating cutting-edge technologies, advancing analytical and data-driven tools, and developing advanced learning methodologies (McAfee & Brynjolfsson, 2012; Waller & Fawcett, 2013; Chen, Chiang, & Storey, 2012). The application of these tools will also depend on the level of investment made in training individuals as well as organizations to manage knowledge with purpose.

Technology is the driving force behind the creation and transformation of Big Data and data science. Today, there are many tools available for working with Big Data and data intensive operations, but the most important and significant challenge facing the industry is the lack of expertise and the knowledge needed for implementation. We must question, how good are these tools and who has the skillset to operate these tools? But most importantly, who has the knowledge and expertise to interpret the results and transform the information into actionable knowledge?

The answer to these questions depends on two types of knowledge-oriented components that must work together in tandem. The first component is knowledge in the form of data and information (explicit knowledge) and the second is knowledge in the form of skills and competencies (tacit knowledge), both of which are needed for research, development, and implementation. In the current era of Big Data and data science-driven initiatives, many CEOs will readily attest that their competitive advantage is the expertise of their employees, or human capital, but this in itself presents a challenge as this type of knowledge is not easily captured, stored, or transferred (Nonaka & Takeuchi, 1995; Liebowitz, 1999; McAfee & Brynjolfsson, 2012). There is a body of knowledge in literature concerning the relationship between explicit, or documented knowledge, in comparison to tacit knowledge, which presents itself in the form of skills and competencies that one would consider to be almost intuitive or nonverbal (Hedlund, 1994; Orlikowski & Baroudi, 1991). However, simply being in agreement that both of these components are necessary in the furthering of knowledge will most likely not substantiate actual growth, but rather requires an active state of ownership to transform tacit competencies to documented and shareable knowledge.

The Wider Concept of Knowledge Management

Knowledge management (KM) is defined here as an interdisciplinary approach to dealing with all aspects of knowledge processes and practices. Knowledge processes include activities such as knowledge creation, acquisition, discovery, organization, retention, and transfer. Knowledge practices include best practices, lessons learned, communities of practice, learning organizations, knowledge sharing, mentoring, shadowing, and apprenticeship. KM is concerned with the big picture and the ability to connect the dots (Alavi & Leidner, 2001). KM introduces a new dimension, the need for managing and transforming tacit knowledge to documented knowledge, which can be brought about through the improvement of communication, information and knowledge transfer, and collaboration.

One aspect of KM is creating an environment for the transfer of knowledge to take place (Al-Hawamdeh, 2002; Hedlund, 1994; Tuomi, 1999). KM can be viewed as the interactions of four major different components, including people, technology, information, and processes (Figure 5.1). When each of these four components is joined with one another, different components emerge. For example, the by-product of the interaction of information and people is research and intelligence activities. Information generation and gathering is a human activity that predates civilization. When people and processes interact, it fosters information growth in both administrative and business environments, as well as the activities that govern many of the norms and practices within organizations. It fosters cultures that are unique to a group of people, an organization, and society as a whole. Combining technology-oriented processes results in the creation of

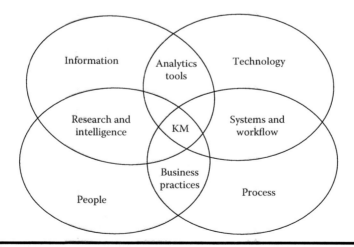

Figure 5.1 Knowledge management as interactions of four major components: people, technology, information, and processes.

new systems and workflows, with much of the automation coming about through normalizing a sequence of events that have proven to be efficient and effective. The intersection of information and technology has been the cornerstone in the way that data and information are generated and used. It can also be viewed as the tool needed in this case for data and test analytics. At the middle of this intersection lie KM processes and practices needed to create the paramount transformation of tacit knowledge to explicit knowledge. For KM to work, the various components must be put together to enable us see the big picture and be able to connect the dots.

Another aspect of KM is the ability to process and integrate information from multiple sources and in varying formats, inclusive of information in such magnitude that we now call it "Big Data," as well as one's reliance on data-driven analytics from differing perspectives (Lamont, 2012). This goal is quite similar, if not synonymous, with the idea of data science, but what is data science? Many individuals grapple with defining data science based on its interdisciplinary nature, as it involves the application of statistics, collection and storage of Big Data, creation of visualizations, and so forth, but at the same time each of these items can also be housed in information science, computer science, business, or even graphic design and art.

Data science uses automated tools and mechanics to extract and relay information, often from very large amounts of data, to assist individuals and organizations in analyzing their own data for better decision-making processes. These tools have been applied to almost every realm of business and industry, as powerful insights can be derived to then generate critical knowledge and action plans. Impacting scientific infrastructures, workflows, processes, scholarly communication, and certainly one's health and well-being, the application of measurements to overwhelming amounts of information is at the core of data science. It's important to keep in mind that data itself isn't that important without a context in which to apply it, as each organization has different needs and curiosities, meaning that a wide variety of tools are now available to accommodate these demands for more information.

The Shift in Data Practices and Access

It could be argued that possibly the greatest achievement in the digital revolution is that an individual now has the ability to interact with data under their own terms without relying on other parties to mediate access to information. Access to one's own analytics has given both individuals and companies the ability to use data to measure their performance and productivity and to determine the return on investment (ROI) of various business activities. There are wide ranges of analytical tools that have become synonymous with formal data-driven roles, but in reality, almost all businesses and individuals rely on analytics in their daily routines in one way or another.

The stock market is a prime example of business activities that rely heavily on analytics and analytics-driven tools. Traditionally one could not trade successfully without relying on the knowledge and expertise of a stock broker who would have access to information relating to trading stocks. All of the information regarding the health of the companies that were being traded and changes within the market, such as an initial public offering (IPO), mergers, or bankruptcy, was handled and disseminated through a broker. With information and data about the stock market quickly becoming available online in the late 1990s, the end user suddenly had the ability to become an actor in what is today is termed as the knowledge economy. The knowledge economy is a concept that revolutionizes the way we do business and demolished the traditional geographical market boundaries. It's difficult to imagine a time period where we were so dependent on local markets and how we've quickly become accustomed to the culture of instant access to information. Whether it's retrieving information via Google, the buying and selling of products and services online, or conducting one's banking online in real time, this dependency on instant access to information is a by-product of the increased values of knowledge, the knowledge economy, and knowledge management.

Smartphones have revolutionized the way we access information and have given us the ability to conduct searches and perform online transactions 24 × 7, removing the dependency on middlemen, increasing autonomy, and reducing the associated waiting time of many transactional processes. This technology has not only reduced one's waiting time but has also removed many of the physical barriers that were once in place, thereby increasing productivity. Online banking through mobile applications is a good example, as it minimized the need to go to a physical location to deposit checks, and provides real-time transaction history, including deposits, withdrawals, and pending items, so that the consumer has the most accurate view of his or her account. It not only helps the consumer feel more informed but also assists in establishing a trusting relationship between the consumer and the bank provider. The elements of trust and privacy were two of the main concerns to many consumers in the past, especially in the context of online transactions, and while this issue still remains a concern, as more and more people use online transaction processes, one's increased access to information minimizes those fears and fosters a trusting relationship between the consumer and the brand (Rittinghouse & Ransome, 2016; Chiregi & Navimipour, 2017).

Access to information in the medical field has been long seen as an issue of a great concern. However, the medical field has seen tremendous change in the last decade, with almost all hospitals and doctors' offices transitioning from paper charts to digital electronic records. This shift to a digital format not only minimizes errors in record keeping, but it also allows the patient to access their own records and take charge of their care under their own terms. Interacting with one's clinic as well as a pharmacy in an online environment has become common practice. Online patient portals are becoming commonplace, especially for specialists who see the same patients on a somewhat frequent basis.

This portal not only allows users to schedule appointments and find helpful resources, but it can also work as a receptacle for storing test results, prescription dosage amounts, and can even facilitate as a record of one's entire medical history. Research pertaining to patient portal use shows a significant increase in dialog between the patient and medical staff, increased attendance in follow-up visits, and increased recovery, with patients attributing these successes to the feeling of being autonomous and in control of their level of care and communication (Zarcadoolas, Vaughon, Czaja, Levy, & Rockoff, 2013). Studies show that when people feel autonomous and are made to believe that they are in control of their choices and actions, especially when they have a medical condition that needs their continued attention, that their adherence to a regimented protocol improves as their access to relevant information is increased (Kipping, Stuckey, Hernandez, Nguyen, & Riahi, 2016).

The notion of increased access to data and information is supported by the advancement of technologies, such as the advancement of data analytics tools, which enhances one's knowledge and awareness. The ability to access information at the right time within a particular situation, without the concern of one's privacy being violated, empowers people to make better decisions and to take control of their lives.

Data Science as a New Paradigm

In 2009, Microsoft Corporation's Vice President, Tony Hey, delivered one of the first presentations of its kind, in which he referred to the current era as an age of data science, of conducting "data-intensive scientific discovery" (Hey, Tansley, & Tolle, 2009). Hey considered data science to be the "fourth paradigm," or a measurable point in history in which science was taking a considerable change in direction. He speculated that the advent of data-driven science would also usher in new theories and models, vastly impacting both the hard and social sciences, and that there is a justified need for this area of expertise. Hey proposed three main activities of data-intensive science to include the capture, curation, and analysis of data. The capturing of data happens in many different ways, as one needs to already have an understanding of the future curation needs, as this determines the collection process (Hey et al., 2009). For example, the act of capturing data from a live experiment in a lab looks very different than capturing Tweets from Twitter using an API. The curation of data determines its longevity and future existence, and data that are not curated is almost guaranteed to be lost. The analysis of data is a multistep ongoing process, as the insights gained from the data change with each implementation. The process is inclusive of building and utilizing databases, cleaning and modeling the data, and data visualization.

The data-driven fourth paradigm raises questions about security, privacy, and storage of data and its resulting information. People with a knowledge of cyber security, cloud computing, and database management are highly sought after by employers, as they want to be reassured that the business's information will not be compromised. Companies today value intellectual capital including the employees' expertise, customer profiles, and organizational memories as some of their greatest competitive advantages (Duhigg, 2012). It's commonplace in online sales and marketing for organizations to seek clients' historical information, an action that is usually included in the barely legible fine print of the contract that you agree to when making any large purchases, such as a new car, or even signing up for a credit card (Campbell & Carlson, 2002). The compilation of this information alone isn't of great value until one puts forth the time-intensive effort of analyzing the data, with the purpose of making sense and finding patterns among the records. The activity of analysis is often very time-intensive, especially for a scientist or researcher who does not have the skillset for the available technologies. We detail some of the currently popular tools later in this chapter that have become mainstream in the realm of data science, particularly ones that provide increased levels of analysis applicable to the needs of all organizations, even those who do not see themselves in the business of data.

Big Data Cost and Anticipated Value

The terms "Big Data" and "data science" are often used interchangeably to refer to the increased interest and anticipated value of digital information; however, Big Data is just one component of data science. Big Data refers to structured and unstructured datasets that are large and complex (Provost & Fawcett, 2013), with one of the main concerns of big data infrastructure being the capability to collect and analyze large datasets beyond the traditional method of sampling and extrapolating (Walker, 2014). For data to be "big" it must meet at least four different characteristics, known as the 4 Vs: volume, velocity, variety, and variability (Lamont, 2012; Marz & Warren, 2015; McAfee & Brynjolfsson, 2012; Russom, 2011).

Volume refers to the size of data an organization is collecting on a continuous basis from difference sources, such as patient information, patient records, data generated from various diagnostic machines, imaging and ultrasound devices, as well as one's drug history and interactions within one's patient portal (Zarcadoolas et al., 2013).

Velocity refers to the data streams and the speed at which data are generated, transferred, and processed. Imagine the speed at which data are generated and transferred using social media platforms such as Instagram and Twitter. Both of these social media sites have the option of using a hashtag to include one's post within a larger but similar conversation. In the last decade, when events have taken place that drew great attention, whether it be a presidential election or a state of emergency, a word or phrase is turned into a hashtag, and within minutes millions of people are connected in conversation through the use of this type of tagging. It's this level of speed that constitutes velocity.

Variety is pretty easy to define, as it's the inclusion of data in many different formats, from traditionally structured data, such as numeric data, to qualitative data, such as videos and audio that add complexity and robustness to one's knowledge of a subject.

Variability refers to the volatility in data and datasets. It is different from variety in the sense that variability deals with the changing nature of the datasets due to uncontrolled variables. We see changes in data over a period of time when analyzing large datasets such as stock market information. The types of data collected and used in stock market-forecasting changes daily, and is based on a wide range of variables that include, human, social, political, and economical value. The variability of that data sometimes makes it complex and difficult to predict or forecast.

Scalability is another important area that's necessary when dealing with Big Data. Scalability refers to the capabilities we build to handle Big Data, from computer networks, information systems, databases, analytics tools, and so on. A tool's scalability means that it can handle changing amounts of data depending on the task, giving flexibility to the user to apply one technology to many situations. Some of the popular tools for the handling of Big Data to be discussed in further sections include Cloudera, Hadoop, and Spark.

The value of Big Data is not only in the data itself but rather in what an organization can do with it. Therefore, knowledge discovery from Big Data using advanced data science and data analytics tools is critical to an organization's overall intellectual capital. But without the infrastructure and the technological capabilities to make use of Big Data, questioning how to handle the data could become a burden to an organization. Organizations will have to make significant investments in capturing, processing, and managing large datasets, and the ROI can only be measured by the value of the knowledge that can be harvested from that data. This could be a challenge for an organization given the tacit nature of that knowledge. Therefore, an organization must have an intellectual capital and KM strategy in place that values tacit KM, and should be able to conduct knowledge audits from time to time to justify the organization's ROI.

Information Visualization

Data and information visualization is key to the process of making sense of big data and data analytics. In order for the knowledge discovery from a dataset (large or small) to be complete, there must be a definitive point of application in which the outcomes convey meaningful action. Data itself are granular in the large scheme of knowledge, especially if one hopes that these data will precipitate a change in human behavior. Let's think of this within the scope of an item that we are all familiar with, which is the nutrition facts panel as seen on all prepackaged foods in the supermarket. This label itself is not dramatic or enticing, probably because we are so accustomed to these labels being present that the majority of individuals don't actually reference the data contained within the panel unless there is a health concern (situation) caused by something such as diabetes or a desire to lose weight, and so on.

The panel contains at minimum the amount of calories per recommended serving, and the grams of carbohydrate, fat, and protein, as well as the daily percentages needed of these macronutrients for the average healthy individual. The visual representation of the data is determined by the frame of reference and the understanding of the metrics used within the panel, but in order to convey the intended message to the user, one must first have a frame of reference to make sense of the data (Reber, 1989; Velardo, 2015). Knowledge acquisition and utilization happen when users are able to understand information conveyed to them in an easy manner, but very few packaged products provide visual panels to assist users in having a frame of reference for making decisions, therefore the data included in the panel often goes unreferenced.

While some visualization techniques are simpler to understand, such as the nutrition facts label panel example above, not everything visual is easy to comprehend without a certain degree of training and expertise. Viewing medical images without the proper training needed to contextualize the data to understand the results can be problematic. To perform any level of analysis through visualization, which is reported to be the most personally valued of the five senses in the Western world (Kambaskovic & Wolfe, 2014), one must contextualize the data first and have a degree of familiarity of the attributes of data visualization.

Data visualization is a hot topic at the moment, even commanding entire university courses to be devoted to its discussion, as its amorphous nature includes the joining of both quantitative and qualitative data, while dancing in the realms of computer science, art, graphic design, and psychology (Healy & Moody, 2014; Telea, 2014). It's difficult to assign data visualization to one discipline as the recently built tools rely on the finesse of computer scientists that have the skillsets to build a program or software that renders an image based on a programmed code, such as the outputs of Python or R, but also has a holistic presence and relies on the expertise of artists and graphic designers. The visualization tool Tableau is especially popular in the business sector, providing aesthetically pleasing graphs and diagrams appropriate for many different types of users, from analyzing a client's data to actually presenting the final analysis to the client, as its functionality and use of color and dimension are almost limitless. The selection of colors used to relay information shouldn't be considered secondary to the importance of the information, as we have been conditioned to associate colors with particular meanings or feelings, and these associations can sometimes be more blatant than the information itself if not appropriately assigned.

In the United States, people associate the color red with having a meaning of centrality or warning, with a heat map usually assigning the deepest shade of red to the area of most importance. When looking at a weather heat map for a storm, the location with the most severe conditions is almost always assigned red, with the intensity fading to orange and yellow as you view the perimeter of the storm. The color red has also been used for stoplight labeling, as a definitive warning to not proceed, and some manufacturers have even played with using this three-color system (inclusive of yellow and green) to assist consumers in more easily understanding the nutritional value of prepackaged items, such as desserts having a red label, meaning to take caution, while milk and eggs have a green label, which gives permission to proceed in purchasing and consuming. Regarding the theme of nutrition, these color-coded labels have proven to be especially helpful for individuals in making more health-conscious food choices, especially for those who have a limited education (including youth and adults) and lack certain literacy skills needed to analyze the data included on the nutrition facts label panel or list of ingredients on the back of the packaged item (Sonnenberg et al., 2013).

Producing visualizations is especially popular as one can summarize the importance of the information presented in a significantly faster amount of time than it would have taken each viewer to actually comb through the data itself in search of its meaning (Fayyad, Wierse, & Grinstein, 2002). Visualizations by industry's definition are simply pictorial or graphical images created for the purpose of presenting data in a summarized format, and range anywhere from a simple bar graph or pie chart to a 3D color-saturated heat map with moving elements (https://www.sas.com/en_us/insights/big-data/data-visualization. html). With this definition in mind, we challenge the reader to consider all of the visualizations that one encounters on a daily basis, as we take many items for granted in our quick analyses that represent quite complex amounts of information. One of the most popular visualizations that summarizes a large amount of data into one single image, is Napoleon's successive losses as his French troops crossed Russia from 1812–1813, as drawn by Charles Minard in 1869 (Tufte, 2001; Friendly, 2002). Minard is considered the first to use graphics in combination with engineering and statistics, specifically numeric data as represented on graphical maps.

The map below combines six different types of data: the number of soldiers within the troops; the direction of the travel (it changes as they advance and then retreat); the distance traveled across Russia; the temperatures encountered; latitude and longitude; and the location of each report. The thick beige line represents the amount of soldiers within the troop at the start of their journey, which is seen to diminish as the line extends from left to right. The black line then represents the amount of soldiers that were still alive as they then began the second half of their journey, which was to return to the western-most point of the map, as seen when the black line diminishes from right to left. It's evident that based on the scale of these two lines, the majority of the troops did not complete the entire march across Russia and back again. This image is often included within the coursework of visualization classes, as it compels the viewer to consider all of the different aspects and the amount of information included in such a simple graphic.

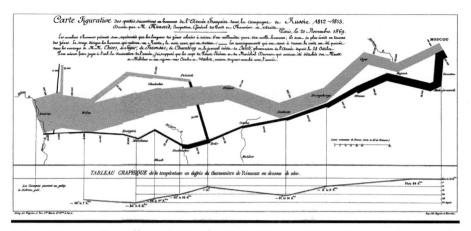

Minard's carte figurative of Napoleon's 1812 campaign. (From Tufte, E., *The Visual Display of Quantitative Information,* **2nd ed., Graphics Press, Cheshire, CT, 2001.)**

Infographics, or a graphic made with the goal of relaying a large amount of information, usually presented in an aesthetically pleasing format, have become increasingly popular over the last few years. Many businesses have jumped on board in utilizing this modern format of relaying large amounts of information to their consumers, that most likely took the team of data collectors a significantly longer amount of time to dig up to create the graphic, than the actual production of the graphic itself. This type of content has proven to be especially useful when it comes to search engine optimization (SEO), as users process visual information much more quickly than text and are infinitely more likely to share visual content across their social media platforms (Stark, Habib, & al Smadi, 2016). However, consumers should be cautious of the quality and accuracy of the information provided.

Tufte (2001) warns of the "lie factor" often included in graphical imagery, which is the value used to describe the relationship between the size of effect shown in a graphic and the size of effect shown in the data, as this is often misinterpreted when purely shown in graphical form. Research pertaining to one's perception of the scientific validity of health claims made on medicine packaging shows significance in the pairing of the textual claim with a data-driven graph or image that supports the theme of the claim (Tal & Wansink, 2016). We do need to be wary of these visual techniques though as they are not always founded on actual data or statistics, but instead lend a scientific tone to the manufacturer's claim, such as including an image of a chemical formula, even if this is not relevant to one's information needs, such as when shopping for Tylenol or aspirin.

Data Analytics Tools

Considering the magnitude of data currently collected and made available, it's practically impossible to manually process that much data in the hope of finding trends or certain patterns that might guide the decision-making process. It is much easier to read every answer collected by a 20-question survey with 100 respondents than it is to comprehend the same survey with 10,000 or 1 million respondents. To make the point here, as the amount of data increases there is a need for an automated means of analyzing data. Proprietary hardware and software provided by some of the big players in the data-wrangling arena include but are certainly not limited to IBM's Watson Analytics, SAS Enterprise Miner (SAS EM), and Tableau (Lamont, 2012).

We discuss these three tools as they are marketed towards both business and academic sectors, as well as introduce popular open-source coding languages that can be used as alternatives. An entire book could be devoted to listing all of the technologies and tools that have been created to assist in the mining and analysis of data. That being said, we believe that anyone with an appreciation for data analytics or a desire to improve upon their future knowledge acquisition, should be familiar with the handful of currently popular tools that we detail in this section.

Watson Analytics, not to be confused with the supercomputer that became the ultimate winner of the popular TV show Jeopardy! in 2011, is a suite of solutions offered to businesses of all sizes by the powerhouse IBM. Trained to use artificial intelligence (AI), Watson Analytics relies on a cognitive computing system and is adept enough to sense, predict, and even infer solutions based on the imported data and given conditions (Guidi, Miniati, Mazzola, & Iadanza, 2016). What differentiates this product from its competition is its functionality to automatically analyze datasets, measure and examine the quality of the data, and recommend the optimal statistical solution to the query. Built using natural language processing (NLP), users can request and receive answers in almost real time in the format of exploratory, predictive, and visual analytics (O'Leary, 2017; Hoyt, Snider, Thompson, & Mantravadi, 2016). The amateur user shouldn't be intimidated by the idea of learning to use a tool such as the one described above. The advent of digital analytics-driven technology fosters the mindset that products created must be flexible and approachable, growing with the user, and not outpacing or lagging behind (Raskin, 2000).

The tool is available through partnership with IBM free of charge to academic institutions who do not seek to commercialize their findings, including both faculty and students. Watson Analytics can be used as a resource in the classroom in which students can try to familiarize themselves with its functionality. As Watson Analytics only came to the market in 2015, we are now beginning to see academic case studies being published pertaining to the real-life application of this tool (Guidi et al., 2016). For example, the exploratory and visual analysis functions were recently harnessed in the medical field to assist in the identification of heart failures, through the measurement and analysis of electrocardiograph signals.

SAS EM is powerful data mining software, and the makers claim that it offers more predictive modeling techniques than any other currently available data mining or data analytics tool in the market. Within the interface, models are built with a visually pleasing process flow diagram environment, giving users the ability to test the accuracy of their predictions using graph-oriented displays. The appeal to users in both business and academic sectors is that the software is scalable, meaning that it can be utilized by an individual as well as an entire team of people, can operate on datasets of varying sizes, and that it is available in a cloud-based environment. Built on the SEMMA methodology, which stands for sampling, exploring, modifying, modeling, and assessing, SAS EM provides users the ability to find patterns and relationships in Big Data. With user-friendliness always being a concern, SAS EM offers a clean interface simple enough to not overwhelm an individual with minimal statistical knowledge, but complex enough to still be of use to someone well-versed in modeling, designing decision trees, neural networks, and performing either linear or logistic regression (Matignon, 2007).

In 2008, researchers used the neural network method of SAS EM to diagnose heart disease, based on the patient profiles of individuals with and without heart disease, and produced an 89% accuracy in their predictions (Das, Turkoglu, & Sengur, 2009). This ensemble-based methodology was made possible by splitting the dataset into two parts, one containing 70% of the data for testing, and the remaining 30% for validation of the proposed system, which is an integral part of neural network modeling, as it's impossible to confirm the accuracy of a model if it's not applied to data similar to what was used in its creation.

Visualizations have come a long way in a very short amount of time, and combine the resources of art and graphic design with analytics, to create concise images that represent much more complex amounts of information (Fox & Hendler, 2011). For example, almost all of the Fortune 500 companies utilize infographics, which are visual images that typically include text, charts, and diagrams. Tableau uses drag and drop dimensions and offers users simple data mining capabilities that have historically only been available to those with a complex coding skillset (Tableau. com, 2017). Tableau gives users the ability to create time series-driven data, such as a moving depiction of the path of Hurricane Katrina across the Gulf of Mexico in 2005. Based on the data provided, the system was not showing only the movement but also the change in intensity of the storm through the use of color as well as varying degrees of opacity. Users can easily use the software interface to create visualizations and save and share the results in a sharable format, such as a PDF, JPG, or PNG.

To spark interest and plant a seed of inspiration, Tableau's website offers thousands of examples of visualizations that have been built using real-world data examples. To incentivize educators to use Tableau within their classrooms, instructors as well as students are given a free annual license, and instructors have the added benefit of access to course materials created by Tableau that can be integrated within their own course research objectives. Librarians at Ohio State University

(OSU) utilized Tableau to visualize library data on three very different datasets, all of which produced results that better enabled them to meet the needs of their university patrons. The datasets illustrated the library's potential to promote library collections in tandem with marketing the library's programming, how combining data from multiple sources could provide support for the prioritization of digitizing special collections, and lastly the ability to better understand user satisfaction from survey data through the Tableau's filtering functions (Murphy, 2015).

While these tools are considered to be cutting edge, their accessibility comes at a cost, most commonly seen as an annual licensing fee either by person or institution. For parties who desire the same results, such as text analysis and data mining, but don't have the resources to purchase the aforementioned software, there are a number of free alternatives. Users should be reminded however that anything advertised as being "free" usually has hidden costs, an idea popularized with the acronym TANSTAAFL, meaning "There ain't no such thing as a free lunch" (Grossmann, Steger, & Trimborn, 2013). Within the context of tools and technologies, this cost is invariably the user's time and attention required in the beginning, to learn how to navigate the tool and generate the desired results. A prime example of this would be the open source and very popular coding languages of R and Python. Python is a multifaceted language and offers everything from data cleaning to graph-based visualizations, with over 40% of data scientists reporting it as their "go-to" for problem solving, per the 2013 survey by O'Reilly (http://www.oreilly.com/data/free/stratasurvey.csp). It's really no wonder why this language has been so widely accepted, as it supports all kinds of data formats, from CSV to SQL tables, and the online community is robust in offering examples and inquiry support (Perkel, 2015). R on the other hand, while also open source, is considered by many users to have a steeper learning curve than Python, due to its complicated syntax, but it is built with the data scientist in mind, as its language is focused on solving statistical problems and generating scientific visualizations (Hardin et al., 2015; Edwards, Tilden, & Allevato, 2014).

The results provided by writing code from either of these languages can assist the user and spark hidden discovery, but to arrive at this point, the user will first need to spend a great deal of time acquainting themselves with the syntax for both of these languages. Almost all undergraduate computer science programs offer courses in programming languages, with Python currently being the most popular introductory language offered at colleges within the United States, compared to business schools, which advertise their access to the proprietary tools mentioned prior (Brunner & Kim, 2016). As discussed more deeply in other sections of this chapter, there is an apparent disconnect in the data analytics-driven tools introduced to students within the collegiate environment depending on their registered course of study, which lends these areas of expertise to being siloes rather than shared for the mutual benefits of all parties who have the same interests (Thompson, 2017).

Additional platforms and tools for the mining of large datasets and creating models include Cloudera, and Apache's Hadoop and Spark. Hadoop has made its name in the software arena in two ways, through its unconventional storage capabilities (splitting files into large blocks that are then distributed across multiple storage systems, with the data being processed in parallel across multiple nodes), as well as its lightning-fast processing procedures that are built upon the MapReduce framework (Padhy, 2013). Similar to Hadoop, Spark can handle large datasets and its speed is marketed as being up to 10 times faster than that of Hadoop. Both of these tools have a learning curve, but are considered to be some of the more user-friendly options for tackling gigantic datasets using open-source packages. These tools are useful to companies that have large and complex datasets, and can also be used in tandem, as they are not mutually exclusive. It depends on the size of the dataset and the needs of the end users as to which option is more appropriate at that time (Lin, Wang, & Wu, 2013).

A precursor to the age of Big Data, was the advent of e-commerce, or the selling of products and services online, which we can see as early as the late 1990s (Kohavi, 2001), and it was in this online environment that the transactional process was made completely trackable. Google has also made this tracking process too easy to not take advantage of, especially for small businesses with limited budgets, through its free Google Analytics dashboard (Clifton, 2012). To benefit from the bevy of data available for collection through Google Analytics, snippets of code are placed on each page of the specified website, and within seconds, the user can monitor the engagement, activity, and trends forming based on visitors to the site (Fang, 2007). These analytics are available in real time in an interactive dashboard that is extremely user-friendly, and can even be integrated through APIs to populate other third party dashboards. While there are many other offerings in the Google suite of services, we focused on this one item due to its market share, which is estimated to having been used in more than 90% of the websites available today.

In 2008, Google launched an effort to help predict trends in illness, specifically the flu, based on users' search queries. The Center for Disease Control (CDC) has a lag time of up to three weeks in collecting data and then making the formal announcement of an epidemic, but Google was able to predict trends in almost real time through their data collection mechanisms (Lazer, Kennedy, King, & Vespignani, 2014). When initially introduced, the idea that illness and disease had the potential to be minimized through pinpointing the volume and location of searches based on similar symptoms, the announcement was met with great fanfare and excitement. This venture gave hope to not only the medical communities in the United States, but to much poorer countries as well, where medical resources are often limited and time is of the essence.

However, in August 2015, Google quietly terminated this initiative, due to an increasing amount of inaccurate predictions, as Google's model failed to adequately account for changes in user search behavior (Pollett et al., 2016). As users, we've become much more adept and accurate in our search queries, as the popular search engines' algorithms now all rely on aspects of semantic analysis, or the meaning and context behind the entire search phrase, rather than just the meaning of each individual word used within a phrase. This improvement in user search behavior, coupled with the fact that not everyone who was searching using terms that are connected to having the flu, actually had the flu, produced inaccurate results based on the model built. Someone may have searched based on having a simple headache, while another user may have entered a query to learn more about flu season itself, or where to get a flu shot in order to be preventative. Researchers are hopeful that in time the model first introduced will be finetuned and be able to deliver an accurate analysis and prediction based on real-time search data as originally promised.

While not normally called "data analysts," those who monitor any kind of technology that records data to be retrieved and used at a later date, certainly utilize a similar skillset. The world of medicine and medical devices has certainly been positively impacted by the integration of analytics, namely digital analytics, as healthcare professionals now have greater precision and clarity to provide the safest and most efficient care possible (Alyass, Turcotte, & Meyre, 2015). There is a long list of technologies that have transformed the way we conduct business, but the foundational element to all of these tools is that the user is now in control of being able to generate new information that will in turn become knowledge, and through application, the promise of progressive change is possible (Howe et al., 2008).

Every aspect of business has been impacted by increased access to data analytics, even including the wine industry, which has traditionally been perceived as a more manual field, steeped in the romanticized idea of picking and sorting grapes by hand to preserve a certain level of product quality (Dokoozlian, 2016). Many wineries are starting to integrate optical sorting machines in their facilities, the leading branding currently being WECO (http://www.weco.com), which operates based on a camera with LED lights that illuminates each individual grape during the sorting process, measuring the fruit's surface for defects as well as sorting based on the desired size of the grape, and discarding all of the items that don't meet the set standards for optimal wine production, including stray twigs, leafs, and debris.

The system can even be calibrated for color and density, selecting grapes that have particular hues that represent the ripeness or saturation of sugars within the grape, as each wine maker desires different levels of these components. While one may argue that this is a stretch for an example of data analytics, analytics themselves don't have to be a physical printable image or graph, but rather information that is derived from the application of measurements or statistics. These same machines are also used within the farming and food manufacturing channels, revolutionizing entire industries through minimizing manual labor and risk while increasing accuracy for a wide gamut of items that we consume on a daily basis, including but not limited to nuts, olives, berries, and grains. This type of optical sorting has been used for some time now, specifically within the grain industry, as a method of removing grains that are infected or contaminated, minimizing the risk of illness in both humans as well as livestock (Dowell, Boratynski, Ykema, Dowdy, & Staten, 2002). While a data-driven tool may have been originally created to facilitate the needs of a more traditional or scientific disciplines, with time, most industries have also started to integrate and adapt these practices.

Applications and Case Studies

The increased efficiency, transparency, and rapid growth of information can be attributed to advances in science and technology including analytics and KM. For example, it's safe to assume that the majority of World Wide Web (WWW) users have used at least one of Google's products, or are at least familiar with the purpose of their offerings, such as the Google search engine, Google Earth, or Google Maps (Jarvis, 2011). Available for downloading on a smartphone, the majority of the users of this application rely on its precision and accuracy for producing a visualization of a map of the most efficient route to take at that time, inclusive of variables such as traffic or road construction. We take for granted the integration and magnitude of data necessary to facilitate this application running efficiently, which is inclusive of thousands of data points all being evaluated and analyzed in tandem, to then produce multiple options for our desired route—such as tolls or no tolls, or if there is a coffee shop on the drive.

The data collection points utilized by Google are more recently also being used by local governments, specifically in the endeavors of smart cities (Vilajosana et al., 2013). "Smart cities" is a relatively new and still unfamiliar term to many individuals outside of the realms of technology and city governance, but one of the primary concepts is that data is constantly being collected that could be of greater value if made easily available to both city authorities as well as the citizens of each municipality. Communities of varying sizes are advocating for the collection and transparency of this data to be made available, as well as increased analytics generated to provide safer, healthier, and happier environments for their citizens. An example of a large metropolitan that is considered a smart city and provides resources built of off analytics-driven tools to its citizens, is the city of Chicago. Connecting over 250,000 street light fixtures, Chicago's smart light project is believed to be the largest city-wide wireless smart light program in the United States, and is built using LED lights that can be remotely dimmed or brightened as needed.

The network can also remotely monitor lights in need of repair, sending an alert instantaneously after a failure has occurred, and is integrated into Chicago's non-emergency 311 system (http://www.chicagoinfrastructure.org/initiatives/smartlighting). At a smaller scale, the City of Denton, Texas, has revamped their waste-collection services through increased access to municipal data. By adding sensors that monitor the fill-level of each large dumpster located in the downtown area, the city can track in real time the need for increased collection and disposal, and design their workflow routes and employee schedules to meet the needs of the city.

This increased access to data provides not only information that can make the city more efficient in their operations, but also provides a cleaner more sanitary environment for its citizens who frequent the area. Their future "smart city" projects include citizen access to utility consumption across the city, including water, gas, and electricity, to have an increased awareness of the resources used as well as to be able to be more efficient in their consumption, such as not overwatering public green spaces in the summer or reducing the electricity used in unoccupied buildings.

Even if you don't live in a smart city (yet), you probably go to the grocery store, or have a membership to one of the bulk-item retailers, such as Costco or Sam's Club, and the chances are pretty good that you have a loyalty card tied to one of these merchants, either physically presented or keyed-in using your name or phone number at the point of transaction. Users are generally incentivized to use these cards to receive discounts, and at the same time the retailer collects data on which items were purchased together and at what frequency, and so on. This process has become the norm and most of us no longer give the sharing of this information a second thought, but what if the data contained within these historical records could actually be of a much more beneficial use than originally estimated?

A prime example of harnessing this innocent, some would even say unnecessary, data for knowledge utilization and the betterment of humanity, is the research conducted at the University of Oxford, with a keen interest in both Big Data and dementia (Deetjen, Meyer, & Schroeder, 2015). Dementia is a good example of a chronic disease that often causes an individual to exhibit particular behaviors for quite a long period of time before diagnosis, and as many chronic diseases, it is speculated to be heavily influenced by diet.

The researchers posit that medical professionals should utilize grocery store loyalty card data to have more information regarding the prior dietary choices of individuals who have a current diagnosis of dementia, as well as to be proactive in spotting the early signs of this disease in its first stages. One of the behaviors exhibited in the early stages of dementia is that of forgetfulness or decreased memory, and this is especially prevalent when grocery shopping, with some people buying the exact same item over and over again with increased frequency. Multiple grocery store brands have already volunteered to share this data with the medical profession, and if approved by the legislation, the release of this longitudinal data for analysis could prove to be quite life-altering.

Another topic of recent debate is the 2016 United States presidential election. The outcome of the election surprised both analysts and statisticians who analyzed and presented data, which led the majority to believe that the Democratic candidate would win by a landslide. A popular website predicting the outcome of the election was http://www.FiveThirtyEight.com, hosted by Nate Silver, who used vast amounts of data collected, paired with rigorous methodology, to provide a quantitative analysis of the discussions and opinions heard across a broad audience. They correctly predicted the 2008 election won by Barack Obama, with his final presidential election forecast accurately predicting the winner of 49 of the 50 states, missing only Indiana, as well as the District of Columbia. He again shocked the nation by accurately predicting the outcomes of all 50 states in 2012, including the 9 swing states, but then in 2016 the model failed. They predicted that Hillary Clinton would win by a 71% majority vote, giving Donald Trump only 29% of the vote.

The data analyzed included a wide range of polling methods, a similar approach to the data collected for the two previous elections. A reason given for this error in accuracy was not the method used, but rather in the reliance on the final data collected, an example that the variability of Big Data could change the predictive outcome. Social media platforms played a large role in this false feeling of security, as users' posts so often lack independence in this age of digital transparency, and while a single Tweet can be viewed hundreds of thousands of times, it really only represents the sentiment and opinion of a single person (Rossini et al., 2017). The election has brought increased attention to the information gathered, and subsequent knowledge formed, by users who rely solely on data presented online, specifically from social media sites such as Facebook and Twitter.

Social media as a channel for communication has reshaped the way that we form our social circles, with conversations never actually dying as online banter and behavioral data are constantly tracked and stored. Businesses are encouraged to take advantage of the immense amount of data that we have chosen to share about ourselves through our social media profiles to improve their marketing strategies, with the most popular and robust platform currently being Facebook. The behavioral tracking and data collection efforts of Facebook include behaviors online and offline, making the connection between user profiles by using credit card information as well as email addresses and phone numbers (Markovikj, Gievska, Kosinski, & Stillwell, 2013). Advertisements can then be tailored and shown just to the intended online audience, saving the business both time and financial resources, while maximizing their ROI.

Researchers and academics are also benefiting from stored information as it provides a more concrete method of discovering communicative patterns exhibited across networks, including informal peer-to-peer interactions as well as business to business (B2B) or business to consumer (B2C) data. Twitter data has been especially useful for having a deeper understanding of users' thoughts and opinions, through the inclusion of hashtags, which note the meaning of the conversation and connect users of the same verbiage to one another. For example, if I tweeted about the 2016 presidential election and used one of the president campaign hashtags, such as #MAGA (Make America Great Again), my Tweet would then be added to the list of all other Tweets that used the same hashtag, essentially creating a conversation around this topic. It is also possible to download all of the Tweets using this hashtag and perform a sentiment analysis to determine the positive or negative feelings associated with this topic.

Twitter's API allows for batch downloads of Tweets, currently up to 10,000 or more Tweets at once, and includes the user's location as well as the dissemination and breadth of share across the Twitter network. Technologies are abundant in fulfilling the need to visualize this type of data. Within the context of the aforementioned example of downloading Twitter data, NodeXL could be an appropriate tool for visualization as it provides a visual representation of the network breadth and sharing of a Tweet or hashtag topic, and is user-friendly as it's an Excel plugin, found at http://www. NodeXL.com. This plugin also provides a list of the top 10 influencers within these conversations based on their centrality, including likes and retweets, as well as the top URLs included in the Tweets, making it much more evident as to who is leading and participating in certain conversations.

Emerging Career Opportunities

One of the notable effects of the data science and data analytics revolution is that new career opportunities are emerging. The job of "data scientist" was termed the sexiest job of the twenty-first century, although it is certainly not the easiest, as employers want their data analytics-focused applicants to wield a set of skills much akin to that of a Swiss army knife (Davenport & Patil, 2012; De Mauro, Greco, Grimaldi, & Nobili, 2016, 2017; Gardiner, Aasheim, Rutner, & Williams, 2017). Not only must a prospective applicant first be able to wrangle large amounts of data, often in multiple formats that don't integrate with one another easily, but after the task of cleaning, mining, and modeling the data, they must then be able to glean the important pieces of information hidden within the data and infer its application including meaningful presentation of the results.

Since this is a leading area of job growth, universities are racing to develop coursework and programs to attract and educate potential future data scientists. Given that data science and data analytics bridge many academic fields, such as computer science, engineering, information science, business, and even art and design, most of the programs offer an adopted interdisciplinary approach. Some of the universities created interdisciplinary programs where students are taking coursework across a multitude of departments while others house the entire program within just one department. Given that this area is still considered to be new in the grand scheme of academia, there is not yet a benchmark of classes or clear curriculum (Tang & Sae-Lim, 2016; Lyon & Mattern, 2017; Tenopir et al., 2016), and as a result, graduates from these programs will have a varying degree of competence in many areas often associated with the title of "data scientist."

Some of the academic institutions that offer programs in data science and data analytics include Massachusetts Institute of Technology (Master of Business Analytics), Carnegie Mellon (Master of Information Systems Management: Business Intelligence & Data Analytics), University of Texas at Austin (Master of Science in Business Analytics, Master of Science in Computer Science), University of Chicago (Master of Science in Analytics), Rutgers University (Master of Science in Data Analytics), Southern Methodist University (Master of Science in Business Analytics), and University of North Texas (Master of Science in Data Science; Master of Science in Advanced Analytics). We compare the programs offered at these institutions to find commonalities as well as what differentiates, depending on if a student enrolls in a more business versus computer science-oriented program of study.

The similar concepts taught across the programs include students gaining a working knowledge of database management, data mining, data visualization, the application of tools used to collect, store, transform, model, and analyze large amounts of data (tools specifically mentioned within their curriculum descriptions include but are not limited to SAS EM, SPSS, Hadoop, and Tableau), marketing analytics, and the ability to apply a wide range of statistics to perform scientific analysis. Most of the programs offer both full-time and part-time options. They acknowledge that many of their students will be at least partially employed during the time that they complete their master's degree, and need the flexibility in the amount of coursework taken each semester. Many of the degrees also include completing an internship or team-based project that is then presented to a panel of executives, greatly emphasizing the importance of communication and team-oriented skills within the workplace.

Positions that one can apply for upon earning their Master of Business or Science degree with an emphasis on data analytics, are on the rise as more and more companies are creating the roles of data scientist, manager, analyst, consultant, or software or project engineer, in order to stay competitive (De Mauro et al., 2016, 2017; Gardiner et al., 2017). Upon review of some of the job postings that included the keywords "data analytics," "data analyst," and "data scientist" on three popular recruitment sites Glassdoor.com, LinkedIn.com, and Indeed.com, we identified a list of the common minimal qualifications desired by employers for their future hires:

- Bachelor's degree in mathematics, engineering, computer science, or statistics, with 3 or more years of work experience or academic research.
- Proficient in more than one statistically driven coding language, specifically R, Python, and SQL.
- Experience with modeling and visualization software, specifically SPSS, SAS, and Tableau.
- Ability to work individually as well as be a team player.
- Strong written and verbal communication skills, comfortable giving technical presentations.
- Strong attention to detail, ability to solve complex problems, has a strong sense of curiosity.

Optional but preferred skills that help candidates stand out from the crowd:

- Master's or PhD in the aforementioned disciplines.
- Proficient in core programming languages, including C/C++, Java, Scala, and Ruby.
- Experience using Apache's Hadoop and or Spark to handle and process very large datasets.
- Machine learning techniques, such as logistic regression, decision trees, random forests, neural networks, and naïve Bayes.

Access to particular tools learned in academia differ greatly by employer and it's important that students realize that although they may be proficient in one software, that the possibility is almost guaranteed that they will need to learn new tools and new systems upon employment. For example, it's rare that a small employer who is in the beginning stages of handling large datasets would have the budget for a yearly Tableau or SAS license, and instead prefer relying on free or budget-friendly resources. In this case, the open-source items discussed prior would most likely be the more attractive option; therefore, a data analyst must be agile and flexible in learning new tools that are similar to ones having been taught within the classroom. As is with any technology, the industry that makes these data-driven tools is quickly advancing and products are constantly being upgraded. This means that a data analyst must be open to learning and keeping up with the new market trends (McAfee & Brynjolfsson, 2012).

Regarding the desired experience, students should utilize internships and become involved in research projects that utilize real datasets that would be of interest to future employers, to establish a portfolio of work experience (http://data.gov). Certifications can also add credibility and enhance one's resume, especially for programs or software that have multiple levels of proficiency. For example, a leading online instructor, Lynda. com, offers courses and tutorials in learning Python, while Tableau holds a tighter rein and true to its proprietary nature, is the only one to have ownership to offer the two-level certification for its software. Both of Google's Analytics and AdWords platforms, which measure one's proficiency in analyzing a website's online presence, health, and the process of running an online advertisement campaign, have multiple proficiency exams, but to stay active and up-to-date, these must be retaken every 18–24 months. And while no one really enjoys having to be recertified on a continual basis, Google has incentivized users to stay current by eliminating the certification fee, making all of the exams now free. Coursera.com offers short online classes that cover many of the concepts considered to be part of data analytics, with the lessons often being offered free of charge, giving the users the option to pay for the course if they want to take the exams and earn a certificate of completion.

Data-related internships, boot camps, hackathons, and incubators are other options to gain relevant real-world experience outside of the classroom (Anslow, Brosz, Maurer, & Boyes, 2016). For example, the Data Incubator program is funded by Cornell's data science training organization (http://thedataincubator. com), and offers a free advanced 8-week fellowship for PhD students and graduates who are about to start searching for industry positions. Companies that have partnered with the Data Incubator for their hiring needs include LinkedIn, Genentech, Capital One, and Pfizer, with alumni now employed by companies such as Verizon, Facebook, KPMG, Cloudera, and Uptake. This fellowship is typically 8 weeks in length, with time requirements similar to that of a Monday to Friday, 8 a.m. to 5 p.m. job, and are physically held in the New York City, San Francisco, Seattle, Washington DC, and Boston metros. There is also an online option that typically requires participants to dedicate 2–3 months of time, but at a part-time commitment. Technical training curriculum in the fellowship includes software engineering and numerical computations, NLP, statistics, data visualization, database management, and parallelization. The soft skills gained from the training include ramping up one's communication skills, as academics and those in industry communicate in very different ways, along with face-to-face networking, and practice interviews to assist fellows in being the most prepared when applying for jobs upon completion of the program.

Conclusion

This chapter served as a review of some of the pertinent areas of analytics (data science, data analytics, and Big Data) and related them to the wider concept of KM. The value and importance of these concepts is linked directly to the increased shift in the economy and the increased emphasis on knowledge and intellectual capital as a key driver of the knowledge-based economy. The currently available tools and the applications that we've outlined will soon be considered ancient history, but the fact remains that the acquisition of knowledge and its systemic management will continue to challenge both scientists and practitioners. We would be remiss to not stress the increasing importance of research and academia and their role in creating more consistent theories, models, and frameworks to guide the development and implementation of data science and data analytics. It is also important to create a curriculum to support knowledge for students who desire to be part of the ever-present data and digital revolution. The task of managing knowledge effectively is built on the collective wisdom and knowledge of each of the contributing people within the organization.

References

Alavi, M., & Leidner, D. E. (2001). Knowledge management and knowledge management systems: Conceptual foundations and research issues. *MIS Quarterly, 25,* 107–136.

Al-Hawamdeh, S. (2002). Knowledge management: Re-thinking information management and facing the challenge of managing tacit knowledge. *Information Research, 8*(1), 143.

Alyass, A., Turcotte, M., & Meyre, D. (2015). From big data analysis to personalized medicine for all: Challenges and opportunities. *BMC Medical Genomics, 8*(1), 33.

Anslow, C., Brosz, J., Maurer, F., & Boyes, M. (2016, February). Datathons: An experience report of data hackathons for data science education. In *Proceedings of the 47th ACM Technical Symposium on Computing Science Education* (pp. 615–620). ACM.

Brunner, R. J., & Kim, E. J. (2016). Teaching Data Science. *Procedia Computer Science, 80,* 1947–1956.

Campbell, J. E., & Carlson, M. (2002). Panopticon. com: Online surveillance and the commodification of privacy. *Journal of Broadcasting & Electronic Media, 46*(4), 586–606.

Chen, H., Chiang, R. H., & Storey, V. C. (2012). Business intelligence and analytics: From big data to big impact. *MIS Quarterly, 36*(4), 1165–1188.

Chiregi, M., & Navimipour, N. J. (2017). A comprehensive study of the trust evaluation mechanisms in the cloud computing. *Journal of Service Science Research, 9*(1), 1–30.

Clifton, B. (2012). *Advanced web metrics with Google Analytics.* Hoboken, NJ: John Wiley & Sons.

Das, R., Turkoglu, I., & Sengur, A. (2009). Effective diagnosis of heart disease through neural networks ensembles. *Expert Systems with Applications, 36*(4), 7675–7680.

Davenport, T., & Patil, D. (2012, October). Data Scientist: The Sexiest Job of the 21st Century. Retrieved from hbr.org/2012/10/data-scientist-the-sexiest-job-of-the-21st-century.

De Mauro, A., Greco, M., Grimaldi, M., & Nobili, G. (2016). Beyond data scientists: A review of big data skills and job families. *Proceedings of IFKAD,* 1844–1857.

De Mauro, A., Greco, M., Grimaldi, M., & Ritala, P. (2017). Human resources for big data professions: A systematic classification of job roles and required skill sets. *Information Processing & Management*, 54(3).

Deetjen, U., Meyer, E.T., & Schroeder, R. (2015). *Big data for advancing dementia research: An evaluation of data sharing practices in research on age-related neurodegenerative diseases.* OECD Digital Economy Papers, No. 246, Paris: OECD Publishing.

Dokoozlian, N. (2016, February). Big data and the productivity challenge for wine grapes. In *Agricultural Outlook Forum 2016* (No. 236854). United States Department of Agriculture.

Dowell, F. E., Boratynski, T. N., Ykema, R. E., Dowdy, A. K., & Staten, R. T. (2002). Use of optical sorting to detect wheat kernels infected with Tilletia indica. *Plant Disease*, 86(9), 1011–1013.

Duhigg, C. (2012). How companies learn your secrets. *The New York Times*, 16, 2012.

Edwards, S. H., Tilden, D. S., & Allevato, A. (2014, March). Pythy: Improving the introductory python programming experience. In *Proceedings of the 45th ACM technical symposium on Computer science education* (pp. 641–646). New York, NY: ACM.

Fang, W. (2007). Using Google analytics for improving library website content and design: A case study. *Library Philosophy and Practice* (e-journal), 121.

Fayyad, U. M., Wierse, A., & Grinstein, G. G. (Eds.). (2002). *Information visualization in data mining and knowledge discovery.* San Francisco, CA: Morgan Kaufmann.

Fox, P., & Hendler, J. (2011). Changing the equation on scientific data visualization. *Science*, 331(6018), 705–708.

Friendly, M. (2002). Visions and re-visions of Charles Joseph Minard. *Journal of Educational and Behavioral Statistics*, 27(1), 31–51.

Gardiner, A., Aasheim, C., Rutner, P., & Williams, S. (2017). Skill requirements in big data: A content analysis of job advertisements. *Journal of Computer Information Systems*, 35, 1–11.

Grossmann, V., Steger, T. M., & Trimborn, T. (2013). The macroeconomics of TANSTAAFL. *Journal of Macroeconomics*, 38, 76–85.

Guidi, G., Miniati, R., Mazzola, M., & Iadanza, E. (2016). Case study: IBM Watson analytics cloud platform as analytics-as-a-service system for heart failure early detection. *Future Internet*, 8(3), 32.

Hardin, J., Hoerl, R., Horton, N. J., Nolan, D., Baumer, B., Hall-Holt, O., ... & Ward, M. D. (2015). Data science in statistics curricula: Preparing students to "think with data". *The American Statistician*, 69(4), 343–353.

Healy, K., & Moody, J. (2014). Data visualization in sociology. *Annual Review of Sociology*, 40, 105–128.

Hedlund, G. (1994). A model of knowledge management and the N-form corporation. *Strategic Management Journal*, 15(S2), 73–90.

Hey, A. J., Tansley, S., & Tolle, K. M. (2009). *The fourth paradigm: Data-intensive scientific discovery* (1st ed.). Redmond, WA: Microsoft Research.

Howe, D., Costanzo, M., Fey, P., Gojobori, T., Hannick, L., Hide, W., ... & Twigger, S. (2008). Big data: The future of biocuration. *Nature*, 455(7209), 47–50.

Hoyt, R. E., Snider, D., Thompson, C., & Mantravadi, S. (2016). IBM Watson analytics: Automating visualization, descriptive, and predictive statistics. *JMIR Public Health and Surveillance*, 2(2), e157.

Jarvis, J. (2011). *What would Google do?: Reverse-engineering the fastest growing company in the history of the world.* New York, NY: Harper Business.

Kambaskovic, D., & Wolfe, C. T. (2014). The senses in philosophy and science: From the nobility of sight to the materialism of touch. *A Cultural History of the Senses in the Renaissance.* Bloomsbury, London, 107–125.

Kipping, S., Stuckey, M. I., Hernandez, A., Nguyen, T., & Riahi, S. (2016). A web-based patient portal for mental health care: Benefits evaluation. *Journal of Medical Internet Research*, *18*(11), e294.

Kohavi, R. (2001, August). Mining e-commerce data: The good, the bad, and the ugly. *Proceedings of the seventh ACM SIGKDD international conference on Knowledge discovery and data mining* (pp. 8–13). San Francisco, CA: ACM.

Lamont, J. (2012). Big data has big implications for knowledge management. *KM World*, *21*(4), 8–11.

Lazer, D., Kennedy, R., King, G., & Vespignani, A. (2014). The parable of Google Flu: Traps in big data analysis. *Science*, *343*(6176), 1203–1205.

Liebowitz, J. (Ed.). (1999). *Knowledge management handbook.* Boca Raton, FL: CRC Press.

Lin, X., Wang, P., & Wu, B. (2013, November). Log analysis in cloud computing environment with Hadoop and Spark. *Broadband network & multimedia technology (IC-BNMT), 2013 5th IEEE international conference on* (pp. 273–276). IEEE.

Lyon, L., & Mattern, E. (2017). Education for real-world data science roles (Part 2): A translational approach to curriculum development. *International Journal of Digital Curation*, *11*(2), 13–26.

Markovikj, D., Gievska, S., Kosinski, M., & Stillwell, D. (2013, June). Mining Facebook data for predictive personality modeling. *Proceedings of the 7th international AAAI conference on Weblogs and Social Media (ICWSM 2013)* (pp. 23–26). Boston, MA.

Marz, N., & Warren, J. (2015). *Big data: Principles and best practices of scalable realtime data systems.* Shelter Island, New York: Manning Publications.

Matignon, R. (2007). *Data mining using SAS enterprise miner* (Vol. 638). Hoboken, NJ: John Wiley & Sons.

McAfee, A., & Brynjolfsson, E. (2012). Big data: The management revolution. *Harvard Business Review*, *90*(10), 60–68.

Murphy, S. A. (2015). How data visualization supports academic library assessment: Three examples from the Ohio State University Libraries using Tableau. *College & Research Libraries News*, *76*(9), 482–486.

Nonaka, I., & Takeuchi, H. (1995). *The Knowledge-creating company: How Japanese companies create the dynamics of innovation.* Oxford, UK: Oxford University Press.

O'Leary, D. E. (2017). Emerging white-collar robotics: The case of watson analytics. *IEEE Intelligent Systems*, *32*(2), 63–67.

Orlikowski, W. J., & Baroudi, J. J. (1991). Studying information technology in organizations: Research approaches and assumptions. *Information Systems Research*, *2*(1), 1–28.

Padhy, R. P. (2013). Big data processing with Hadoop-MapReduce in cloud systems. *International Journal of Cloud Computing and Services Science*, *2*(1), 16.

Perkel, J. M. (2015). Pick up python. *Nature*, *518*(7537), 125.

Pollett, S., Boscardin, W. J., Azziz-Baumgartner, E., Tinoco, Y. O., Soto, G., Romero, C., ... & Rutherford, G. W. (2016). Evaluating Google flu trends in Latin America: Important lessons for the next phase of digital disease detection. *Clinical Infectious Diseases*, *64*(1), ciw657.

Provost, F., & Fawcett, T. (2013). Data science and its relationship to big data and data-driven decision making. *Big Data*, *1*(1), 51–59.

Raskin, J. (2000). *The humane interface: New directions for designing interactive systems.* Boston, MA: Addison-Wesley.

Reber, A. S. (1989). Implicit learning and tacit knowledge. *Journal of Experimental Psychology: General 118*(3), 219.

Rittinghouse, J. W., & Ransome, J. F. (2016). *Cloud computing: Implementation, management, and security.* Boca Raton, FL: CRC Press.

Rossini, P. G., Hemsley, J., Tanupabrungsun, S., Zhang, F., Robinson, J., & Stromer-Galley, J. (July, 2017). Social media, US presidential campaigns, and public opinion polls: Disentangling effects. *Proceedings of the 8th International Conference on Social Media & Society* (p. 56). New York, NY: ACM.

Russom, P. (2011). Big data analytics. *TDWI best practices report, fourth quarter, 19,* 40.

Sonnenberg, L., Gelsomin, E., Levy, D. E., Riis, J., Barraclough, S., & Thorndike, A. N. (2013). A traffic light food labeling intervention increases consumer awareness of health and healthy choices at the point-of-purchase. *Preventive Medicine, 57*(4), 253–257.

Stark, H., Habib, A., & al Smadi, D. (2016). Network engagement behaviors of three online diet and exercise programs. *Proceedings from the Document Academy, 3*(2), 17.

Tableau.com. (2017). www.tableau.com, accessed 2017.

Tal, A., & Wansink, B. (2016). Blinded with science: Trivial graphs and formulas increase ad persuasiveness and belief in product efficacy. *Public Understanding of Science, 25*(1), 117–125.

Tang, R., & Sae-Lim, W. (2016). Data science programs in US higher education: An exploratory content analysis of program description, curriculum structure, and course focus. *Education for Information, 32*(3), 269–290.

Telea, A. C. (2014). *Data visualization: principles and practice.* Boca Raton, FL: CRC Press.

Tenopir, C., Allard, S., Sinha, P., Pollock, D., Newman, J., Dalton, E., ... & Baird, L. (2016). Data management education from the perspective of science educators. *International Journal of Digital Curation, 11*(1), 232–251.

Thompson, G. (2017). Coding comes of age: Coding is gradually making its way from club to curriculum, thanks largely to the nationwide science, technology, engineering and mathematics phenomenon embraced by so many american schools. *The Journal (Technological Horizons In Education), 44*(1), 28.

Tufte, E. (2001) *The visual display of quantitative information* (2nd ed.). Cheshire, CT: Graphics Press.

Tuomi, I. (1999). *Corporate knowledge: Theory and practice of intelligent organizations* (pp. 323–326). Helsinki, Finland: Metaxis.

Velardo, S. (2015). The nuances of health literacy, nutrition literacy, and food literacy. *Journal of Nutrition Education and Behavior, 47*(4), 385–389.e1.

Vilajosana, I., Llosa, J., Martinez, B., Domingo-Prieto, M., Angles, A., & Vilajosana, X. (2013). Bootstrapping smart cities through a self-sustainable model based on big data flows. *IEEE Communications Magazine, 51*(6), 128–134.

Walker, S. (2014). Big data: A revolution that will transform how we live, work, and think. *International Journal of Advertising, 33*(1), 181–183.

Waller, M. A., & Fawcett, S. E. (2013). Data science, predictive analytics, and big data: A revolution that will transform supply chain design and management. *Journal of Business Logistics, 34*(2), 77–84.

Zarcadoolas, C., Vaughon, W. L., Czaja, S. J., Levy, J., & Rockoff, M. L. (2013). Consumers' perceptions of patient-accessible electronic medical records. *Journal of Medical Internet Research, 15*(8), 284–300.

Chapter 6

Fundamentals of Data Science for Future Data Scientists

Jiangping Chen, Brenda Reyes Ayala,
Duha Alsmadi, and Guonan Wang

Contents

Data, Data Types, and Big Data

Data has been frequently discussed along with two other concepts: information and knowledge. However, the concept of data seems less ambiguous than that of information or knowledge. Data are considered to be symbols or raw facts that have not yet been processed (Ackoff, 1989; Coronel, Morris, & Rob, 2012, p. 5). In contrast, information has a number of different definitions, such as information as a thing, something informative, a process, or equivalent to knowledge (Buckland, 1991; Losee, 1997; Saracevic, 1999; Madden, 2000). Knowledge also does not have an agreed-upon definition. Davenport and Prusak (1998) viewed knowledge as "a fluid mix of framed experience, values, contextual information, and expert insight that provides a framework for evaluating and incorporating new experiences and information" (Davenport and Prusak, 1998, p. 5).

The three concepts of data, information, and knowledge are related. They are products of human intellectual activity at different cognitive levels. Data usually need to be processed and organized to become informative for human beings, while knowledge is obtained by making use of data and information through experiences. There is a fine line between data and information: when we talk about digital information or digital data, they are sometimes used interchangeably.

To facilitate our discussion, we follow the idea that data are symbols or raw facts that are stored in electronic media. They can be produced or collected manually by humans or automatically by instruments. The following are examples of data in different forms and formats:

■ *Scientific data*: Scientific data are data collected by scientists or scientific instruments during observations or experiments. Examples of scientific data include astronomical data collected by telescopes, patients' vitals collected by heartbeat monitors, and laboratory data collected by biologists or physicists.

- *Transaction data*: Transaction data are data collected by companies to record their business activities, such as sales records, product specifics, and customer profiles.
- *Data that are of public interest*: Data that reflects environmental status, human activities, and social change, such as geographical data and weather forecasts.
- *Web data*: Web data are documents, web pages, images, and videos published or released on the Internet. These data also record human activities and knowledge. They can be raw facts for certain tasks or decision-making processes.
- *Social media data*: Social media data are generated by Internet users or mobile users, such as Tweets or postings on social media sites, or comments to others' postings.

Data can also be classified based on different facets depending on different perspectives. For example, scientific data can be numeric, textual, or graphical. Or, we can categorize data as structured, semistructured, or unstructured based on whether they have been organized.

Data type is an important concept in this regard. It has been a basic concept in computer science. A computer programmer knows that one needs to determine the data type of a particular variable before applying appropriate operations to it. Computer languages can usually handle different types of data. For example, Python, one of the most popular computer programming languages for data analysis, contains simple data types including numerical data type, string, and bytes, and collection data types such as tuple, list, set, and dictionary. Data might need to be changed to a different type to be processed properly.

Most data we talk about in this chapter are actually digital data in electronic form. Data are generated every day and at every moment. One of the characteristics of data is that it can be copied easily and precisely, or transferred from one medium to another with great speed and accuracy. However, that is not always true, especially when we start to deal with Big Data. Big Data refer to digital data with three Versus: (1) volume, (2) variety, and (3) velocity (Laney, 2011). Volume refers to the situation that data grows at a rapid rate, variety refers to the many data types in which data exist, and velocity refers to the speed of data delivery and processing. Dealing with Big Data requires theories, methods, technologies, and tools, which lead to the emergence of data science, a new discipline that focuses on data processing, which will be discussed in the remaining sections.

Data are an important resource for organizations and individuals who use them to make decisions. When Big Data becomes publicly available, no one can ignore the huge potential impact of Big Data on business, education, and personal lives. Therefore, data science as a new discipline has drawn great attention from governments, industry, and educational institutions.

Data Science and Data Scientists

Dr. William S. Cleveland formally defined data science in 2001 as it is used today in his article "Data Science: An Action Plan for Expanding the Technical Areas of the Field of Statistics" (Cleveland, 2001). Since then, the concept of data science has been further developed and has been increasingly linked to data analytics and Big Data. Companies have realized the value of data science and its usefulness in discovering knowledge and helping with decision-making. The need for data science workers, or data scientists, has increased tremendously in recent years (Leopold, 2017).

Defining Data Science: Different Perspectives

Providing a precise definition for a discipline is important, especially for students who are eager to acquire important knowledge and skills for future job seeking. This section summarizes the definitions of data science and analyzes the features of this discipline. We also provide our own definition based on our understanding of this important field.

The following are prominent data science definitions we found in the literature:

- Data science as an expansion of the technical areas of the field of statistics. (Cleveland, 2001)
- "Data science involves principles, processes, and techniques for understanding phenomena via the (automated) analysis of data." (Provost & Fawcett, 2013)
- "Data science is the study of the generalizable extraction of knowledge from data." (Dhar, 2013)
- "An extension of information systems research." (Agarwal & Dhar, 2014)
- "This coupling of scientific discovery and practice involves the collection, management, processing, analysis, visualization, and interpretation of vast amounts of heterogeneous data associated with a diverse array of scientific, translational, and interdisciplinary applications." (Donoho, 2015)
- "Data science is now widely accepted as the fourth mode of scientific discovery, on par with theory, physical experimentation and computational analysis. Techniques based on Big Data are showing promise not only in scientific research, but also in education, health, policy, and business." (Michigan Institute for Data Science, 2017)
- "Data science enables the creation of data products." (Loukides, 2011, p. 1)
- "Data science is the study of where information comes from, what it represents, and how it can be turned into a valuable resource in the creation of business and IT strategies." (Banafa, 2014)
- "Data science is not only a synthetic concept to unify statistics, data analysis and their related methods but also comprises its results. It includes three phases, design for data, collection of data, and analysis on data." (Hayashi, 1998, p. 41)

- "Data science is a combination of statistics, computer science, and information design." (Shum et al., 2013)
- "Data science is a new trans-disciplinary field that builds on and synthesizes a number of relevant disciplines and bodies of knowledge, including statistics, informatics, computing, communication, management, and sociology, to study data following data thinking." (Cao, 2017)

The above definitions reflect current understanding of this discipline. Together they help to describe the characteristics of data science, which can be summarized into the following aspects:

- The center of data science is data. Especially, Big Data becomes the subject that is investigated.
- The purpose of data science is to obtain information or knowledge from data. The information will help to make better decisions, and the knowledge may help an organization, a state, a country, or the whole of humanity to better understand the development or change of the nature or the society.
- Data science is a multidisciplinary field that has applied theories and technologies from a number of disciplines and areas such as mathematics, statistics, computer sciences, information systems, and information science. It is expected that data science will bring change to impact these disciplines.

We define data science as an interdisciplinary field that explores scientific methodology and computational technology about data, including data management, access, analysis, and evaluation for the benefits of human beings.

In our definition, data are at the center of data science. Any activities around data should be considered within the scope of this field. In other words, data science not only deals with data mining or data analysis, but also explores technologies and approaches for other important activities including data management, data access, and evaluation. We avoid listing related disciplines in the definition because we believe data science has impact and promises new opportunities to every scientific and industry field.

As an emerging discipline, data science is still in its evolving stage. More research and investigation are needed to understand data science problems, challenges, theories, and methodologies. Interesting questions in data science to explore can be: what are its other major concepts in addition to data? What are the problems data science should attempt to tackle? And do we have enough theories and frameworks to guide future practice in data science? As specified by Provost & Fawcett (2013), in order for data science to flourish as a separate field, we should think beyond the specific algorithms, techniques, and tools that we are currently using and work on developing the core principles and concepts that underlie related techniques.

Most Related Disciplines and Fields for Data Science

Data science has been considered an interdisciplinary field since its emergence. It was originally developed within the statistics and mathematics community (Cao, 2017). At its current stage, data science is more closely related to several disciplines than others, which is evident from the definitions in the literature provided in the previous section. Even though we believe data science is relevant to every discipline, some disciplines or fields have more impact on data science than others. These disciplines include mathematics (especially probability and statistics), information science, computer science, knowledge management, management information systems, and decision science. The data science programs at the doctoral and master's levels (see Tables 6.3 and 6.4) confirm the above list, as most of the data science programs were established in colleges, schools, or departments of information science, computer science, statistics, and business management.

Many studies have explored the connections between data science and mathematics, computer science, and management information systems (Cleveland, 2001; Shum et al., 2013; Agarwal & Dhar, 2014). Mathematics including statistics provides numerical reasoning and methodologies for data analysis and processing. Some mathematical concepts and theorems such as those in calculus, linear algebra, and statistics needed to be mastered by potential data science workers; computer science provides programming languages, database systems, and algorithms that a data science worker needs to manipulate numerical and textual data; management information systems, on the other hand, have explored data mining and data analytics for years.

However, we found that the connection of data science with information science was largely neglected which may negatively impact the development of data science as a discipline. The impression of the public on information science is actually library science, which has focused more on classic information resource management, library management, and reference services in libraries. Even though information science has been closely related to library science in history, its landscape has dramatically changed since the advent of the Internet. Information science can definitely provide useful theories and guidance to the development of data science. For example, one of the important concepts to explore in data science may be the data lifecycle. There are many ways to describe and depict data lifecycle (Chang, 2012). For data management, the data lifecycle can include creating data, processing data, analyzing data, presenting data, giving access to data, and reusing data (Boston University Libraries, n.d.). Or it can be a series of processes to create, store, use, share, archive, and destroy data (Spirion, n.d.). Explanations of the different stages included in the lifecycle can be different. But some processes, such as archiving, storing, and accessing have been well explored for information in information science.

Data science is also closely related to knowledge management. Hawamdeh (2003) describes the field of knowledge management and its relationship with

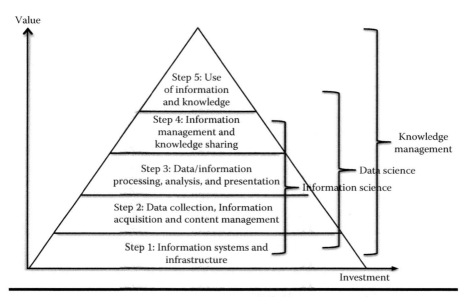

Figure 6.1 Data science in the context of knowledge process. (Adapted from Hawamdeh, S., *Knowledge Management: Cultivating Knowledge Professionals,* Chandos Publishing, Oxford, UK, 2003.)

other disciplines for the purpose of educating knowledge professionals (p. 168). He considered knowledge management a multidisciplinary subject. Knowledge management professionals need to acquire information technology and information science skills, which serve as foundation for higher-level knowledge work such as knowledge management and sharing. We expand on his framework by including an additional step, shown in Figure 6.1, called data processing, analysis, and processing, between the original step 2 (information acquisition and content management) and step 3 (information and knowledge sharing). In this consideration, information science, data science, and knowledge management are closely related and overlapping disciplines or areas. Typically, data science centers on activities in step 2, but should also include activities and tasks in steps 2, 4, and 5.

Data Scientists: The Professions of Doing Data Science

The increasing interest in data science calls for more data science workers. Several reports indicated that data scientists, which is very frequently called the "sexiest job of the twenty-first century" (Davenport & Patil, 2012), are in high demand. McKinsey & Company (2011) pointed to a fact that the United States is facing a shortage of 140,000–190,000 analytical and managerial skills necessary to handle business processes related to critical decision-making and Big Data. Additionally, companies in other sectors have started recruiting data scientists as a crucial component to conduct their business strategy. For example, Greylock Partners, a leading venture capital firm in Silicon Valley that was an early investor in such well-known companies such as Facebook and LinkedIn, is now showing more concern about expanding its team-based portfolio and recruiting more talented people in data science and analytics fields (Davenport & Patil, 2012).

A data scientist is usually described as "a hybrid of data hacker, analyst, communicator, and trusted adviser" (Davenport & Patil, 2012, p. 73). Data scientists' jobs "involve data collecting, cleaning, formatting, building statistical and mathematical models, and communicate professionally with their partners and clients" (Carter & Sholler, 2016). According to Stodder (2015), data scientists begin by analyzing business problems, and then providing strategic and meaningful insights about probable actions and future plans. This strategy involves identifying business drivers, establishing a creative team which is characterized by excellent technical and communication skills, and using the latest analytics and visualization techniques (Stodder, 2015). Provost and Fawcett (2013) specified that successful data scientists "must be able to view business problems from a data perspective."

The literature also describes knowledge and skills that are necessary for data scientists. For example, Dhar (2013) believed that "a data scientist requires an integrated skill set spanning mathematics, machine learning, artificial intelligence, statistics, databases, and optimization, along with a deep understanding of the craft of problem formulation to engineer effective solutions." He considered four basic types of courses that contribute to the knowledge and skills for a qualified data scientist: machine learning, computer science (data structure, algorithms, and systems), correlation and causation, and problem formulation.

Data scientist can be a general term for people working in data science. Different titles can be found from job postings and literature for related positions. It is important to realize that data scientists may vary in their background, experiences, skills, and analytical workflows. In semistructured interviews of 35 data analysts, Kandel, Paepcke, Hellerstein, & Heer (2012) indicated that analysts fall into three categories which are hacker, scripter, and application user, each with different levels of technical skills and tasks that can be performed. They, however, generally share five common tasks that support daily insights, which are discovery, wrangling, profiling, modeling, and reporting.

To verify what we have learned from the literature and to understand the increase in the demand for data scientists (Leopold, 2017), we conducted a small-scale job analysis. The analysis is reported in the next section.

Data Science and Data Analytics Jobs: An Analysis

One of the goals of this chapter is to explore how higher education could produce qualified data science workers, or data scientists. We therefore conducted a preliminary job analysis to understand current job markets for data science workers.

This section will present the purposes of the analysis, the research questions, the process of data collection and analysis, and the results.

Purposes of Analysis and Research Questions

This analysis aimed to understand the current job market for data scientists. Such understanding would provide us with guidelines to designing data science courses and programs at undergraduate and graduate levels. Specifically, we would like to find answers to the following questions:

1. What characterizes the employers who hire data scientists?
2. What were the job titles used when employers are looking for qualified data science workers?
3. What are the general qualifications, knowledge, tools, or skills required by employers?

As mentioned earlier, our analysis also served as a validation of knowledge, skills, and competences for qualified data scientists addressed in the literature.

Data Collection

We recruited the students of a graduate-level information science class at the University of North Texas to collect the data. INFO 5717 is a course in which students learn to build web-based database systems. The students were assigned to five teams with three members in each team. They were asked to collect, process, and upload at least 60 job postings pertaining to data analysts, information management specialists, data scientists, and business intelligence analysts to a MySQL database.

Prior to data collection, we provided a list of metadata elements to guide the process. Students were required to collect information for at least the following elements for each job posting:

- Title of the job
- Start date and end date for the position
- Name of the employer
- Salary
- URL of the job posting at the employer's website
- Position requirements
- Position responsibilities

Students were instructed to collect the above data on the Internet using methods they considered appropriate. The data was collected from job aggregator websites such as Linkedin.com, indeed.com, dice.com, and glassdoor.com in November 2016. Additionally, some of the teams also collected data pertaining to the employers such as the URL of the official employer website, contact email, and the mailing address. The data was initially saved as text files.

We also used this project as an opportunity to train students for future data science-related skills. Therefore we asked students to perform a series of tasks to process the collected data: each student team was required to design a job database using MySQL, containing one or two database tables to store the job posting and the employer information. Using HTML and PHP scripts, students created simple web entry forms to insert the job and employer information into their databases. As part of their course requirements, students developed web pages to manipulate or send queries to retrieve information from the data. They also developed authentication and session-control mechanisms to secure and use their database. In total, the teams collected 364 job postings.

Data Cleanup and Integration

After the students completed the classwork, we exported the job and employer data from five databases to Excel spreadsheets for analysis. We found that even though we gave students metadata specifications, the database tables from each team were not completely aligned—some had more elements, or the same elements were called different names. Employer information was more problematic because we did not specify what exactly information should be collected about employers except that we needed to have the name and URL of the employer for each job posting. Furthermore, students might have collected the same job posting (duplication) as the teams worked independently of each other. Some of the collected records had some missing values such as job location and salaries because the information pertaining to the corresponding job was not stated clearly in the original announcements.

We conducted manual data cleanup to fix the above data issues. Some minor cases of missing salary values were handled by using online popular salaries estimate tools such as the job search engine indeed.com (Doherty, 2010). The manual data cleanup was possible because we had only 364 job postings. After removing the duplicates and consolidating the data elements, we obtained 298 usable observations for analysis.

Tools for Data Analysis

For this small dataset, we mainly used Excel to sort the different elements, to generate frequency counts and percentages, and to produce diagrams. Additionally, SAS Enterprise Miner (SASEM) 14.1 was utilized to perform cluster analysis (SAS Institute, 2014) on qualifications and required experience as presented in the job postings. We also wrote a simple program using the R language to extract the most frequent keywords and produced a visual representation for the results called a word cloud (Heimerl, Lohmann, Lange, & Ertl, 2014). Manual data transformation and adjustment were performed to use these tools for the purpose of answering the research questions proposed in Section 3.1.

Results and Discussion

In this section, we present the results of our analysis and answer our research questions.

Characteristics of the Employers

From the 298 valid job postings, 236 company names were identified by the students. Among them we found some well-known companies such as Apple, Blue Cross and Blue Shield, Capital One, Bank of America, Ericsson, Google, IBM, Infosys, Nike, Samsung, Sprint, Verizon, Walmart, Wells Fargo, and Xerox. These employers covered different industrial fields such as information technology, healthcare, insurance, finance, retail, biotechnology, business services, manufacturing, and media sectors.

With respect to job postings' geographical concentration, Texas leads the United States with more than 24% of the jobs posting, and next comes Illinois and California with 17% and 15% respectively (Figure 6.2). The collected data also shows that both

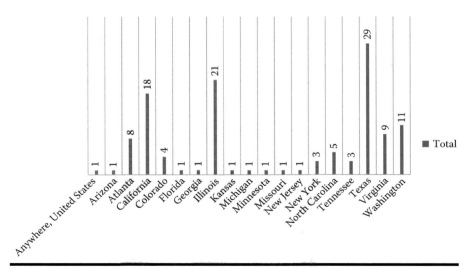

Figure 6.2 Job distributions among states.

graduates and undergraduates were possible candidates for data science jobs. The preferred levels of experience ranged from two to more than 10 years. These results portrayed an image of the job market for data science.

Job Titles

A frequency analysis of the 298 job postings showed that the top job titles used by the employers were data scientist, data analyst, business intelligence analyst or intelligence analyst, information management specialist or analyst, and data engineer. For data scientist and data analysts, the job title could specify different levels, such as lead or principal, senior, junior, or intern. Some job titles, such as data analytics scientist and data science quantitative analyst, were a mixture of data analyst, data science, or data scientist. Other titles included Big Data architect, data integrity specialist, data informatics scientist, marketing analyst, MIS specialist, project delivery specialist, and consultant on data analytics (Table 6.1).

Word Cloud and Clusters on Qualifications and Responsibilities

We used the R programming language to visually summarize the most frequent terms that occurred in two fields: qualifications and responsibilities. The word cloud in Figure 6.3 shows that terms like "experience," "business," "analyst," "analysis," "management," and "skills" stood out for their frequent occurrences, as depicted in Figure 6.1. This was supported by many field researchers, career practitioners, and job posters who confirmed that effective data scientists and analysts need to have a substantial level of managerial and business experience in addition to sophisticated skills in programming, development, design, and statistics (Davenport & Patil, 2012; Kandel et al., 2012).

Table 6.1   Job Titles

Job Titles	Frequency
Data Scientist	95 (31.9%)
Data Analyst	95 (31.9%)
Business Intelligence Analyst (Intelligence Analyst)	50 (16.8%)
Information Management Specialist/Analyst	13 (4.4%)
Data Engineer	10 (3.3%)
Research Analyst	5 (1.7%)
Other	30 (10%)
Total	298

Figure 6.3   Most frequent terms in qualifications and responsibilities.

We also conducted cluster analysis on these two fields. The purpose of the cluster analysis was to identify concepts that are implied in the dataset. The process starts by exploring the textual data content, using the results to group the data into meaningful clusters, and reporting the essential concepts found in these clusters (SAS Institute, 2014). Using the SASEM cluster analysis function, which is based on the mutual information weighting method, we obtained five clusters as presented in Table 6.2. According to the terms in each cluster, we could name cluster 1 as project management; cluster 2 as machine learning and algorithmic skills; cluster 3 as statistical models and business analytics; cluster 4 as database management and systems support, and cluster 5 as communication skills. Among them, cluster 2 is the largest cluster with a relative frequency of 47% among all the terms in these two fields.

The text filter node functionality in SAS enabled us to invoke a filter viewer to explore interactive relationships among terms (SAS Institute, 2014). To preview the

Table 6.2   The Clusters of the Job Postings

Cluster ID	Cluster Name	Descriptive Terms	Frequency (Percentage)
1	Project management	Stakeholder, project, presentation, recommendation, communicate, plan, understand, fast-pace, requirement, manage, team multiple, issue, and improvement	46 (15%)
2	Machine learning and algorithmic skills	Statistical, technique, machine, model, test, learning, system, report, create, design, requirement multiple, and implement	144 (47%)
3	Statistic models and business analytics	Sale, market, customer analytics, help analytic, build, insight, service, opportunity, engineer key, solution, and team	60 (19%)
4	Database management and systems support	Request, assign, perform, process, procedure information, maintain, database, improvement, relate, assist, support, issue, and service	37 (12%)
5	Communication skills	Ability, experience, skill strong, fast-pace highly, demonstrate, communication, write multiple, service analytical, and customer information	21 (7%)

concepts that are highly correlated with the term data, a concepts link map was generated, as depicted in Figure 6.4. The five clusters in Table 6.2 become the six nodes (two notes on analysis/analyze) with links to other outer notes.

The word cloud and cluster analysis showed a broader view of the terms from the job market perspective. Also, further exploration of the concepts link map revealed that data scientists need to have substantial skills in Hadoop, Hive, Pig, Python, Machine Learning, C++ and Java programming, R, and SQL, in addition to being proficient with statistical and modeling packages such as SPSS and SASEM.

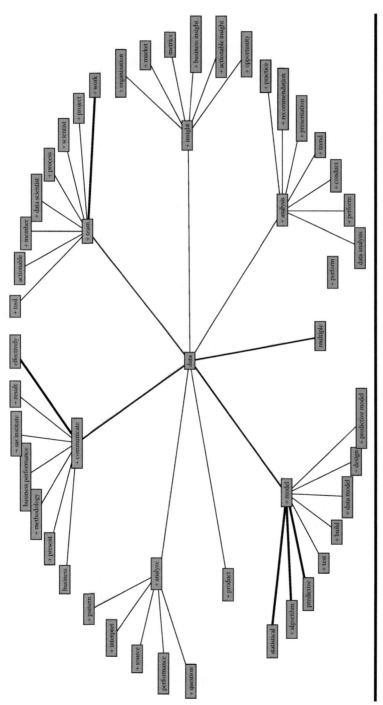

Figure 6.4 The concepts link map of the job postings.

Summary of the Job Posting Analysis

This small-scale study covers many activities a data analyst may perform according to Donoho (2015). We realized that a data science worker needs to know how to organize and clean up data, how to analyze data, and how to create data visualizations using different tools. It would be very beneficial if the data analyst also possesses basic research skills, such as how to collect data and how to evaluate it. Data analysis also involves team-based work of data scientists who use effective tools to build statistical and predictive models to produce business and actionable insights within their organizations, which should be also built in data science curriculum.

Our job analysis above provides insights into this new data science discipline and its career opportunities. These insights can guide educators to develop appropriate data science programs and courses that meet the short-term and long-term needs of the digital economy. Next, we present our thoughts on data science education based on this job analysis and an overview of existing data science programs.

Data Science Education: Current Data Science Programs and Design Considerations

The increasing demand for data science workers has motivated the rapid growth of data science-related programs, as presented in this section. Universities and colleges have been actively developing new data science programs and courses at different levels to meet the needs of the job market on qualified data science workers. It is necessary to review and summarize what has been established so far to develop high-quality data science programs.

Data Science Programs Overview

As we have discussed earlier, data science is an interdisciplinary field. It integrates and applies knowledge and techniques in multiple disciplines and fields. Data science programs and courses have been established to meet the needs for training effective data scientists. These programs offer a variety of courses to train students with the research and professional skills to succeed in leading edge organizations. This section summarizes existing data science programs and trainings.

PhD Programs in Data Science

According to a data science salary study, 88% of data scientists have advanced degrees and 46% have a PhD (Burtch Works Executive Recruiting, 2017). As of June 2017, the website MastersInDataScience.org lists 18 universities offering doctorate-level programs in data science and data analytics, as showed in Table 6.3.

The majority of these PhD programs require students to have undertaken technical coursework or earned a technical degree in computer science, mathematics,

Table 6.3 PhD Programs in Data Science

University	School or Department	Degree, Online or Campus, and URL
Colorado Technical University	Computer Science Department	Doctor of Computer Science with a concentration in Big Data Analytics, online, http://www.coloradotech.edu/degrees/doctorates/computer-science/big-data-analytics
Indiana University Bloomington	School of Informatics and Computing	PhD minor in Data Science, campus or online, http://www.soic.indiana.edu/graduate/degrees/data-science-minor.html
Brown University	Computer Science Department	PhD in Computer Science—concentration in Data Science, campus, http://cs.brown.edu/~kraskat/phd.html
Chapman University	Schmid College of Science and Technology	PhD in Computational and Data Sciences, campus, http://www.chapman.edu/scst/graduate/phd-computational-science.aspx
Georgia State University	Department of Computer Science	PhD (Bioinformatics Concentration), campus, http://cs.gsu.edu/graduate/doctor-philosophy/ph-d-bioinformatics-concentration-degree-requirements/
Indiana University-Purdue University-Indianapolis	School of Informatics and Computing	PhD in Data Science, campus, https://soic.iupui.edu/hcc/graduate/data-science-phd/

(Continued)

Table 6.3 (*Continued*)   PhD Programs in Data Science

University	School or Department	Degree, Online or Campus, and URL
Kennesaw State University	Department of Statistics and Analytical Sciences	PhD in Analytics and Data Science, campus, http://csm.kennesaw.edu/datascience/
New York University	Tandon School of Engineering	PhD in Computer Science with Specialization in Visualization, Databases, and Big Data, campus, http://engineering.nyu.edu/academics/programs/computer-science-phd
Newcastle University	School of Computer Science	EPSRC CDT in Cloud Computing for Big Data, campus, http://www.bigdata-cdt.ac.uk/
Oregon Health and Science University	Department of Medical Informatics and Clinical Epidemiology	PhD in Biomedical Informatics—Bioinformatics and Computational Biology Track, Clinical Informatics Track, campus, http://www.ohsu.edu/xd/education/schools/school-of-medicine/departments/clinical-departments/dmice/educational-programs/clinical-informatics.cfm
University of Cincinnati	College of Medicine	PhD in Biostatistics—Big Data Track, campus, http://med.uc.edu/eh/divisions/bio/programs/phd-biostatistics-big-data-track
University of Maryland-College Park	College of Information Studies	PhD in Information Studies—Concentration in Big Data/Data Science, campus, https://ischool.umd.edu/phd-admissions

(Continued)

Table 6.3 (Continued) PhD Programs in Data Science

University	School or Department	Degree, Online or Campus, and URL
University of Massachusetts-Boston	College of Management	PhD in Business Administration—information systems for data science track, campus, https://www.umb.edu/academics/cm/business_administration_phd/mis
University of Michigan-Ann Arbor	Department of Statistics	PhD in Statistics, campus, http://lsa.umich.edu/stats/graduate-students/graduate-degree-programs/phdprograms.html
University of Pittsburgh	Graduate School of Public	PhD in biostatistics, campus, http://www.publichealth.pitt.edu/biostatistics/academics/phd
University of Southern California	Marshall School of Business	PhD in Data Sciences and Operations, campus, https://www.marshall.usc.edu/index.php/departments/data-sciences-and-operations
University of Washington-Seattle	eScience Institute	PhD in Big Data and Data Science, campus, http://escience.washington.edu/education/phd/igert-data-science-phd-program/
Worcester Polytechnic Institute	College of Arts and Sciences	PhD in Data Sciences, campus, https://www.wpi.edu/academics/study/data-science-phd

or statistics as a prerequisite for entrance. These programs are on-campus and require students to be present in class to complete their degrees; only a couple of programs offer online classes but still have the requirement of attending campus activities to obtain the degree.

Masters Programs in Data Science

The same data science salary study also mentioned that 59% of data scientists who are early in their career hold a master's degree, a significant increase from 48% in 2015 (Burtch Works Executive Recruiting, 2017). As of June 2017, the site MastersInDataScience.org lists 23 schools offering master's degree programs in data science, as showed in Table 6.4.

These master's degree programs have a duration from nine months to about two years, depending on curriculum requirements and the part-time or full-time status of the students. Students usually begin by taking required courses specified in the curriculum in the first semester. Once they obtain basic knowledge and become familiar with technologies in data science, such as programming, databases, and statistics, they proceed to do practicums, application, and capstone projects. Some of the schools provide internship opportunities and practicums so students can work with partner companies in industry or government agencies during their study.

Graduate Certificate Programs in Data Science

Certificate programs provide a flexible way for people to study data science. A certificate program generally takes less time and costs less money than a program intended for earning a degree. It is an attractive option for people who are interested in pursuing a career in this field. The website MastersInDataScience.org has a list of 92 universities offering graduate certificate programs in data science and related specialties.

Massive Open Online Courses

Massive open online courses (MOOCs) are online courses with open (mostly free) access via the Internet. These courses are similar to university courses, but do not tend to offer academic credit. They allow students to choose their own academic path and complete the courses on their own schedule. To study data science or data analysis, students may choose courses from MOOCs, such as Coursera, and construct a learning plan primarily from the following sample areas:

- *Programming languages*: The most popular languages currently used are Python and R for data science courses.
- *Statistics and probability*: There are broad theories that are used for making informed choices in analyzing data and for drawing conclusions about scientific truths from data.

Table 6.4 Master Programs in Data Science

University	School/College	Degree, Online Option, URL
Southern Methodist University	Dedman College of Humanities and Sciences, Lyle School of Engineering and Meadows School of the Arts	Master of Science in Data Science, online option, https://requestinfo.datascience.smu.edu/index10.html
University of California, Berkeley	School of Information	Master of Information and Data Science, online option, https://requestinfo.datascience.berkeley.edu/index3.html
Arizona State University	W.P. Carey School of Business	Master of Science in Business Analytics, online option, https://programs.wpcarey.asu.edu/masters-programs/business-analytics
Carnegie Mellon University	School of Computer Science	Master of Computational Data Science, no online option, https://mcds.cs.cmu.edu/
Columbia University	Data Science Institute	Master of Science in Data Science, no online option, http://datascience.columbia.edu/master-of-science-in-data-science
Cornell University	Department of Statistical Science	Master of Professional Studies in Applied Statistics (Option II: Data Science), no online option, http://stat.cornell.edu/academics/mps

(Continued)

Table 6.4 (*Continued*) Master Programs in Data Science

University	*School/College*	*Degree, Online Option, URL*
Georgia Tech	College of Computing, College of Engineering, and Scheller College of Business	Master of Science in Analytics, no online option, https://analytics.gatech.edu/
Illinois Institute of Technology	College of Science	Master of Data Science, online option, http://science.iit.edu/programs/graduate/master-data-science
Indiana University, Bloomington	School of Informatics and Computing	Master of Science in Data Science, online option, http://www.soic.indiana.edu/graduate/degrees/data-science/index.html
New York University	Center for Data Science	Master of Science in Data Science, no online option, http://cds.nyu.edu/academics/ms-in-data-science/
North Carolina State University	Institute for Advanced Analytics	Master of Science in Analytics, no online option, http://analytics.ncsu.edu/
Northwestern University	McCormick School of Engineering and Applied Science, and School of Continuing Studies	Master of Science in Analytics, and Master of Science in Predictive Analytics, http://www.mccormick.northwestern.edu/analytics/

(Continued)

Table 6.4 (*Continued*) Master Programs in Data Science

University	School/College	Degree, Online Option, URL
Rutgers, The State University of New Jersey	Computer Science Department	Master of Science in Data Sciences, no online option, https://msds-cs.rutgers.edu/msds/aboutpage
Stanford University	Department of Statistics	Master of Science in Statistics: Data Science, no online option, https://statistics.stanford.edu/academics/ms-statistics-data-science
Texas A & M University	Department of Statistics	Master of Science in Analytics, online option, http://analytics.stat.tamu.edu/
University of California, San Diego	Departments of Computer Science and Engineering	Master of Advanced Study in Data Science and Engineering, no online option, http://jacobsschool.ucsd.edu/mas/dse/
University of Illinois at Urbana-Champaign	Department of Computer Science, Department of Statistics, School of Information Sciences and Coursera	Professional Master of Computer Science in Data Science, online option, https://online.illinois.edu/mcs-ds
University of Minnesota—Twin Cities	College of Science and Engineering, College of Liberal Arts and School of Public Health	Master of Science in Data Science, no online option, https://datascience.umn.edu/

(Continued)

Table 6.4 (*Continued*) Master Programs in Data Science

University	School/College	Degree, Online Option, URL
University of San Francisco	College of Arts and Sciences	Master of Science in Analytics, no online option, https://www.usfca.edu/arts-sciences/graduate-programs/analytics
University of Southern California	Viterbi School of Engineering	Master of Science in Computer Science—Data Science, online option, https://gapp.usc.edu/graduate-programs/masters/computer-science/data-science
University of Virginia	Data Science Institute	Master of Science in Data Science, no online option, https://dsi.virginia.edu/academics
University of Washington—Seattle Campus	Interdisciplinary	Master of Science in Data Science, no online option, https://www.datasciencemasters.uw.edu/
University of Wisconsin	Department of Statistics, and UW Extension	Master of Science in Statistics—Data Science, and Online Master of Science in Data Science, https://advanceyourcareer.wisc.edu/degrees/data-science/

Source: http://www.mastersindatascience.org.

- *Data collection*: Extracting data from the web, from APIs, from databases, and from other sources.
- *Data cleaning and managing*: Manipulating and organizing the data collected to make them useful for data analysis tasks.
- *Exploratory data analysis*: Exploring data to understand the data's underlying structure and summarizing the important characteristics of a dataset.
- *Data visualization*: Using plotting systems to construct data graphics, analyzing data in graphical format, and reporting data analysis results.
- *Machine learning*: Building prediction functions and models by using supervised learning, unsupervised learning, and reinforcement learning.

Bootcamps

Bootcamp programs are non-traditional educational paths. Compared with traditional degrees, these programs are intense and have faster routes to the workplace. It is another education option for considering a career as a data scientist. Data science bootcamps provided by DataScience.Community (Datascience.Community, n.d.) has a list of bootcamps available for data science and data engineering.

In summary, data science programs at undergraduate, master, and doctoral level institutions have been developed in the United States. These programs provide a wide range of choices to students who want to obtain knowledge and skills in data science. Still, more data science programs are being developed. It may be the time to explore the characteristics of a competitive data science program.

Data Science Program: An Integrated Design

Based on our understanding of data science and its related concepts, the knowledge and skills required for data science workers, and current data science programs, we believe that a high-quality data science program should provide courses and training in the following areas:

- *Fundamental concepts, disciplines, and the profession*: The program should offer one or two courses to introduce the student to basic concepts, disciplines, and the profession of data science. These courses set up a solid foundation for students to learn more advanced concepts and applications. They may teach not only concepts and characteristics of data, information, knowledge, and the data lifecycle, but also related mathematical concepts, functions, models, and theorems. Students should develop an affection for data and be willing to do data science.
- *Statistical analysis and research methodology*: The program should offer multiple courses to teach data collection and data analysis skills, which are usually taught in master or doctoral level research methodology classes.

- *Programming languages, algorithms, and tools*: Courses such as computer algorithms, programming languages, data mining, database design, and machine learning help students to apply computational techniques for data collecting, cleaning, analysis, prediction, and presentation.
- *Business logic and soft skills*: Students need to understand the purpose of data science and they need to learn how to communicate and collaborate with different business units such as sales, marketing, and production. Also, the need to learn to effectively present data analysis results is also important.
- *Practicum or a capstone project*: Data science students should do at least one project that integrates what they have learned to address a challenging real-world data science problem. The ideal case would be to have the student work in a company or an organization to participate in real data science projects that may help the organization to make better business decisions.

Each student can choose different courses or focus on different areas to improve their knowledge and skills based on their backgrounds and personal interests. For example, a student with a business background may want to take mainly courses on data management and programming languages, while, an information science student should study business logic in addition to statistical analysis and data visualization through coursework or the practicum.

A course may also teach a student knowledge and skills in different areas. For example, a database course may teach students data collection, cleaning, analysis, and report writing through its term project. The instructor can also design projects consisting of research values or reflecting real industry needs.

In general, a flexible curriculum and close connections with profit and non-profit organizations provide students with much flexibility and opportunities to become successful data workers.

Summary and Conclusions

This chapter explored the concept and characteristics of data. It reviewed different perspectives from the literature on data science and provided the authors' definition of this emerging field. The interdisciplinary nature of data science was discussed. We then presented a preliminary analysis of about 300 job postings on data scientists and related positions. Based on the literature review and the job posting analysis, we summarized the knowledge and skills required for different levels of data science workers. Furthermore, we reviewed current data science programs, focusing on the shared values and commonalities of these programs. This chapter concluded with a list of curriculum considerations for designing a data science program.

References

Ackoff, R. L. (1989). From data to wisdom. *Journal of Applies Systems Analysis, 16,* 3–9.

Agarwal, R., & Dhar, V. (2014). Editorial—Big data, data science, and analytics: The opportunity and challenge for IS research. *Information Systems Research, 25*(3), 443–448.

Banafa, A. (2014). *What is data science?* Available at: http://works.bepress.com/ahmed-banafa/15/.

Boston University Libraries. (n.d.). *Data life cycle.* Available at: https://www.bu.edu/datamanagement/background/data-life-cycle/.

Buckland, M. K. (1991). Information as thing. *Journal of the American Society for Information Science, 42*(5), 351–360.

Burtch Works Executive Recruiting. (2017). *Data science salary study.* Available at http://www.burtchworks.com/big-data-analyst-salary/big-data-career-tips/.

Cao, I. (2017). Data science: Challenges and directions. *The Communications of ACM, 60*(8), 59–68.

Carter, D., & Sholler, D. (2016). Data science on the ground, hype, criticism, and everyday work. *Journal of the Association for Information Science & Technology, 67*(10), 2309–2319.

Chang, C. (2012). Data life cycle. Available at: https://blogs.princeton.edu/onpopdata/2012/03/12/data-life-cycle/

Cleveland, W. S. (2001). Data science: An action plan for expanding the technical areas of the field of statistics. *ISI Review, 69,* 21–26.

Coronel, C., Morris, S., and Rob, P. (2012). *Database systems: Design, implementation, and management,* 10th ed. Boston, MA: Course Technology, Cengage Learning.

DataScience.Community. (n.d.). Data science bootcamps. Available at: http://datascience.community/bootcamps.

Davenport, H. T., & Patil, D. J. (2012). Data scientist: The sexiest job of the 21st century. *Harvard Business Review, 90*(5), 70–77.

Davenport, T., & Prusak, L. (1998). *Working knowledge: How organizations manage what they know.* Cambridge, MA: Harvard University Press.

Dhar, V. (2013). Data science and prediction. *Communications of the ACM, 56*(12), 64–73.

Doherty, R. (2010). Getting social with recruitment. *Strategic HR Review, 9*(6), 11–15.

Donoho, D. (2015). 50 years of Data Science. unpublished. Available at: https://dl.dropbox usercontent.com/u/23421017/50YearsDataScience.pdf. University of Michigan's Data Science Initiative.

Hawamdeh, S. (2003). *Knowledge management: Cultivating knowledge professionals.* Oxford, UK: Chandos Publishing.

Hayashi, C. (1998). What is data science? Fundamental concepts and a heuristic example. In C. Hayashi, K. Yajima, H. H. Bock, N. Ohsumi, Y. Tanaka, & Y. Baba (Eds.), *Data science, classification, and related methods.* Springer, Tokyo: Studies in Classification, Data Analysis, and Knowledge Organization.

Heimerl, F., Lohmann, S., Lange, S., & Ertl, T. (2014). Word cloud explorer: Text analytics based on word clouds. *Proceedings of the Annual Hawaii International Conference on System Sciences* (pp. 1833–1842). doi:10.1109/HICSS.2014.231.

Kandel, S., Paepcke, A., Hellerstein, J. M., & Heer, J. (2012). Enterprise data analysis and visualization: An interview study. *IEEE Transactions on Visualization and Computer Graphics, 18*(12), 2917–26. doi:10.1109/TVCG.2012.219.

Laney, D. (2011). 3D data management controlling data volume, velocity, and variety. *META Group: Application Delivery Strategies*, Available at: https://blogs.gartner.com/doug-laney/files/2012/01/ad949-3D-Data-Management-Controlling-Data-Volume-Velocity-and-Variety.pdf.

Leopold, G. (2017). Demand, salaries grow for data scientists. Datanami, January 24, 2017. Available at: https://www.datanami.com/2017/01/24/demand-salaries-grow-data-scientists/.

Losee, R. M. (1997). A discipline independent definition of information. *Journal of the American Society for Information Science, 48*(3), 254–269.

Loukides, M. (2011). *What is Data Science?* Sebastopol, CA: O'Reilly Media.

Madden, A. D. (2000). A definition of information. *Aslib Proceedings, 52*(9), 343–349.

McKinsey & Company. (2011). Big data: The next frontier for innovation, competition, and productivity. *McKinsey Global Institute*, (June), 156. doi:10.1080/01443610903114527.

Michigan Institute for Data Science (MIDAS). (2017). About MIDAS. Available at: http://midas.umich.edu/about/.

Provost, F., & Fawcett, T. (2013). Data science and its relationship to big data and data-driven decision making. *Big Data 1*(1), BD51–BD59.

Saracevic, T. (1999). Information science. *Journal of the American Society for Information Science, 50*(12), 1051–1063.

SAS Institute. (2014). *Getting Started with SAS® Text Miner 13.2*. Cary, NC: SAS.

Shum, S. B., Hall, W., Keynes, M., Baker, R. S. J., Behrens, J. T., Hawksey, M., & Jeffery, N. (2013). Educational data scientists: A scarce breed. Available at: http://simon.buckinghamshum.net/wp content/uploads/2013/03/LAK13Panel-Educ Data Scientists.pdf.

Spirion. (n.d.). Data lifecycle management. Available at: https://www.spirion.com/data-lifecycle-management/.

Stodder, D. (2015). Chasing the data science unicorn. Available at: https://tdwi.org/articles/2015/01/06/chasing-the-data-science-unicorn.aspx.

Chapter 7

Social Media Analytics

Miyoung Chong and Hsia-Ching Chang

Contents

Introduction

Social media analytics (SMA) has become one of the core areas within the extensive field of analytics (Kurniawati, Shanks, & Bekmamedova, 2013). Generally speaking, SMA facilitates proper analytic techniques and tools to analyze user-generated content (UGC) for a particular purpose. The content of social media comes in different forms of repositories (Sinha, Subramanian, Bhattacharya, & Chaudhary, 2012). For instance, the repositories could be blogs (Tumblr), microblogs (Twitter), wikis (Wikipedia), social networking sites (Facebook and LinkedIn), review sites (Yelp), and multimedia sharing sites (YouTube) (Holsapple, Hsiao, & Pakath, 2014).

SMA has involved many techniques and tools to mine a variety of UGC, including sentiment analysis, topic modeling, data mining, trend analysis, and social network analysis. A primary reason for growing interest in SMA is the real time accessibility regarding size and diffusion speed of the UGC on social media (Holsapple et al., 2014). For example, a YouTube video showing a Coke bottle exploding because of Mentos became the most popular video on YouTube and subsequently in news shows in 2009 (Kaplan & Haenlein, 2010). Despite the burgeoning attention to SMA, social media analysts need to deal with arduous tasks in the given context to accomplish a specific analytical objective because the UGC from social media is generally improvised, freeform, and a mixture of relevant and irrelevant resources.

During the 1990s, the Internet and the World Wide Web (WWW) were adopted to facilitate social communication. The development and quick diffusion of Web 2.0 technologies made a revolutionary leap in the social element of the Internet application. Social media users can take advantage of the user-centered platforms with UGC while the various set of possibilities exist to connect these cyberspaces to construct online social networks. Adopting social media platforms serves diverse purposes, such as social interaction, marketing, digital education, disaster management, and civic movement, and by user groups, including business, governments, nongovernment organizations, politicians, and celebrities. For example, Psy's music video, Gangnam Style released on July 15, 2012, became the most watched video on YouTube history with the view of more than 2.8 billion times, which also tremendously contributed to making the singer to become a world star.

Facebook, starting its service in 2004, currently reached more than 2 billion active users worldwide as of June 2017. Launched in 2006, Twitter has obtained 319 million active users as of April 2017 and created 500 million Tweets every day. Though Twitter has a smaller number of active accounts than YouTube, WhatsApp, or Facebook Messenger, Facebook and Twitter have been the most visible social media platforms initiating new functional services by incorporating Web 2.0 components into other web-based applications (Obar & Wildman, 2015). When compared to the lists in 2015 and 2017, overall the popular social media platforms are similar besides the leap of WhatsApp and Facebook Messenger. Figure 7.1 presents

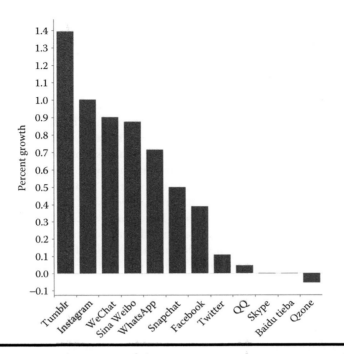

Figure 7.1 The percentage growth between 2015 and 2017 among social media services that have most active users.

the significant growth of users of Tumblr, Instagram, WeChat, Sina Weibo, and WhatsApp between 2015 and 2017.

An official definition of social media requires distinguishing the two interrelated concepts, Web 2.0 and UGC (Kaplan & Haenlein, 2010). The term Web 2.0 was first adopted in 2004 to describe the new web environment where software developers and end users began to mash up the content and applications cooperatively and innovatively. Though Web 2.0 does not indicate any particular technical term related to the WWW, it includes a few essential functionalities for its performances such as Adobe Flash, Really Simple Syndication (RSS), and Asynchronous JavaScript (AJAX) (Kaplan & Haenlein, 2010).

While considering Web 2.0 as the platform of the social media development and technological foundation, UGC refers to the collective outcomes of user interactions with social media. Extensively employed in 2005, UGC represents the diverse formats of content in social media (Kaplan & Haenlein, 2010). The Organization for Economic Cooperation and Development specified three requirements of UGC: public accessibility of the content, creativity, and amateurism (Peña-López, 2007). Drawing upon these descriptions about Web 2.0 and UCG, Kaplan and Haenlein (2010) define social media as a "group of Internet-based applications that build on the ideological and technical foundations of the Web 2.0, and that allow the creation and exchange of User Generated Content" (p. 61). Extending this perspective, Batrinca and Treleaven (2015) delineate social media as "web-based and mobile-based Internet applications that allow the creation, access and exchange of UGC that is ubiquitously accessible" (p. 89).

To characterize social media, Kaplan and Haenlein (2010) categorize social media platforms by two dimensions: social presence and media richness, and self presentation and self disclosure. Table 7.1 presents the six different types of social media. Regarding social presence and media richness, collaborative social media applications, such as Wikipedia, blogs represent the lowest scores because these are usually text-based services and limited to simple trade regarding content sharing.

The middle level presents content communities, such as YouTube, and social networking sites, such as Facebook and Instagram, which allow sharing images and multimedia content as well as text-based communication. The highest level includes virtual game and social worlds, such as World of Warcraft, Second Life, and virtual reality role-playing games, which pursue any possible aspects of direct interactions in a cyber world (Kaplan & Haenlein, 2010). As for self-presentation and self-disclosure, blogs ranked higher than collaborative projects because they are more focused on particular content areas. Likewise, social networking sites present more self-disclosure than content communities. Lastly, cyber social worlds demand further self-disclosure than virtual game worlds because disclosers are strictly bounded by game rules (Kaplan & Haenlein, 2010).

Obar and Wildman (2015) focus on the importance of user profile elements in defining social media. Despite the substantial differences in the options of identifying users and information requested depending on the social media platforms, they usually require creating a user name along with contact details and a profile picture. The requirement enables social media users to make connections and share their content without many doubts because without verifying user information, discovering and linking to other users could be a challenge. Obar and Wildman (2015) describe the functions of social media platforms as connecting the online

Table 7.1 Classification of Social Media by Social Presence/Media Richness and Self-Presentation/Self-Disclosure

Self-presentation/ Self-disclosure		Social Presence/Media Richness		
		Low	Medium	High
	High	Blogs, Quora, Flicker	Facebook, Instagram	Second life
	Low	Wikis	YouTube, Pinterest	World of Warcraft, virtual reality role-playing game

Source: Kaplan, A. M., and Haenlein, M., *Bus. Horiz.*, 53, 59–68, 2010.

social networks. Along this line, boyd and Ellison (2010) considered user profiles as the backbone of social media platforms.

A majority of Americans now state that they obtain news more increasingly via social media (Kohut et al., 2008). According to a recent study, 60% of all Americans use social media to get news (Gottfried & Shearer, 2016), whereas approximately 80% of online Americans are currently Facebook users and among these users, 66% obtain news on the site (Greenwood, Perrin, & Duggan, 2016). On Twitter, around 60% of the users receive news on the site, which results in a bigger percentage with a smaller user foundation (16% of American adults). In addition, research studies have continuously discovered that the more people read news media, the more likely they are civically and politically involved across various measures (Wihbey, 2015).

Historical Perspective of Social Networks and Social Media

Throughout history, humans have developed technologies to communicate with each other. There are plenty of ideas regarding the first advent of social media, but the adoption of the telegraph to exchange messages in 1792 can be one of the earliest examples social media (Edosomwan, Prakasan, Kouame, Watson, & Seymour, 2011). In the late 1800s, drawing from the social contract conceptions of society, German sociologist Ferdinand Tonnies explained that social groups have formed by sharing values or conflicts, and Durkeim, a French sociologist, applied sociological theory to empirical research. People used the radio and telephone for social interaction, although the radio communication was one-sided (Rimskii, 2011).

Social networks have evolved over the decades by employing digital media in the modern day. However, social communication using technology started with the telephone system in the 1950s. Phone phreaks used handcrafted devices to access the telephone network for both exploration and to exploit vulnerabilities allowing free access (Borders, 2009). E-mail was introduced to the public in the 1960s. In 1969, ARPANET, a network of time-sharing computers that established the foundation of the Internet created by a U.S. government agency, Advanced Research Projects Agency, was developed (Edosomwan et al., 2011). In the same year, CompuServe, an early form of cloud computing, was designed to provide time-based services by leasing time on its machine, but this was ineffective because of the high fees for the service (Rimskii, 2011).

During the 1970s, the idea of social media was developed through computer technologies. For example, Multi-User Dungeon (MUD), a real time and online role-playing game organized through interactive fiction, is mainly text-based, where users need to input commands in a natural language. Bulletin board system (BBS) was introduced in 1978, and it allows its users to upload and download software, trade messages, and read news by logging into the system. Bulletin boards were the precursor of the WWW. The Usenet, developed by professors at Duke University in 1980, is an online discussion system where users can post news, messages, and articles (Ritholz, 2010).

During the 1980s, the Whole Earth Lectronic Link (WELL), General Electric Network for Information Exchange (GEnie), Listserv, and IRC were introduced. The WELL started in 1985 as a conversation-based community through BBS and is one of the long-standing online communities. GEnie is an Internet application programmed with ASCII language and was in a competitive relationship with CompuServe. Launched in 1986, Listserv was the first electronic mailing application through software installed on the server, and it enabled automatic and machine-processed electronic emailing distribution, which allowed an e-mail to reach a group of people registered on the mailing list. IRC (Internet relay chat) was created for group conversations as a type of real time chat, online text messaging, concurrent conferencing, and data transferring between two people (Ritholz, 2010).

During the 1990s, many social networking sites, such as SixDegrees.com, MoveOn.org, BlackPlanet.com, and AsianAve.com, were created. These websites were Internet niche social networking sites with which users can interact. Additionally, blogging sites were created, such as Blogger and Epinions where users can find and write review comments about commodities (Edosomwan et al., 2011). In this period, ThirdVoice, a free plug-in user-commenting service on web pages, and Napster, a peer-to-peer music file sharing software application, were generated. Opponents of ThirdVoice criticized that the comments were frequently insulting. Through Napster, users could exchange music files by deviating from the legal distribution methods; this eventually determined the end user to be a violator of copyright laws (Ritholz, 2010).

In 2000, many social networking sites sprang up, including LunarStorm, Cyworld, and Wikipedia. LunarStorm (www. LunarStorm.se), a Swedish commercial virtual site and Europe's first Internet community, was specially designed for teenagers. In 2001, Fotolog, sky blog, and Friendster started their services. In 2003, MySpace, LinkedIn, tribe.net, Last.fm and in 2004, Facebook Harvard, Mixi, and Dogster emerged. In 2005, Yahoo! 360, BlackPlanet, and YouTube evolved (Junco, Heibergert, & Loken, 2011). Though BlackPlanet was created in the 1990s, it became popular and evolved in 2005. Twitter launched its service in 2006, and the number of users rapidly increased because of its microblogging style and celebrity adoption (Jasra, 2010).

2010 was Facebook's year. Facebook took over Google's position as in the biggest website regarding market share in July 2010, and *Time* magazine recognized Facebook CEO Mark Zuckerberg as Man of the Year (Cohen, 2010). In the *Harvard Business Review*, Armano (2014) describes six social media trends for 2010. He points out that there will be more scaled social initiatives beyond one-off marketing or customer relations efforts. For instance, Best Buy's employees were directed to participate in customer support on Twitter through a company-built system that monitors their participation. With approximately two-thirds of organizations prohibiting access to social network applications from corporate-owned devices, while simultaneously the sales of smartphones have skyrocketed, employees are likely to feed their social media cravings on their mobile devices (Armano, 2014). Sharing content through online social networks rather than by email became mainstream for people. For example, the *New York Times* created a content sharing application that enables them to quickly distribute an article across social media networks such as Facebook and Twitter (Armano, 2014).

Communication methods, such as face-to-face meetings, phone calls, and email, have restrictions. For example, people can easily forget and lose the content of a conversation because of memory loss or missing notes. Social media aids interaction and communication among individuals and helps them to reach a larger audience. Adopting social media has effectively raised the channels of communication (Edosomwan et al., 2011) because it has become much easier for people to send timely messages via a Tweet or an instant messenge, which saves time and energy (interacting with people using social media technology). In addition, engaging social media reinforces brand experience, which helps to establish a brand reputation. Customer brand awareness begins with the employees' experiences about the company (Carraher, Parnell, & Spillan, 2009).

Evolution of Analytics

The area of analytics started in the mid-1950s along with new methods and tools that created and captured a significant amount of information and distinguished patterns in it faster than unaided human intelligence ever could. The concepts and techniques of business analytics have continuously changed based on analytical circumstances. The era of Analytics 1.0 can be described as the era of "business intelligence (BI)." In this period, an analytic technique made progress on providing an impartial comprehension of major business developments and provided managers with fact-based understanding for better insights to make marketing decisions. From this period, production data, including processes, sales, and customer care, was recorded, accumulated, and innovative computing techniques played a critical role. For the first time, customized information systems with considerable scale increased the investment to the systems. Analysis process required far more time to have data than to analyze it. The process, which took weeks to months to conduct analysis, was slow and meticulous. Therefore, it was critical to discover the most important questions and concentrate on them. In this era, analytics focused on better decision-making opportunities to enhance predictions on specific primary components.

The first generation of analytics was followed by the era of Analytics 2.0, known as the era of Big Data. Analytics 1.0 continued until the mid-2000s, when internet-based companies, such as Google, Facebook, Twitter, and LinkedIn, started to collect and analyze new types of information (Davenport & Court, 2014). In this Analytics 2.0 phase, the industry adopted new tools to have innovative analytic techniques and attract customers. For instance, LinkedIn built many big data products, such as Network Updates, People You May Know, Groups You May Like, Jobs You May Be Interested In, Companies You May Want to Follow, and Skills and Expertise (Davenport & Court, 2014). This improvement required powerful infrastructure and analytics talents who can develop, obtain, and master different types of analytics tools and technologies. A single server could not perform analysis of big data.

Hadoop, an open-source parallel server software structure for fast batch data processing, was introduced. A NoSQL database system was employed to manage relatively unstructured data. The cloud-computing system was adopted to manage and analyze lots of information. Machine-learning (ML) techniques were also applied to quickly create models from the rapidly-shifting data. Data analysts had to make it organized for statistical analysis with scripting languages, such as Python, Hive, and Pig. Spark, an open source cluster computing system especially for streaming data, and R, a framework for statistical computing and graphics, have become popular as well. The required capabilities for Analytics 2.0 were much higher than those for Analytics 1.0. The quantitative analysts of Analytics 2.0 were named data scientists, who had both computational and analytical skills.

The revolutionary companies in Silicon Valley started to pay attention to analytics to promote shopper-facing products and services. They drew consumers' attention to their websites through certain computational algorithms, recommendations from other users, product reviews, and targeted advertisements, which are driven by analytics drawn from enormous quantities of data.

When analytics entered the 3.0 phase, almost every company in every industry makes decisions on products and services based on data analytics. An enormous amount of data has been created whenever companies produce, deliver, and do marketing for products and services or interact with customers. For example, UPS entirely reorganized its delivery routes by implementing big data analytics. UPS has traced and recorded package shipment and delivery since the 1980s by accumulating information on the 16.3 million (on average) package transactions and movements daily, which obtains 39.5 million daily tracking requests. The telematics sensors installed in more than 46,000 UPS trucks trace metrics such as speed, braking, direction, and training performance. This tracked data demonstrates daily performance as well as information about a substantial reconstruction of delivery routes. UPS's On-Road Integrated Optimization and Navigation (ORION) is a proprietary route optimization software that relies greatly on online map data and optimization algorithms.

This analytics solution will eventually be able to redesign a UPS driver's pickups and distributions in real time. It is also known as "the world's largest operations research project" due to its scale and scope (Davenport & Court, 2014). In 2015, UPS adopted ORION to 70% of U.S. routes identified as part of the initial deployment, which aided UPS to diminish driving miles, fuel consumption, and car emissions. With the full employment of the ORION system in 2016, UPS expected to save $300–$400 million by saving 10 million gallons of fuel, reducing 100 million miles, and reducing 100,000 metric tons of carbon dioxide production annually, which is equal to taking more than 20,000 passenger vehicles off the road each year. ORION costs $250 at full deployment, but ORION has already saved UPS more than $320 million at the end of 2015 (Informs.org, 2017).

The ORION project demonstrates that a company's application of analytics not only improved internal business decisions but also generated values, including environmental protection. This is the core of Analytics 3.0. Companies employed analytics to assimilate information and provide examples while hinting at other possible applications. Internet-based firms, such as Google, Facebook, Amazon, eBay, and Facebook, processing enormous amounts of streaming data, have become frontiers of this approach. These online companies have bloomed mainly by assisting their customers to make effective decisions and actions, and not by providing simple information for them. This strategic shift signifies an advanced function for analytics within institutions. Companies need to identify a division of related tasks and react to new competence, environments, and preferences.

Social Media Analytics

Social media has dramatically changed information creation and sharing. What was once the exclusive domain of traditional news media, television, radio, public institutions, and private entities are willing to reach out to social media. Thanks to the prevalence of social media, information dissemination on a mass scale now flows from a customer's fingertips to the Internet. According to Chung (2016), "Its key capabilities include universal reachability, speed of interaction, multimodal communication, device compatibility, and emotional appeals of SM" (p. 105). Nearly ubiquitous Internet access combined with intuitive web applications has democratized mass media. Social media supports many forms of Internet-fueled communication. Therefore, an individual can rate any customer experience (review websites), maintain an electronic diary (weblog or blog), provide recommendations for product purchase (recommender website), and communicate privately with friends and family or publicly with strangers around the world through different social media platforms.

Defining Social Media Analytics

Regardless of medium, social media outlets capture tremendous amounts of data. The data are a treasure trove awaiting discovery. In recognition of the underlying value embedded in social media data, analysts have begun to apply the power of data analytics to social media data. While sharing analytic techniques with other forms of media, SMA focuses exclusively on various types of data from social media. SMA attempts to unveil patterns of unstructured data by developing and evaluating tools that can capture, analyze, synthesize, and visualize social media data (Zeng, Chen, & Li, 2010). From the perspective of business, Holsapple et al. (2014) define business SMA as "all activities related to gathering relevant social media data, analyzing the gathered data, and disseminating findings as appropriate to support business activities such as intelligence gathering, insight generation, sense-making, problem recognition/opportunity detection, problem solution/opportunity exploitation, and/or decision-making undertaken in response to sensed business needs" (p. 4).

Processes of Social Media Analytics

Regarding the processes of performing SMA, Fan and Gordon (2014) describe SMA as a three-step process progressing mostly linearly (some steps overlap) from capture to understand and ending with presentation. During capture, data are gathered and preprocessed, and pertinent information is extracted. In the understanding phase, data are further analyzed using advanced analytics techniques, such as sentiment analysis, topic modeling, social network analysis, trend analysis, and so on. During the presentation phase, the analysis is summarized and visualized as a series of reports. Bengston et al. (2009) view SMA as a monitoring process consisting of five phases: the definition of the problem, the identification of online news sources that will be used to collect content, the development of search terms with respect to several strategies and the storage of search results, the analysis of textual content as the core phase of the analytics process, and the presentation of gained results.

Another approach is proposed by Bruns and Liang (2012) for the analysis of Twitter data. They suggest a three-step approach including data collection and storage, data analysis, and the publication of results. Similarly, Stavrakantonakis et al. (2012) mention four steps starting with the collection of data, the establishment of a "listening grid" to store gathered data, followed by the actual analysis, and finally the generation of "actionable reports" to support decision-making (p. 54). Likewise, Ruggiero and Vos (2014) present a four-step approach, which are preparation, data collection, data analysis, and reporting. Summarizing the existing procedures of an analytics pipeline, Wittwer, Reinhold, and Alt (2016) suggest four main steps. First, the definition of media sources and appropriate search terms. Second, the pre-processing of data for integrating and aligning the heterogeneous data from different sources as well as the application of various analysis techniques. Third, the presentation and interaction of users with the results. Fourth, the use of results, which is the only manual task (Wittwer et al., 2016).

Social media data are roughly categorized as historical datasets and real-time feeds. Data is further classified as raw, cleaned, and value-added. Historical data are archives of previously captured news feeds, financial and economic data, and social commentary, whereas real-time feeds are continuous data streams taken directly from a social media site, news feed, or financial exchanges (Batrinca & Treleaven, 2015). Social media data is acquired (captured) through a variety of techniques. Some data providers (e.g., Wikipedia) allow a full database download. Other vendors provide controlled access to data via dedicated software tools (e.g., Google Trends). Lastly, data are available through an application programming interface (API). An API exposes data over HTTP for system-to-system data transfer. Examples include Wikipedia's DBpedia, Twitter, and Facebook (Batrinca & Treleaven, 2015).

Raw data are the data as the source creates it. It may contain errors and may have never been fixed. Cleaned data are raw data that has had some amount of pre-processing to remove errors such as typos, wrong facts, outliers, or missing information. Value-added data are cleaned data that have been augmented with additional knowledge gleaned from the analysis (Batrinca & Treleaven, 2015).

As for data formats, data is most commonly encoded into hypertext markup language (HTML), extensible markup language (XML), JavaScript object notation (JSON), or comma-separated value (CSV) files. HTML is a markup language for authoring web pages. It is responsible for delivering page structure directives (tables, paragraphs, and sections) and content (text, multimedia, and images) to a web browser. XML is a markup language used to create structured textual data. XML markup functions as metadata that describes the content contained within the markup. Both XML and HTML wrap content between start and end tags (e.g., </div>). JSON is a lightweight data structure derived from a subset of the JavaScript programming language. It has a simple key-value structure that is simultaneously a machine and human readable format. Because JSON is familiar to C language software developers, it is used in many data feeds. CSV refers to any file that has a single record per line, fields that are separated by a comma, and text that is encoded using ASCII, Unicode, or EBCDIC (Batrinca & Treleaven, 2015).

After data are captured, analysts can proceed to the understanding phase. In this step, the actual value of SMA is revealed. Value is defined through the lens of the analyst. An analyst working under the auspices of a corporation has a different concept of value than an analyst working to support a political candidate. Holsapple et al. (2014) offer the following benefits of business SMA: improved marketing strategy, better customer engagement, better customer service, reputation management, and new business opportunities. Like epinion.com and Amazon, review websites allow customers to post comments about an experience with a product or service. SMA can provide insights regarding these experiences that in turn inform marketing strategy.

Customer engagement improves when businesses target customer values and provide additional channels for two-way communications. Customer service improves as companies become more in tune with the needs of their clients. As an example, SMA identified stock picking experts on the Motley Fools CAPS voting site. Picks by these experts consistently outperformed other voters. These stock picks were provided to Motley Fool customers creating a better product and improving customer service. SMA assists with reputation management by allowing a business to extract sentiment insight from social media content related to the company. Finally, SMA can identify new business opportunities by monitoring distinctive phrases that spread rapidly on social media applications.

These phrases offer insights that can help a firm's decision on providing a new product or what new features to add to an existing product. That is because SMA has played a role in each stage of the product development lifecycle: design development, production, utilization, and disposal (Fan & Gordon, 2014). In the design development phase, trend analysis can identify a change in customer attitude and desires for a product. Application development firms regularly release multiple product versions and then solicit feedback directly from potential end users. Responses then drive the next product feature sets. In extreme cases, end users and software developers cocreate products using social media as a collaboration platform. In the production phase, SMA allows firms to anticipate changing demand by scaling production either up or down. A company can also use social media data to track issues posted by competing companies regarding a supplier, thus allowing the company to avoid the same problem.

Furthermore, SMA is most commonly leveraged during the product adoption phase. Firms closely monitor brand awareness (introduction to the brand), brand engagement (connection to the brand), and word of mouth (customer chatter about a brand). Metrics, like the number of Tweets and followers, indicate an excitement (positive or negative) for a brand. Customer segmentation (grouping customers based on shared attributes) allows a firm to tailor marketing messages for a particular group. In the product disposal stage, SMA can follow, and companies can directly communicate with, consumers on disposal strategy. This is especially useful when product disposal creates an environment risk (e.g., toxic batteries). Because product disposal is often accompanied with product replacement, firms can use SMA to market replacement products selectively.

Similar to people in business, politicians benefit from SMA as well. The benefit, however, is different as politicians have wildly divergent needs. Reputation and impression management are of utmost importance politically. Politicians are interested in how people discuss them, new topics that might trigger crises or sway sentiment against them, and measuring the degree of influence they exert. Additionally, SMA is applied in an exploratory fashion to monitor for "late-breaking" topics thus gaining advanced notification and a longer preparation period (Stieglitz & Dang-Xuan, 2013). Politicians use social media to gain support, encourage civic engagement, and promote themselves. President Obama's 2008 campaign is widely considered the first campaign where social media and SMA had a measurable and significant effect on the election outcome (Grubmüller, Götsch, & Krieger, 2013).

Stieglitz and Dang-Xuan (2012) propose an analytics framework for conducting SMA in the political realm consisting of two major parts: data tracking and monitoring, and data analysis. Data tracking and monitoring are concerned with techniques for acquiring social media data while data analysis pertains to analysis methods and reporting. For data analysis, they suggest three approaches: topic or issue-related, opinion or sentiment-related, and structural. Topic or issue-related analysis or issue management uses text mining techniques to identify topics (issues) that might become a crisis or scandal and damage reputation. Opinion or sentiment-related analysis is an attempt to identify citizen (voter) opinion regarding a topic. Politicians use such information to make informed decisions about upcoming votes, which issues to address or refute, and decision-making. The structural analysis attempts to identify key players in a community network. This information allows a politician to seek favor with influential people. The same approach can get extended to entire communities if it appears that individual communities exert more influence on an issue.

Social Media Analytics Techniques

This section introduces SMA techniques pertaining to different procedures of SMA. Here we reiterate the common processes of SMA specified in relevant studies: identifying data sources, data acquisition, data analysis, and representing results.

Identifying Data Sources

Data sources could be solely textual content or aggregating multiple data sources, including social networking services, blogs, and RSS feeds, supplemented by various types of data, such as geospatial data, video data, and communication data (Batrinca & Treleaven, 2015). Depending on looking backward or forward, we can group social media data into:

- *Historical data*: Recently amassed and stocked social, news, and business-related data.
- *Real-time data*: Streaming data from social media, news agencies, and monetary trade, telecommunication services, and global positioning system (GPS) gadgets.

Data also subdivides into raw, cleaned, and value-added data based on the level of the process. Raw data are primary and unprocessed data directly from the source that includes errors. Cleaned data is processed and modified raw data by removing erroneous parts. Value-added data are cleaned, analyzed, tagged data, or augmented data with information. Analysts need access to historical and real-time social media data, particularly the primary sources, to conduct comprehensive research. In many cases, it is beneficial to combine different data sources with social media data. For example, opinions about negative financial news, such as a downslide of the stock market, might be presented in social media. Thus, when conducting textual data analysis of social media, considering multiple data sources will potentially help detect the context and perform deeper analysis with a better understanding of the data. The aggregation of data sources is certainly the focus of future analytics (Batrinca & Treleaven, 2015).

Data Acquisition

There are four ways to directly acquire data from social media, including scraping, application programmable interfaces (API), RSS feed, and file-based data acquisition. Scraping is gathering social media data from the Internet, and this is usually unstructured text data. Scarping is also known as site scraping or web data extraction, web harvesting, and web data mining. Analysts can collect social media data systematically if social media data repositories guarantee programmable HTTP-based approach to the data through APIs. Facebook, Twitter, and Wikipedia provide access via APIs. An RSS feed is a systematic method of streaming social media data to deliver to its subscribers. An RSS feed is considered semi-structured data. File-based data is usually acquired from spreadsheets and text files.

Data Analysis Techniques

Sentiment analysis, topic modeling, and visual analytics are the primary analysis techniques of social media data. Sentiment analysis, namely opinion mining, is usually employed to determine the attitude and perception of an author concerning an event. For this purpose, natural language processing (NLP), computational linguistics, and text analytics are deployed to identify and extract subjective information from textual data. Two quantitative methods of topic modeling, latent Dirichlet allocation (LDA) and latent semantic analysis (LSA), facilitate unsupervised (unlabeled) text analysis and can also be implemented to analyze social media data.

Sentiment Analysis

Web 2.0 users can form, participate in, and collaborate with virtual online communities. Microblogs, such as Twitter, became user-generated information-abundant resources, because users began sharing information and opinions in diverse online events and domains, ranging from breaking news to celebrity gossip, to product reviews, and to discussions about recent incidents, such as the Orlando massacre in June 2016, the U.S. Presidential election, and hurricanes Harvey and Irma in 2017.

Social media includes a copious amount of sentiment-embodied sentences. Sentiment refers to "a personal belief or judgment that is not founded on proof or certainty" (WordNet 2.1 definitions), which may depict the emotional state of the user, such as happy, sad, angry, or the author's viewpoint on a topic. Sentiment analysis, an important aspect of opinion mining, aims to discover whether the polarity of a textual corpus, a collection of written texts, leans towards positive, negative, or neutral sentiments.

During the past decade, sentiment analysis has been a popular research area in SMA, particularly on Twitter because of the accessibility to diverse fields. Scholars have applied sentiment analysis to business predictions (Liu, Huang, An, & Yu, 2007), politics (Park, Ko, Kim, Liu, & Song, 2011), finances (Dergiades, 2012), and the automatic identification of sentiment on textual corpora (Kontopoulos, Berberidis, Dergiades, & Bassiliades, 2013). Sentiment analysis can be conducted at different levels, such as the document, sentence, word, or feature level. In the case of Twitter analytics, word-level analysis fits better because Twitter limits users to 140 characters of information. Cross-examination of the literature verifies that the evaluation measures of mechanically annotating sentiment at the word level lie in two approaches: "dictionary-based approaches" and "corpus-based approaches" (Kumar & Sebastian, 2012, p. 372). Moreover, for automated sentiment analysis, diverse methods have been adopted to forecast the sentiment of words, articulations, or corpora, and these approaches incorporate NLP and ML algorithms (Kumar & Sebastian, 2012). Recently, researchers analyzed Twitter activities and reported that more than 80 percent of users either post status updates to their followers or spread information regarding their daily experiences (Thelwall, Buckley, & Paltoglou, 2011).

Sentiment analysis is about mining attitudes, emotions, feelings—it is subjective impressions rather than facts. Generally speaking, sentiment analysis aims to determine the attitude expressed through the text with respect to the topic or the overall contextual polarity of a document (Mejova, 2009). Pang and Lee (2008) provide a thorough documentation on the fundamentals of sentiment classification and extraction, including sentiment polarity, degrees of positivity, subjectivity detection, opinion identification, nonfactual information, term presence versus frequency, parts of speech (POS), syntax, negation, topic-oriented features, and term-based features beyond term unigrams. Moreover, sentiment analysis divides into sentiment context, sentiment level, sentiment subjectivity, sentiment orientation (polarity), and sentiment strength depending on specific tasks. Used for discovering the context of the text to identify opinion, sentiment context varies from expert review portals to generic forums where opinions cover a spectrum of topics (Westerski, 2008). Sentiment level means that the text granularity of sentiment analysis can be performed, such as the document, sentence, or feature level. Sentiment subjectivity determines whether a given text conveys an opinion or states a fact. Sentiment orientation (polarity) identifies whether a viewpoint in a text is positive, negative or neutral. Sentiment strength discovers the "degree" of an idea in a given text.

Topic Modeling

According to Blei, Ng, and Jordan (2003), "topics" are probability distributions across all terms in the dictionary. For example, the topic "education" can be associated with terms such as "students," "teachers," "schools," and so on, with high probability, and the likelihood of the rest of the words in the dictionary—which typically will include hundreds or thousands of terms—will be near 0. LDA assumes that the author of a document produces text by following a generative model, according to which, given the document, first a topic is selected from a corresponding conditional multinomial distribution of topics, and then, given the topic, words are chosen from the multinomial distribution of terms that corresponds to that topic.

Since the problem of estimating the parameters of the multinomial (Dirichlet) distributions of documents across topics and topics across terms is intractable, they are estimated using Markov chain Monte Carlo simulations. LDA-based topic modeling has two advantages. First, utilizing clues from the context, the topic models connect words with related meanings and separate uses of words from multiple meanings (McCallum, 2002). Second, since the topic modeling method is performed automatically based on a mathematical algorithm, subjective bias in analyzing data is minimized. A tutorial for LDA is available in ConText (Diesner et al., 2015).

LSA facilitated topic extraction as well as document similarity and was introduced as a novel way of automatic indexing and information retrieval in library systems in the early 1990s (Deerwester, Dumais, Furnas, Landauer, & Harshman, 1990). According to Deerwester et al. (1990), "this approach takes advantage of implicit higher-order structure in the association of terms with documents ('semantic structure') in order to improve the detection of relevant documents on the basis of terms found in queries" (p. 1). LSA has been extended as a theory of meaning where topics are represented in two equivalent forms: as linear combinations of related terms and as linear combinations of relevant documents (Foltz, Kintsch, & Landauer, 1998). As a statistical estimation method, LSA quantifies a collection of records by applying the vector space model (VSM), which arranges text-based data into a term-by-document matrix where term frequencies are recorded. LSA then employs singular value decomposition (SVD), which is an extension of principal component analysis and quantifies patterns of term-document co-occurrence using least squares estimation. After SVD, an analysis similar to what is done in numerical factor analysis can produce interpretable topics (Evangelopoulos, Zhang, & Prybutok, 2012; Sidorova, Evangelopoulos, Valacich, & Ramakrishnan, 2008). Despite the more rigorous statistical estimation in LDA, LSA has higher computational efficiency, provides reproducible results, and is readily available in several implementation packages (Anaya, 2011).

Visual Analytics

Visual analytics (Cook & Thomas, 2005) draws on methods from information visualization (Card et al., 1999) and computational modeling. Visual analytics has considerably contributed to SMA (Diakopoulos et al., 2010; Hassan et al., 2014). The theory of visual analytics employs interactive visualization in processing analytical reasoning instead of the static display as the results of the analysis (Cook & Thomas, 2005). The analytic process starts with a high-level assessment that leads analysts to interesting features of the data. Then the analysts can reconstruct the perspective by filtering or generating brand-new visualizations, which help them explore better via the qualitative data analysis.

Brooker, Barnett, and Cribbin (2016) address the advantage of visual analytic approach to social media data. By integrating data collection and data analysis as a single process, exploring a dataset may inspire innovative ideas that eventually result in new levels of data collection with findings discovered during the continuing iterations (Brooker et al., 2016). Therefore, through visual analytics, data analysts can independently examine their hypotheses and viewpoints to discover phenomena as they develop questions to assess their ability in mapping digital data to traditional social science concepts (Brooker et al., 2016).

Stream Processing

Data analytics of real-time social media characterizes large quantity of temporal data with little latency. This process requires applications that support online analysis of data streams. However, traditional database management systems (DBMSs) lack the pre-established concept of time and cannot manage online data in real time, which results in developing data stream management systems (DSMSs) (Hebrail, 2008). DSMSs can process main memory while saving the data on the system, which enables it to deal with the transient online data streams and to process constant queries of the streaming data (Botan et al., 2010). Commercial DSMSs includes CEP engine (Oracle), StreamBase, and StreamInsight (Microsoft) (Chandramouli et al., 2010). Taking Twitter as an example, Tweets generated from public accounts that represent more than 90% of Twitter accounts including replies and mentions, can be retrieved in JSON format. For example, the Twitter Search API is used to request past data on Twitter and streaming API, filtered by user ID, search keyword, and geospatial location, is used to request a real time stream of Tweets (Batrinca & Treleaven, 2015).

Social Media Analytics Tools

Many opinion mining tools have been employed by businesses and are mainly aimed at sentiment analysis of customer feedback about goods and services. There is an extensive selection of tools for textual analysis, including open-source tools and multi-performance business toolkits and platforms (Batrinca & Treleaven, 2015). This section introduces essential tools and toolkits to collect, clean, and analyze. Popular scientific analytics tools, such as R, MATLAB®, Python, and Mathematica, have been improved to support retrieving and analyzing text. Additionally, network visualization tools, such as NodeXL and UCINET, enable researchers to draw relationships among actors or entities of social media within the network.

Scientific Programming Tools

R is a language and open-source software tool for statistical computing and graphics, which provides a wide variety of statistical techniques, including linear and nonlinear modeling, classical statistical tests, time series analysis, classification, and clustering, and graphical methods (www.r-project.org). MATLAB® is designed for numeric experimental programming, including data processing and data modeling, which enables time series analysis, GUI and array-based statistics. MATLAB® can be employed for an extensive range of applications, and significantly improved processing speed when compared to the traditional programming languages.

Python is a high-level programming language for general purpose programming, and it has gained popularity because the language is comparatively easy to learn and it enables programmers to write clear programs on both a small and large scope. Python can deal with natural language detection, topic extraction, and query matching. It also can be established to conduct sentiment analysis using different classifiers.

Apache Unstructured Information Management Applications (UIMA) is an open-source project that analyzes "Big Data" and discovers information that is relevant to the user. Mathematica is adopted for allegorical programming such as computer algebra.

Network Visualization Tools

NodeXL is an add-in for Excel and supports the analysis and exploration of social media with importing features that can pull data from Twitter, Flickr, YouTube, Facebook, WWW hyperlinks, and wikis (Hansen, Shneiderman, & Smith, 2011). NodeXL allows users to create visual network diagrams of actors (vertices), estimate the network influence (e.g., betweenness centrality or page rank) of a single actor on others, and retrieve information on a large scale (Hansen et al., 2011). NodeXL has a professional version supporting more functions than its free version.

UCINET, developed by Lin Freeman, Martin Everett, and Steve Borgatti, is a software package for the analysis of social network data. UCINET only runs on the Windows operating system and comes with the NetDraw which is a social media data visualization tool using a network format.

Business Applications

Business applications refer to commercial software tools that users can collect, search, and analyze text for business purposes. For example, SAS Sentiment Analysis Manager included in the SAS Text Analytics program allows users to access source content, such as social media outlets, text data inside the organization, websites, and generate reports about customers and competitors in real time (Batrinca & Treleaven, 2015). RapidMiner is also a popular tool providing an open-source community edition as well as a fee-based enterprise edition (Hirudkar & Sherekar, 2013). RapidMiner offers data mining and ML procedures. These procedures include data extraction, transformation, and loading (ETL), data visualization, structuring, assessment, and employment. RapidMiner, written in Java, adopts learning schemes and attribute evaluators using the Weka ML environment, and the R project is employed for statistical modeling schemes in RapidMiner. Similar to SAS Enterprise Miner, IBM SPSS Modeler is one of the most popular applications that support various data analytics tasks.

Social Media Monitoring Tools

In general, organizations adopt social media monitoring tools to conduct sentiment analysis, and to trace and measure what users have been saying about a company, products, services, and certain topics on social media. The examples of social media monitoring tools contain Social Mention which offers social media notifications called social mention alerts. Amplified Analytics emphasizes reviews regarding products and services and marketing-related information. Reputation monitoring tools include Lithium Social Media Monitoring and Trackur, which track conversations and discussions online. Useful and free tools provided by Google include Google Trends, which compares a specific search term or multiple search terms to the total search volume, and Google Alerts, which is an automatic content notification tool established around Google Search (Batrinca & Treleaven, 2015).

Text Analysis Tools

Text analysis tools are mainly designed for NLP and text analytics. Many open source tools were developed by research groups and NGO (nongovernmental organizations) to capture, explore, and analyze opinions (Batrinca & Treleaven, 2015), and examples include Stanford NLP group applications and LingPipe, a package of Java libraries for the linguistic analysis of human language (Teufl, Payer, & Lackner, 2010). Python natural language toolkit (NLTK), a popular open source text analysis application, provides Python modules, linguistic data, and documentation for text analytics. GATE, another example of open source sentiment analysis tools, has been employed for media content analysis, customer experiences analysis, cancer and pharmaceutical research, and web mining. Its website explains that sentiment analysis and polarity categorization are more challenging when compared to traditional document classification.

Lexalytics Sentiment Toolkit is an automatic sentiment analysis application. Though Lexalytics is not capable of data scraping, it is especially powerful when applied to a large number of documents. Other commercial applications for text mining include IBM LanguageWare, SPSS Text Analytics for Surveys, STATISTICA Text Miner, Clarabridge, WordStat, AeroText, and Language Computer Corporation (Batrinca & Treleaven, 2015).

Data Visualization Tools

Combining BI competences, data visualization tools enable diverse users to make sense of Big Data. Through interactive user interfaces, users can conduct exploratory data analysis with visualizations on smartphones and tablet computers. These tools assist users in recognizing patterns and relationships in data. Impromptu visualization on the data can discover trends and outliers, which also can be conducted on the framework of Big Data, including Apache Hadoop or Amazon Kinesis.

Examples of data visualization tools include SAS Visual Analytics, Tableau, Qlik Sense, Microsoft Power BI, and IBM Watson Analytics.

Social Media Management Tools

Social media management tools can be handy especially when users run an agency or manage social media for other businesses. Examples of ways to manage social media include generating content, retargeting content, engaging with users, and sharing schedules and related content. Social media management tools can help to perform these tasks effectively and efficiently. Though many of the tools share similar characteristics, users can choose the best ones that suit their needs. Notable social media management tools include Hootsuite, Buffer, Agora Plus, Sprout Social, CoSchedule, Everypost, Sendible, Social Pilot, MavSocial, and Friends+Me.

Almost all social media management tools are commercial and charge fees for use, and among those examples, Hootsuite is one of the most popular applications. One of the most important features of Hootsuite is that it enables users to assist and plan content, gauge their social ROI, and run advertisements in social media in an all-in-one system. Hootsuite allows users to manage various accounts and keywords, link with more than 35 social networks, and multischedule social media posts.

Representative Fields of Social Media Analytics

Social media reflects dynamic human information interaction, which creates new opportunities to understand people, organizations, and society. More and more researchers and business professionals are discovering innovative methods of automatically gathering, incorporating, and investigating rich data on social media. Therefore, documenting all the cases of social media applications in one section is challenging. As Batrinca and Treleaven (2015) mentioned, three indicative fields of SMA include business, bioscience, and social science. The early adopters of SMA were retail industries and finance. Retail businesses employ social media to reinforce their brand awareness, service enhancement for customers, product improvement, marketing strategies, information procreation, and even fraud detection.

In finance, social media is applied to evaluate market sentiment, and news data is adopted for trading (Batrinca & Treleaven, 2015). For instance, Bollen, Mao, and Zeng (2011) discovered with near 90% accuracy, by evaluating the sentiment of a random sample of Twitter data, that the Dow Jones Industrial Average (DJIA) prices are interrelated with the Twitter sentiment two or three days earlier. Wolfram (2010) employed Twitter data to train a support vector regression (SVR) model in forecasting values of each NASDAQ stock. The experiment resulted in revealing a "considerable advantage" for predicting future values.

In marketing, the UGC of social media is used for detecting online and offline opinions of customers. Anderson and Magruder (2012) applied a regression discontinuity approach to identifying a causal relationship between Yelp ratings and dining reservations in restaurants. The study discovered that the more star ratings the restaurants get, the more reservations they obtained, resulting in a complete booking for the top-rated restaurants. Srinivasan, Rutz, and Pauwels (2016) examined the influence of a mix of marketing activities on users' Internet activity and sales. Applying a vector autoregressive (VAR) model, the researchers investigated the relationship between price and marketing channels, TV advertisements, the number of commercial search clicks and website visits, and Facebook activity on sales. In addition, paid search was influenced by TV advertisements and the number of Facebook likes, while it affects marketing channels, website visits, likes on Facebook, and sales (Moe, Netzer, & Schweidel, 2017).

In the field of biosciences, social media is used to quit smoking, deal with obesity, and supervise diseases by gathering related data on peers for behavioral change initiatives. Penn State University biologists (Salathe et al., 2012) created novel applications and methods to trace the transmission of contagious diseases through the data from news websites, blogs, and social media.

In computational social science, social media has been adopted to observe public responses to political issues, public agenda, events, and leadership. As an illustration, Lerman, Gilder, Dredze, and Pereira (2008) applied computational linguistics to forecast the news impact on the public regarding political candidates. Others explored how Twitter is used within the election context to predict election results (DiGrazia, McKelvey, Bollen, & Rojas, 2013) by discovering candidates' patterns of political practice (Bruns & Highfield, 2013). These applications focused more on the behaviors of the candidates and paid less attention to the behavior of the public. Social media provides a new channel for political candidates to get closer to the public, and these public spheres also open communication channels for the online audience to connect with each other and get involved with antagonistic politics.

Conclusions

Social media is a fundamentally measurable, information communication technology that extends web-based communications. Web 2.0 is the second transformative phase of the WWW and the social nature of Web 2.0 makes it possible to foster social media platforms, such as Facebook, Twitter, LinkedIn, Reddit, YouTube, and wikis. These social media platforms provide open environments for collective intelligence that creates collaborative content, and this cogenerated content increases its value with increased adoptions. The convenient access to APIs of the social media platforms has resulted in an explosion of social data creation and the use of SMA.

Besides identifying definitions and processes of SMA, this chapter focused on introducing major techniques and tools for SMA. Presented techniques include social media data scraping, sentiment analysis, topic modeling, visual analytics, and streaming processing. Representative fields of social media analytics are business, bioscience, and computational social science. One of the critical issues regarding SMA is that social media platforms are increasingly limiting access to their data to make profits from their content. Data scientists and researchers have to find ways to collect a large scale of social media data for research purposes with reasonable costs. Otherwise, computational social science could be the privilege of big organizations, resourceful government agencies, and elite scholars that can afford costly social media data and therefore the studies they conducted would hardly be evaluated or replicated.

References

Anaya, L. H. (2011). *Comparing latent dirichlet allocation and latent semantic analysis as classifiers* (Doctoral dissertation). University of North Texas, Denton, TX.

Anderson, M., & Magruder, J. (2012). Learning from the crowd: Regression discontinuity estimates of the effects of an online review database. *The Economic Journal, 122*(563), 957–989.

Armano, D. (2014, July 23). Six social media trends for 2010. Retrieved July 7, 2017, from https://hbr.org/2009/11/six-social-media-trends

Batrinca, B., & Treleaven, P. C. (2015). Social media analytics: A survey of techniques, tools and platforms. *AI & Society, 30*(1), 89–116.

Bengston, D. N., Fan, D. P., Reed, P., & Goldhor-Wilcock, A. (2009). Rapid issue tracking: A method for taking the pulse of the public discussion of environmental policy. *Environmental Communication, 3*(3), 367–385.

Blei, D. M., Ng, A. Y., & Jordan, M. I. (2003). Latent dirichlet allocation. *Journal of Machine Learning Research, 3*, 993–1022.

Bollen, J., Mao, H., & Zeng, X. (2011). Twitter mood predicts the stock market. *Journal of Computational Science, 2*(1), 1–8.

Borders, B. (2009). A brief history of social media. Retrieved December 5, 2010, from http://socialmediarockstar.com/history-of-social-media.

Botan, I., Derakhshan, R., Dindar, N., Haas, L., Miller, R. J., & Tatbul, N. (2010). SECRET: A model for analysis of the execution semantics of stream processing systems. *Proceedings of the VLDB Endowment, 3*(1–2), 232–243.

boyd, D., & Ellison, N. (2010). Social network sites: Definition, history, and scholarship. *IEEE Engineering Management Review, 3*(38), 16–31.

Brooker, P., Barnett, J., & Cribbin, T. (2016). Doing social media analytics. *Big Data & Society, 3*(2), 2053951716658060.

Bruns, A., & Highfield, T. (2013). Political networks on twitter: Tweeting the queensland state election. *Information, Communication & Society, 16*(5), 667–691.

Bruns, A., & Liang, Y. E. (2012). Tools and methods for capturing Twitter data during natural disasters. *First Monday, 17*(4), 1–8.

Card, S. K., Mackinlay, J. D., & Shneiderman, B. (Eds.). (1999). *Readings in information visualization: Using vision to think*. San Francisco, CA: Morgan Kaufmann.

Carraher, S. M., Parnell, J., & Spillan, J. (2009). Customer service-orientation of small retail business owners in Austria, the Czech Republic, Hungary, Latvia, Slovakia, and Slovenia. *Baltic Journal of Management, 4*(3), 251–268.

Chandramouli, B., Ali, M., Goldstein, J., Sezgin, B., & Raman, B. S. (2010). Data stream management systems for computational finance. *Computer, 43*(12), 45–52.

Chung, W. (2016). Social media analytics: Security and privacy issues. *Journal of Information Privacy and Security, 12*(3), 105–106.

Cohen, A. H. (2010, December 27). 10 social media 2010 highlights. Retrieved July 8, 2017, from https://www.clickz.com/10-social-media-2010-highlights-data-included/53386/

Cook, K. A., & Thomas, J. J. (2005). *Illuminating the path: The research and development agenda for visual analytics.* Los Alamitos, CA: IEEE Computer Society.

Davenport, T. H., & Court, D. B. (2014, November 5). Analytics 3.0. Retrieved August 26, 2017, from https://hbr.org/2013/12/analytics-30

Deerwester, S., Dumais, S. T., Furnas, G. W., Landauer, T. K., & Harshman, R. (1990). Indexing by latent semantic analysis. *Journal of the American Society for Information Science, 41*(6), 391.

Dergiades, T. (2012). Do investors' sentiment dynamics affect stock returns? Evidence from the US economy. *Economics Letters, 116*(3), 404–407.

Diakopoulos, N., Naaman, M., & Kivran-Swaine, F. (2010, October). Diamonds in the rough: Social media visual analytics for journalistic inquiry. In *Visual analytics science and technology (VAST), 2010 IEEE symposium on* (pp. 115–122). IEEE.

Diesner, J., Aleyasen, A., Chin, C., Mishra, S., Soltani, K., & Tao, L. (2015). ConText: Network construction from texts [Software]. Retrieved from http://context.lis.illinois.edu/

DiGrazia, J., McKelvey, K., Bollen, J., & Rojas, F. (2013). More tweets, more votes: Social media as a quantitative indicator of political behavior. *PloS One, 8*(11), e79449.

Edosomwan, S., Prakasan, S. K., Kouame, D., Watson, J., & Seymour, T. (2011). The history of social media and its impact on business. *Journal of Applied Management and Entrepreneurship, 16*(3), 79.

Evangelopoulos, N., Zhang, X., & Prybutok, V. (2012). Latent semantic analysis: Five methodological recommendations. *European Journal of Information Systems, 21*(1), 70–86.

Fan, W., & Gordon, M. D. (2014). The power of social media analytics. *Communications of the ACM, 57*(6), 74–81.

Foltz, P. W., Kintsch, W., & Landauer, T. K. (1998). The measurement of textual coherence with latent semantic analysis. *Discourse Processes, 25*(2–3), 285–307.

Gottfried, J., & Shearer, E. (2016, May 26). News use across social media platforms 2016. *Pew Research Center.*

Greenwood, S., Perrin, A., & Duggan, M. (2016, November 11). Social media update 2016: Facebook usage and engagement is on the rise, while adoption of other platforms holds steady. *Pew Research Center.*

Grubmüller, V., Götsch, K., & Krieger, B. (2013). Social media analytics for future oriented policy making. *European Journal of Futures Research, 1*(1), 20.

Hansen, D., Shneiderman, B., & Smith, M. A. (2011). *Analyzing social media networks with NodeXL: Insights from a connected world.* Burlington, MA: Morgan Kaufmann.

Hassan, S., Sanger, J., & Pernul, G. (2014, January). SoDA: Dynamic visual analytics of big social data. In *Big data and smart computing (BIGCOMP), 2014 international conference on* (pp. 183–188). Piscataway, NJ: IEEE.

Hebrail, G. (2008). *Data stream management and mining.* Mining Massive Data Sets for Security, IOS Press, 89–102.

Hirudkar, A. M., & Sherekar, S. S. (2013). Comparative analysis of data mining tools and techniques for evaluating performance of database system. *International Journal of Computational Science and Applications, 6*(2), 232–237.

Holsapple, C., Hsiao, S. H., & Pakath, R. (2014). *Business social media analytics: Definition, benefits, and challenges.* Twentieth Americas Conference on Information Systems, Savannah, GA.

Informs.org. UPS On-road integrated optimization and navigation (ORION) project. Retrieved September 6, 2017, from https://www.informs.org/Impact/O.R.-Analytics-Success-Stories/UPS-On-Road-Integrated-Optimization-and-Navigation-ORION-Project

Jasra, M. (2010, November 24). The history of social media [Infographic]. Retrieved December 4, 2010, from Web Analytics World: http://www.webanalyticsworld.net/2010/11/history-of-social-media-infographic.html

Junco, R., Heibergert, G., & Loken, E. (2011). The effect of Twitter on college student engagement and grades. *Journal of Computer Assisted Learning, 27,* 119–132.

Kaplan, A. M., & Haenlein, M. (2010). Users of the world, unite! The challenges and opportunities of Social Media. *Business Horizons, 53*(1), 59–68.

Kevthefont (2010). Curse of the Nike advert-it was written in the future. *Bukisa,* p. 1.

Kohut, A., Keeter, S., Doherty, C., & Dimock, M. (2008). Social networking and online videos take off: Internet's broader role in campaign 2008. *TPR Center, The PEW Research Center.*

Kontopoulos, E., Berberidis, C., Dergiades, T., & Bassiliades, N. (2013). Ontology-based sentiment analysis of Twitter posts. *Expert Systems with Applications, 40*(10), 4065–4074.

Kumar, A., & Sebastian, T. M. (2012). Sentiment analysis on Twitter. *IJCSI International Journal of Computer Science Issues, 9*(4), 372.

Kurniawati, K., Shanks, G. G., & Bekmamedova, N. (2013). The business impact of social media analytics. *ECIS, 13,* 13.

Lerman, K., Gilder, A., Dredze, M., & Pereira, F. (2008, August). Reading the markets: Forecasting public opinion of political candidates by news analysis. *Proceedings of the 22nd International Conference on Computational Linguistics, 1,* 473–480.

Liu, Y., Huang, X., An, A., & Yu, X. (2007). ARSA: A sentiment-aware model for predicting sales performance using blogs. In *Proceedings of the 30th annual international ACM SIGIR conference on research and development in information retrieval* (pp. 607–614). Amsterdam, the Netherlands, July 23–27.

McCallum, A. K. (2002). MALLET: A machine learning for language toolkit: Topic modeling. Retrieved November 4, 2016, from http://mallet.cs.umass.edu/topics.php

Mejova, Y. (2009). Sentiment analysis: An overview. Comprehensive exam paper. Retrieved February 3, 2010, from http://www.cs.uiowa.edu/~ymejova/publications/CompsYelenaMejova.pdf

Moe, W. W., Netzer, O., & Schweidel, D. A. (2017). Social media analytics. In B. Wierenga and R. van der Lans (Eds.), *Handbook of marketing decision models* (pp. 483–504). Cham, Switzerland: Springer. https://www.springerprofessional.de/en/handbook-of-marketing-decision-models/13301802

Obar, J. A., & Wildman, S. S. (2015). Social media definition and the governance challenge: An introduction to the special issue. *Telecommunications Policy, 39*(9), 745–750.

Pang, B., & Lee, L. (2008). Opinion mining and sentiment analysis. *Foundations and Trends® in Information Retrieval, 2*(1–2), 1–135.

Park, S., Ko, M., Kim, J., Liu, Y., & Song, J. (2011). The politics of comments: Predicting political orientation of news stories with commenters sentiment patterns. In *Proceedings of the ACM 2011 conference on computer supported cooperative work (CSCW '11)* (pp. 113–122). New York, NY: ACM.

Peña-López, I. (2007). *Participative web and user-created content: Web 2.0, Wikis, and social networking.* Paris: OECD. Retrieved October 24, 2007, from http://213.253.134.43/oecd/pdfs/browseit/9307031E.pdf

Rimskii, V. (2011). The influence of the Internet on active social involvement and the formation and development of identities. *Russian Social Science Review, 52*(1), 79–101.

Ritholz, B. (2010). History of social media. Retrieved December 5, 2010, from http://www.ritholtz.com/blog/2010/12/history-of-social-media/

Ruggiero, A., & Vos, M. (2014). Social media monitoring for crisis communication: Process, methods and trends in the scientific literature. *Online Journal of Communication and Media Technologies, 4*(1), 105.

Salathe, M., Bengtsson, L., Bodnar, T. J., Brewer, D. D., Brownstein, J. S., Buckee, C., … Vespignani, A. (2012). Digital epidemiology. *PLoS Computational Biology, 8*(7), e1002616.

Sidorova, A., Evangelopoulos, N., Valacich, J.S., & Ramakrishnan, T. (2008). Uncovering the intellectual core of the information systems discipline. *MIS Quarterly, 32*(3), 467–482, A1–A20.

Sinha, V., Subramanian, K. S., Bhattacharya, S., & Chaudhary, K. (2012). The contemporary framework on social media analytics as an emerging tool for behavior informatics, HR analytics and business process. *Management: Journal of Contemporary Management Issues, 17*(2), 65–84.

Srinivasan, S., Rutz, O. J., & Pauwels, K. (2016). Paths to and off purchase: Quantifying the impact of traditional marketing and online consumer activity. *Journal of the Academy of Marketing Science, 44*(4), 440–453.

Stavrakantonakis, I., Gagiu, A. E., Kasper, H., Toma, I., & Thalhammer, A. (2012). An approach for evaluation of social media monitoring tools. *Common Value Management, 52*(1), 52–64.

Stieglitz, S., & Dang-Xuan, L. (2013). Social media and political communication: A social media analytics framework. *Social Network Analysis and Mining, 3*(4), 1277–1291.

Teufl, P., Payer, U., & Lackner, G. (2010, September). From NLP (natural language processing) to MLP (machine language processing). In *International conference on mathematical methods, models, and architectures for computer network security* (pp. 256–269). Berlin: Springer.

Thelwall, M., Buckley, K., & Paltoglou, G. (2011). Sentiment in Twitter events. *Journal of the American Society for Information Science and Technology, 62*(2), 406–418.

Westerski, A. (2008). *Sentiment analysis: Introduction and the state of the art overview* (pp. 211–218). Madrid, Spain: Universidad Politecnica de Madrid.

Wihbey, J. (2015). How does social media use influence political participation and civic engagement? A meta-analysis. Journalist Resource. Retrieved February 17 2016.

Wittwer, M., Reinhold, O., & Alt, R. (2016). Social media analytics in social CRM-towards a research agenda. In *Bled eConference*, p. 32.

Wolfram, M. S. A. (2010). *Modelling the stock market using Twitter.* Scotland, UK: University of Edinburgh.

Zeng, D., Chen, H., Lusch, R., & Li, S. H. (2010). Social media analytics and intelligence. *IEEE Intelligent Systems, 25*(6), 13–16.

Chapter 8

Transactional Value Analytics in Organizational Development

Christian Stary

Contents

Introduction

Stakeholders are increasingly involved in organizational change and development, and thus, value creation processes (Tantalo & Priem, 2016). However, stakeholders whose values are incongruent with the values of their organization are likely to be disengaged in work (Rich, Lepine, & Crawford, 2010), and finally, organizational development processes (Vogel, Rodell, & Lynch, 2016). Understanding value congruence, that is, the relationship between stakeholder values and values created on the organizational level, is therefore key to developing organizations (Edwards & Cable, 2009), and a prerequisite to developing a culture for high performance (Posner & Schmidt, 1993).

To avoid the emergence of behavior (patterns), for example, pretending to fit into an organization of work (Hewlin, Dumas, & Burnett, 2017), when stakeholders feel that their values do not match those of the organization, a deeper analysis of values of self-knowledge can help. It allows addressing relationships between value congruence, perceived organizational support, core self-evaluation, task performance, and organizational citizenship behavior (Strube, 2012). In the following, self-knowledge analytics based on individual stakeholder transactions as an integral part of an enriched value network analysis (VNA) approach (Allee, 2008), is introduced. Thereby, interactions between stakeholder roles relevant for operation are analyzed. They are represented in diagrammatic networks (termed holomaps) and can be tangible or intangible. Tangible interactions encode role-specific deliverables that are created by persons acting in a certain role and need to be exchanged for task completion. Intangible interactions encode deliverables facilitating business operations that are not part of formal role specifications. However, they are likely to influence the successful completion of business operations.

Since VNA targets organizational change as considered possible by stakeholders, individual value systems need to be analyzed before changes can become effective. This organizational transaction analysis ensures interactional coherence based on individual value systems. In business-to-customer relationships, value cocreation is well established (Payne, Storbacka, & Frow, 2008; Vargo, Maglio, & Akaka, 2008). The subjective analytics in value alignment help to identify individual capabilities and needs, which otherwise might be overlooked or unrecognized (Jankowicz, 2001). In large organizations, software repositories may help in understanding and evaluating requirements and encoding needs for change (Selby, 2009). Dashboards have been designed not only to display complex information and analytics but also to keep indicators with respect to volatility. The latter requires stakeholder judgement, also termed as a user- or demand-driven approach in analytics (Lau, Yang-Turner, & Karacapilidis, 2014), which focuses on participative elicitation processes, recognizing that any data analysis needs to make sense for the involved stakeholders.

Studies in that context reveal that decision making on variants or proposals not only requires individual intellectual capabilities but also collective knowledge and active sharing of expertise (Lau et al., 2014). In particular, in complex work settings, people with diverse expert knowledge need to work together toward a meaningful interpretation of interactional change (Treem & Leonardi, 2017). Thus, any consensus-building mechanism needs to accommodate input from multiple human experts effectively. This work proposes to enrich VNA methodologically by using repertory grids (Easterby-Smith, 1976; Fransella & Bannister, 1977; Senior, 1996; Boyle, 2005) to embody individual value systems into organizationally relevant ones. This enrichment is intended to facilitate decision making in organizational change management processes.

The chapter starts with an introduction to the methodological challenge by providing the relevant background on VNA. It identifies those elements in VNA that may cause fragmentation of value-driven alignment of change proposals, which should become part of a more structured consensus-building process, in particular when business processes should be developed on that ground (Stary, 2014). Several interviews have been conducted with experts dealing with aligning value systems in stakeholder-driven change management, to identify suitable candidates to enable value-system-based consensus building. A report on the interviews and the evaluated candidates is given. Based on the results, the consensus-building process can be supported by individually-generated repertory grids, which are introduced subsequently. They allow individuals in certain roles to reflect on their value system when acting in that role. In this way, individual stakeholders can rethink organizational behavior changes offered by or to other stakeholders and assess them according to their role-specific value system. An exemplary case demonstrates the feasibility of this methodological enrichment and reveals first insights from a typical use scenario, namely organizational sales development. We conclude the chapter by summarizing the achievements and providing topics for further studies.

Value Network Analysis

Since its introduction, VNA has targeted a key question of business, namely "How is value created through converting intangible assets to tangible ones, and thus negotiable forms of value?" (Allee, 2008). The method aims at developing organizations beyond the value chain, since traditional value-chain models represent a linear, if not mechanistic, view of business and its operation. Complex constellations of values, however, require analyzing business relationships by taking into account the role of intangible value exchange as a foundation for value creation. Value exchange needs to be analyzed before changing business transactions in practice. In particular, complex relationships require preprocessing from a value-based perspective, as they influence effectiveness and efficiency, and cause possible friction in operational processes (Allee, 2008).

VNA is meant to be a development instrument beyond engineering, as it aims to understand organizational dynamics, and thus to manage structural knowledge from a value-seeking perspective, for individuals and the organization as a whole. However, it is based on several fundamental principles and assumptions (Allee, 1997, 2003, 2008)[1]:

■ Participants of an organization and organizationally-relevant stakeholders participate in a value network by converting what they know, both individually and collectively, into tangible and intangible values that they contribute to the network, and thus to the organization.

■ Participants accrue value from their participation by converting value inputs into positive increases of their tangible and intangible assets, in ways that will allow them to continue producing value outputs in the future.

■ In such a network, each participant contributes and receives value in ways that sustain both their own success and the success of the value network as a whole. This mutual dependency is a condition sine qua non. Once active participants either withdraw or are expelled, the overall system becomes unstable and may collapse, and needs to be reconfigured.

■ Value networks require trusting relationships and a high level of integrity and transparency on the part of all participants. Then, insights can be gained into interactions by identifying and analyzing not only the patterns of exchange, but rather the impact of value transactions, exchanges, and flows, and thus the dynamics of creating and leveraging value.

■ A single transaction is only meaningful in relation to the system as a whole. It is set by role carriers who utilize incoming deliverables from other role carriers (inputs) and can assess their value, and they realize value that is manifest by generating output.

As stakeholders—in roles relevant for business—are responsible for their relations with others, the organization itself needs to be conceptualized as a highly dynamic and complex setting.

Organizations as Self-Adapting Complex Systems

Organizations as sociotechnical systems are complex living systems and as such are self-regulating and self-managing (Allee, 2003). When managing organizations from the outside in, by predetermining structures, systems, rules, and

---

[1] For the sake of preciseness, the VNA description closely follows the original texts provided by Allee (2003).

formal reporting relationships, inherent dynamics may get overlayed and become completely controlled externally. In rapidly changing settings, self-organization of concerned stakeholders is an effective way to handle requirements for change. However, for self-organization to happen, stakeholders need to have access to relevant information and, more importantly, an understanding of the organization and the situation as a whole. Both are required to make informed decisions and initiate socially effective action. Since the behavior of autonomous stakeholders cannot be predicted fully, organizations need rules to guide behavior management according to the understanding of stakeholders and their capabilities to change their behavior.

This need can be adeptly demonstrated in customer service. When there are too many rules, customers are locked into a bureaucracy that seems unresponsive to their needs. When there are too few rules, inconsistency and chaos are likely in complex business cases. Stakeholders need to develop guiding principles that effectively support them in organizational design and the respective decision making through information and technology provision (Firestone & McElroy, 2003). These principles need to tackle both tangible and intangible relationships and stakeholder interaction. They should be qualified to reflect on their tangible and intangible exchanges, and finally negotiate their own "protocols," that is, activities with those with whom they interact (Allee, 2008).

Although no one person or group of people can manage a complex system, the stakeholders can self-organize their inputs and outputs and negotiate exchanges with others in the organizational system as necessary. Modeling work and business relations as dynamic patterns of tangible and intangible exchanges help stakeholders to identify individually consistent roles and understand the system. They also allow them to make it transparent and therefore communicable with coworkers, business partners, and the economic ecosystems of which they are part. According to Allee (2003), all organizational layers are concerned with the following:

- *Operationally*: Stakeholders need to understand how digital networks and technologies support creating, organizing, and accessing the everyday knowledge they need to complete their tasks and make informed decisions.
- *Tactically*: They need to understand how social webs such as knowledge networks and communities of practice help create, diffuse, and leverage knowledge and innovation.
- *Strategically*: Stakeholders need to develop an understanding of their organization as a participant in (multiple) business networks, where intangibles are important for building relationships and work interactions.

Above all, stakeholders need to accept the dual nature in interaction in networked ecosystems through tangibles and intangibles in order to learn how to engage in conversations that matter.

Patterns of Interaction as Analytical Design Elements

VNA builds upon organizations as self-adapting complex systems, which are modeled from that perspective by the following:

1. Identifying patterns of interactions representing tangible and intangible relations between stakeholder roles.
2. Describing these patterns in a structured way, recognizing:
   a. Sources and sinks of information exchanged between stakeholder roles.
   b. The impact of received information and objects.
   c. The capabilities of produced and delivered assets.
3. Elaborating critical processes or exchanges, and thus proposing changes from both a cognitive perspective and the flow of energy and matter.

In line with the living system perspective, VNA assumes that the basic pattern of organizing a business is that of a network of tangible and intangibles exchanges. Tangible exchanges correspond to flows of energy and matter, whereas intangible exchanges point to cognitive processes. Describing a specific set of participating stakeholders and exchanges allows a detailed description of the structure of any specific organization or a network of organizations.

Although VNA considers the act of exchange to be a fundamental activity, it goes beyond traditional economic understanding of stakeholder interactions. Exchange includes goods, services, and revenue, but also considers the transaction between stakeholders as representative of organizational intelligence, thus as a cognitive interaction process. Transactions ensure successful task accomplishment and business through cognitively reflected exchanges of information and knowledge sharing, opening pathways for informed decision making. Hence, exchanges not only have value per se, but encode the currently available collective intelligence (determining the current economic success).

Tangible and Intangible Transactions

In VNA, knowledge, either implicit or explicit (Dalkir, 2011), and intangible exchanges are different from tangible ones, and as such they need to be specifically treated in relation to their characteristics. Tangible exchanges include goods, services, and revenue, in particular physical objects, contracts, invoices, return receipts of orders, requests for proposals, confirmations, and payments. They also include knowledge, products, or services that directly generate revenue, or that are expected (contractual) and paid for as part of a service or good.

Intangible exchanges comprise knowledge and benefits. Intangible knowledge and information exchanges occur and support the core product and service value chain, but are not contractual. Intangibles are extras that stakeholders in a certain role provide to others to help keep business operations running. For instance, a service organization asks sales experts to volunteer time and knowledge on organizational development in exchange for an intangible benefit of prestige by affiliation.

Stakeholders involved in intangible transactions help to build relationships by exchanging strategic information, planning knowledge, process knowledge, and technical know-how, and in this way they share collaborative design work, performing joint planning activities and contributing to policy development. Intangibles, like other assets, are increased and leveraged through deliberate actions. They affect business relationships, human competence, internal structure, and social culture. VNA considers intangibles as assets and negotiables that can actually be delivered by stakeholders engaged in a knowledge exchange. They can be held accountable for the effective execution of that exchange, as they are able to articulate them accordingly when following the VNA's structured procedure.

Although there are various attempts to develop new measures and analytical approaches for calculating knowledge assets and for understanding intangible value creation, traditional scorecards need to move beyond considering people as liabilities, resources, or investments. Responsible stakeholders need to understand how intangibles create value and, most importantly, how intangibles go to market as negotiables in economic exchanges. As a prerequisite, they need to understand how intangibles act as deliverables in key transactions with respect to a given business model.

Value Network Representation of Organizations

Value networks represent organizations as a web of relationships that generates tangible and intangible values through transactions between two or more roles. These roles stem from any public or private organization or sector and stand for individuals, groups, or entire organizations. The network, instead of representing hierarchical positions, structures the dynamics of processing and delivering tangibles and intangibles. Although the roles need to be related to the organization at hand, suppliers, partners, and consumers—regardless of their physical location—need to become part of the network once they generate value or receive transactional deliverables.

When modeling an organization as a value network, several assumptions apply (Allee, 2008):

- An exchange of value is supported by some mechanism or medium that enables the transaction to happen. As organizations can also be considered sociotechnical systems, typical enablers are information and communication technologies. For instance, a sales briefing is scheduled by utilizing some specific web application, such as doodle.com.
- There is provided value, which in relation to the briefing is based on a tangible exchange of inputs of customer service, and response to inquiries between organizers and participants. The intangibles are targeted news and offerings

as well as updates on services and customer status (knowledge), and a sense of community (benefit).

■ There is return value, which is efficiency in terms of short handling time of customer requests as tangible, and informed customer request and feedback on latest developments (knowledge) and customer loyalty (benefits) as intangibles.

Value exchanges are modeled in a special type of concept map (Novak & Cañas, 2006), termed a holomap. Concept maps have turned out an effective means to articulate and represent knowledge (cf. Trochim & McLindon, 2017). They have been in use since the 1980s for acquiring mental models while graphically generating a coherent conceptual model, both supporting individuals, and groups in sharing and planning processes (cf. Goldman & Kane, 2014). Recently, concept maps served as a baseline for value analytics (cf. Ferretti, 2016), and have been used effectively for congruence analysis of individual cognitive styles (Stoyanov et al., 2017). Their ease of use allows triggering learning processes in distributed settings, for example, supported by web technologies (cf. Wang et al., 2017).

The VNA mapping from the observed reality to a role-specific concept map (holomap) is based on the following elements:

■ Ovals represent the functional roles of stakeholders, termed participants of the value network, that is, the nodes of the network.
■ Participants send or extend deliverables to other participants. One-directional arrows represent the direction in which the deliverables are moving during a specific transaction. The label on the arrow denotes the deliverable.

When modelers create holomaps, they think of participants as persons they know carrying out one or more roles in the organizational system at hand. Holomapping is based on the assumption that only individuals or groups of people have the power to initiate action, engage in interactions, add value, and make decisions. Hence, VNA participants can be individuals, small groups or teams, business units, whole organizations, collectives such as business networks or industry sectors, communities, or even nation-states. VNA does not consider databases, software, or other technology to be a participant. It is the decision-making capability about which activities to engage in that qualifies only humans as VNA participants.

Transactions or activities are represented by an arrow that originates with one participant and ends with another. The arrow represents movement and denotes the direction of addressing a participant. In contrast with participants, which tend to be stable over time, transactions are temporary and transitory in nature. They have a beginning, a middle, and an end point.

Deliverables are those entities that move from one participant to another. A deliverable can be physical or tangible, like a document or a physical object, but it can also be nonphysical, such as a message or request that may only be delivered verbally. It can also be an intangible deliverable of knowledge about something, or a favor.

In VNA, an exchange only occurs when a transaction results in a particular deliverable coming back. A gap is when something is provided without anything being received in return. However, focusing on the exchange as the molecular element of value creation is a generic concept that enables capturing a variety of organizations as value networks. Tangible and intangible exchanges establish patterns typical of business relationships. In many cases, tangible exchanges comprise exchanges of matter and energy (goods and money), while the intangible exchanges capture cognitive and emotive exchanges such as favors and benefits.

In the following, we provide an example of a VNA case in the Sales and Presales group of a service company providing innovative instruments (methods and technologies) for knowledge acquisition and sharing. To understand the overall patterns of exchange, and to determine the impact of tangible and intangible inputs for each participant, the company decided to perform a VNA. This should not only help to analyze the state of affairs, but also leverage potential changes for each participant. The Sales and Presales group aims to improve their ability to utilize operations and customer feedback in further developing their services.

The first step participants need to consider in the modeling process involves all the roles, organizational units, or work groups, both internally and externally, that are considered of relevance in the activities of the Sales and Presales group. In this case, four groups (participants) inside the organization are identified, namely Sales and Presales, Product Development, Customer Service, and freelance interviewers outside the organization. They represent the nodes in the holomap in Figure 8.1.

For modeling, first the stakeholders need to think about tangible exchanges that take place between the participants. What are the transactions adding value? What are the tangible deliverables in the work system? Figure 8.1 shows tangible deliverables such as product information, feedback from market, requests, and updates. For these cases, the transaction and communication channel is considered a tangible deliverable because it either comprises core data relevant for operating the business, or affects essential relations to organizational units for product and (customer) knowledge management.

Intangible transactions or exchanges are modeled the same way. To distinguish the intangible deliverables from the tangible deliverables, modelers use a different line style (dotted line in Figure 8.1). For the service provider at hand, intangibles are incomplete information, involving an order handling report, customer report, customer preparation data, and so on, which various stakeholders make available through reporting and active sharing of knowledge (Figure 8.1). They are considered intangible because there is no direct monetary income related to them. They are neither contracted by the provider nor expected by the recipients. They are extra offerings to participants to keep the operation running, and product development informed, mainly based on informal learnings and experiential knowledge.

As shown in Figure 8.1, several tangible exchanges occur. For example, the Product Development group provides product announcements to the Presales group in exchange for feedback from the market, whereas the Sales group provides requests

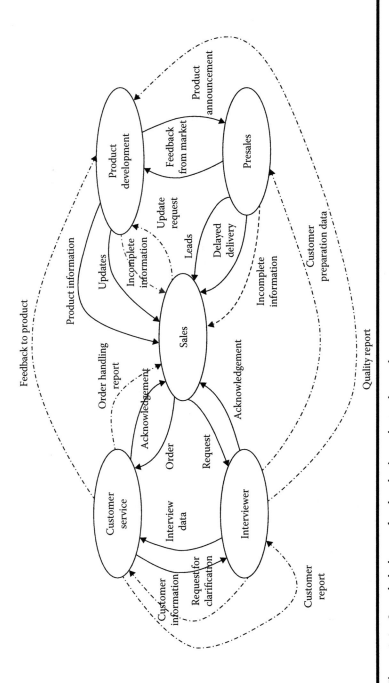

Figure 8.1 Sample holomap for developing sales and presales.

to interviewers who acknowledge them. In addition, several intangible exchanges occur, such as interviewers provide customer information to the Customer Service group in exchange for customer reports. The latter complements the formal, role-specific exchange specified through the pair "request for clarification—interview data." It documents the intention to provide a comprehensive picture of customers to build trustful relationships with customers (representing the benefits of the exchange).

However, several one-sided transactions, with respect to tangibles and intangibles, become evident, as also shown in the holomap in Figure 8.1. For instance, concerning intangibles, the interviewers provide customer preparation data and quality reports to the Presales and Product Development groups respectively, without any intangible return. Concerning tangibles, the Product Development group provides both product information and updates to the Sales group without any return, for example.

Once all exchanges and deliverables are captured in the holomap, a diagram of how the business is perceived from a stakeholder perspective is established. The value network view of an organization helps understand the role of knowledge and intangibles in value creation. The modeling process allows capturing strategically critical intangible exchanges from a stakeholder perspective, thus enabling further targeting opportunities for value creation. This issue is addressed through analyzing the value network as represented by the holomap in the next subsection.

Analyzing the Value Network

According to Allee (2003, 2008), analyzing the capabilities of a value network and developing opportunities for constructive change requires three different types of analysis. Transactional value analytics is based on elicited transactions between people acting in certain roles and explores patterns in terms of reciprocity and the perceived value of generated deliverables transmitted to other stakeholders, and received deliverables from other stakeholders. Based on this data, change proposals can be developed in terms of offering novel deliverables to other stakeholders, either changing their type from intangible to tangible, generating novel ones, or including further stakeholders not involved in interaction from the perspective of a certain role carrier so far.

The first analysis is about assessing the structure and dynamics of the represented system as a whole. The second and third analyses focus on each participant's role in the interactions of the value network:

1. *Exchange analysis*: What is the overall pattern of exchanges in the represented system?
2. *Impact analysis*: What impact does each value input have on the participants?
3. *Value creation analysis*: How can an organization create, extend, and leverage value, either through adding value by participants, extending value to other participants, or converting one type of value to another?

The exchange analysis investigates the overall patterns of exchanges addressing a variety of structural issues. When starting to identify missing relations for operating the business or links requiring a rationale, potential breakdowns in flow that can be critical for the business can be determined. In that context, the coherence of relations and resulting flows of how value is generated in the network can be evaluated. By checking the amount of tangibles and intangibles, a possible dominance of a particular type of exchange can be determined. The overall pattern of reciprocity reveals involvement data of the participants (as perceived by the respective modeler). Extensive sources and sinks of interactions should be noted as potential for optimizing the entire network, avoiding specific participants benefiting at the expense of others.

In the holomap in Figure 8.1, several patterns can be noticed. The Sales and Presales groups gain knowledge about the product from the Product Development group, but that knowledge results in dead ends and so is not passed on explicitly to the Customer Service group and interviewers. In fact, there is no significant product knowledge exchange between Sales, Presales, Customer Service, and interviewers. Another pattern that shows up is that knowledge from the Customer Service group to the Product Development group only flows one way. The Customer Service group delivers feedback on a product to the Product Development group as intangible, but does not have any channel for a two-way communication about product-related data with the Product Development group.

Allee (2003, 2008) states there would be no straightforward interpretation for what certain patterns mean, although they become apparent through the exchange analysis. Patterns trigger discussions among those participating in identified patterns, in particular when missing steps in a core process or cultural issues regarding the organization pop up. According to practical experience, it is the set of upcoming questions that lead to informed impact and value creation analysis, and thus change.

The impact analysis focuses on received inputs that trigger some response by the receiver. The analysis considers activities and effort for handling the input and for leveraging the value received. The concerned stakeholders need to estimate the tangible and intangible costs (or risks) and gains for each input. They need to describe how a certain activity is generated in response to a certain (tangible or intangible) input. Then, they are asked to describe the estimated increase or decrease of tangible assets in terms of overall cost or risk-benefit analysis (Table 8.1).

Table 8.1 shows the value inputs for the Sales group. Each of the received inputs is explored according to the various activities to be set, as well as the costs and benefits it brings. In this way, each Sales participant can address the value impact for the whole organization. Each input presented in the holomap (Figure 8.1) is described using a single line as it may affect essential tangible and intangible inputs to the Sales and Presales groups. For instance, the fourth-line item shows incomplete delivery of information from the Presales group to the Sales group. This results in a low value since the required knowledge on sales transactions was not available

Table 8.1 Sample Impact Analysis of a Sales Participant

Transactions				Impact Analysis						
					Impact on Intangible Assets (Positive/Negative)					
Deliverable	From	To	Activities That are Generated	Impact on Financial Resources (Positive/Negative)	Human competence	Internal structure	Business relationships	Overall Cost/Risk for This Input	Overall Benefit of This Input	Perceived Value in View of Recipient
Order handling report (Intangible)	Customer service	Sales	Evaluation of report	Extra time and effort to be calculated	Knowledge on calculation schema	Contact to presales and accounting	Knowledge on competition and market	H = High M = Medium L = Low; H: 2 hours/ M: recalculation required	H = High M = Medium L = Low; M: Documentation of order handling	+2, +1 Neutral −1, −2; Neutral
Incomplete information (Intangible)	Product development	Sales	Additional information from product development to be collected	Effort to be spent for information collection	Knowledge on product	Expert interviews	Technical skills and knowledge	H: 3 hours/ H: Availability of experts required	M: Completeness of product information	Neutral
Delayed delivery (Tangible)	Presales	Sales	Delay in customer service to be communicated	Loss of service time	–	Presales contact	Presales reminder	H: Extension of loss of service time/ H: Loss of loyalty	L	−2
Incomplete information (Intangible)	Presales	Sales	Additional information from presales to be collected	Effort to be spent for information collection	Knowledge on sales transaction	Presales contact	Appointment	M: 1 hour/ M: Availability of presales required	M: Completeness of product information	−1

(Continued)

Table 8.1 (*Continued*) Sample Impact Analysis of a Sales Participant

Transactions				Impact Analysis						
Deliverable	From	To	Activities That are Generated	Impact on Financial Resources (Positive/Negative)	Impact on Intangible Assets (Positive/Negative)			Overall Cost/Risk for This Input	Overall Benefit of This Input	Perceived Value in View of Recipient
Acknowledgement (Tangible)	Customer service	Sales	Recognition	–	Customer service satisfied	Order process	Customer contacted	–	H: Order processed	+2
Acknowledgement (Tangible)	Interviewer	Sales	Recognition	–	Interviews enabled	Order process	Customer contacted	–	H: Order processed	+2
Product information (Tangible)	Product development	Sales	Material to be studied	Time and effort to be studied	Knowledge acquisition	Informed external contacts	Qualification	H: 2 hours per feature/ M: Request for clarification	H: Informed collaboration	+1
Updates (Tangible)	Product development	Sales	Feature(s) to be studied	Time and effort for studying	Knowledge acquisition	Informed method application	Qualification	H: 1 hour per feature/ M: Request for clarification	H: Informed collaboration	+2
Leads (Tangible)	Presales	Sales	Update of CRM-database	Time and effort for update	Knowledge on opportunities	CRM-timeliness	Data accuracy	M: 5 min per record/ M: requires clarification	H: Documentation of opportunities	+2

in time. The Presales group needs to be contacted for availability to overcome short-comings in customer relationship management.

The value creation analysis not only reflects the situation as it is but supports the proposal of changes to the way a participant is committed to the delivery. Although the structure of this analysis is similar to the impact analysis, it instead focuses on a participant's capability regarding how to extend value to other participants represented in the holomap. It analyzes the tangible and intangible costs (or risks) and gains for each value output. Each value output could add new tangible or intangible value and thus extend value to other participants represented in the holo-map. By assessing each value output, a participant needs to determine the activities, resources, and processes required, as well as the costs and anticipated benefits of each value-creating activity.

In Table 8.2 a sample analysis is given for the Sales group. An important value creation for the Sales group is to add value to product features received from the Product Development group by providing timely information for interviewers and the Customer Service group (in terms of what is actually ordered). Their role in contributing to timely information in cooperation with the Product Development group is actually very small, being of both low cost for them and also low benefit in case of not actively requesting updates. Hence, they could be an active agent to extend value through targeted requests and other efforts to reach the Product Development group. They would need to engage in a value conversion process in which they convert one relevant type of value to another with respect to a working product lifecycle. Engaging in increase value creation actually explores new strategic possibilities and changes costs and benefits for the organization. It is the moment where a member of the organization develops an offer to others by individual behavior change instead of elaborating the expected behavior of others.

Looking at the value outputs that go to the Product Development group, the Sales group can open a tangible communication channel to support the process used by the Product Development group, and an intangible transaction of feeding back product-relevant information. In this way, the Sales group can leverage their intangible value outputs (information about product use) into more advanced product development that could be turned into a revenue stream. In the example, the communication channels were mostly one way, conveying product knowledge from the Product Development group to the Sales group. With a focus on converting that expected communication channel with another type of value gain, the Sales group enables communication after gaining feedback for the Product Development group. Both converting an intangible value to gain a tangible value, and establishing an intangible deliverable with product feedback, supports the strategic intent of a rapid response to the changing needs of customers.

Although value creation analysis can become very rich due to anticipated changes in the flow of deliverables, the participants need to understand what impact a particular output has on the participant who receives it. From an organizational development perspective, a participant could maximize the effectiveness

Table 8.2 Sample Value Creation Analysis

Transactions			Perceived Value		Value Creation Analysis	
Deliverables	*From*	*To*	*Recipient Highly Values This Deliverable. Strongly Agree(+2) Agree(+1) Neutral(0) Disagree(−1) Strongly Disagree(−2)*	*Tangible Asset Utilization Is: H = High M = Medium L = Low*	*What Are The Tangible Costs? (Financial And Physical Resources)*	*How High Is The Risk Factor In Providing This Output? H = High M = Medium L = Low*
Update request (Intangible)	Sales	Product development	−1	L	Processing time by request	L
Request (Tangible)	Sales	Interviewer	+2	H	Interview	M
Order (Tangible)	Sales	Customer service	+2	H	Processing time per order	M

(Continued)

Table 8.2 (Continued) Sample Value Creation Analysis

Intangible asset utilization is: H = high M = medium L = low (Human competence Internal structures Business relationships)	IS	BR	What are other intangible costs or benefits? (Industry, society, environment)			How do we add to, enhance, or extend value?	What is the overall combined cost/risk for this input?	What is the overall benefit for us in providing this input?
			Industry	Society	Env			
HC	M	M	Benefit-timely information	—	—	Formal procedure	Formulating request/—	Trigger to product development
M								
	H	H	Customer care	—	—	—	—	High customer satisfaction
H								
	H	H	Order management	—	—	CRM-Database update	Delivery of update/completeness of data	Enriched CRM-Database
H								

and efficiency of a certain business operation when following the created value. The overall cost benefit analysis could result in excellent data. However, a closer analysis of the proposed value creation could lead to inconveniences of the involved participants, in particular business partners and customers (Augl & Stary, 2015, 2017). For instance, collecting context data from interviewers when applying the product in the field could easily lead to rejecting the proposed value creation once they consider a request for preparing such a report as a negative value input.

Before proposed value creations can become effective in business operations, they need to be acknowledged by the involved participants as well as by responsible management. To allow the constructive elaboration of value creations affecting the collective, each member of the organization should be empowered with an instrument enabling him or her to reflect on individual values. Applying such an instrument requires stepping out of the VNA logic while providing a baseline for discussing proposals for change. Once value-creation proposals are congruent with individual value systems, the resulting business operation becomes coherent for the involved stakeholders.

How to Ensure Coherent Value Creation

To facilitate the alignment of proposed value creations with individual value systems, repertory grids are suggested in this section. They allow inconsistencies of value creations with individual value systems to be overcome, and are selected based on several expert semi-structured interviews addressing value management from a stakeholder perspective. Accordingly, in the first subsection the identification of a proper method is described, before the repertory grids are detailed and exemplified to demonstrate their analytical support capability in the context of the VNA.

Eliciting Methodological Knowledge

In search of a method allowing individual value systems to be externalized, 14 organizational analysts involved in value management were invited to a semi-structured interview. The overarching question they were asked was "How far can organizational stakeholders be supported with methods when needing to align individual with collective values or value systems?"

Each of the 14 analysts is active in Germany and Austria and have the following profile:

- Long-term experience in practical value management, particularly in method-driven approaches (e.g., utilizing Bohm's Dialogue)
- Interest in theory-driven work praxis (e.g., system thinking)
- Experience in international organizational development projects

The analysts were interviewed using the following sequence of items:

1. What is your particular work background with respect to value management?
2. What are your qualifications and background in organizational design, learning, and development by means of knowledge management techniques?
3. Which methods do you actually use with respect to value management?
4. Which methods do stakeholders need most when adjusting individual with collective values or value systems?
5. Which method do you consider most effective when adjusting individual value systems with collective ones? Kindly refer to:
   a. Repertory grid
   b. Critical incident technique
   c. Storytelling
   d. Another method you consider most effective based on your experience

The methods were picked from a selected collection of practically tested methods for gaining and analyzing stakeholder knowledge (Stary, Maroscher, & Stary, 2012). The repertory grid technique (Fransella & Bannister, 1977; Fromm, 1995; Goffin, 2002) supports an understanding of how individuals perceive a certain phenomenon in a structured way, namely by identifying the most relevant characteristics of the carriers of those characteristics (e.g., roles, persons, objects, situations) on an individual scale. The technique enables not only identifying attributes that relate to the carriers but also to elicit explanations of these characteristics. Repertory grids have been introduced by Kelly (1955) based on the personal construct theory in psychology, and have been used successfully in a variety of contexts including expert, business, and work knowledge elicitation (Ford, Perty, Adams-Webber, & Chang, 1991; Gaines & Shaw, 1992; Hemmecke & Stary, 2006; Stewart, Stewart, & Fonda, 1981), strategic management (Hodgkinson, Wright, & Paroutis, 2016), supplier-manufacturer relationship management (Goffin, Lemke, & Szwejczewski, 2006), learning support (Stary, 2007), attitude studies (Honey, 1979), team performance (Senior, 1996; Senior & Swailes, 2004), project management (Song & Gale, 2008), and consumer behavior (Kawaf & Tagg, 2017).

The critical incident technique (Flanagan, 1954) is an exploratory qualitative method that has been shown to be both reliable and valid in generating a comprehensive and detailed description of a content domain. The technique consists of asking eyewitnesses for factual accounts of behaviors. The critical incident technique consists of a set of procedures for collecting direct observations of human behavior in such a way as to facilitate their potential usefulness in solving practical problems and developing psychological principles. By "incident," what is meant is any observable human activity that is sufficiently complete in itself to permit inferences and predictions to be made about the person performing the act. To be critical, an incident must occur in a situation where the purpose or intent of the act seems fairly clear to the observer and where its consequences are sufficiently definite to leave little doubt concerning its effects. In this way, role-specific behavior can be valued by observers, including their transactions with other stakeholders.

The critical incident technique outlines procedures for collecting observed incidents that have special significance and meeting systematically defined criteria, for example, behavior considered to be most effective or efficient in a certain situation (Butterfield, Borgen, Amundson, & Maglio, 2005). The many and varied applications of the technique comprise counseling (Hrovat & Luke, 2015), tourism (Callan, 1998), service design (Gremler, 2004), organizational merger (Durand, 2016), and healthcare (Eriksson, Wikström, Fridlund, Årestedt, & Broström, 2016). In contrast to repertory grids, which are grounded on introspection, the critical incident technique relies on making observations about other people, thus valuing the behavior of others. In doing so, people make notes when observing others' behavior to reconstruct the meaning later.

Storytelling has evolved from systemic representations of narratives (Livo & Rietz, 1986), today termed mainly digital storytelling, as personal stories are kept in digital form to preserve relevant happenings and share via the web (Lundby, 2008). A story represents the world around us as perceived by specific individuals (storyteller as creator and filter). Depending on its purpose, a story can be more like an objective or fact-driven report or a dedicated trigger for change addressing persons in certain social or functional roles (Denning, 2001). Digital storytelling, due to its narrative and social structure, is used to capture not only individual characteristics (Cavarero, 2014) but also workplace specifics (Swap, Leonard, Shields, & Abrams, 2001) and organizational issues (Boje, 1991). Similar to repertory grids, stories are created from introspection, although they result in a linear text in contrast to a matrix representing individual value systems.

The interviews with the value management experts lasted around 30 minutes each. Since they had various backgrounds (marketing, information systems, psychology, linguistics etc.) and diverse work experience, most of the analysts could identify methods not included in the selection offered in item 5 of the interview guide. However, they agreed that repertory grids are the most effective, since they constitute a noninvasive method used to elicit value systems as the externalization procedure (see next section) does not direct the way of answering the respective questions or thinking. A personal construct can be understood as the underlying mechanism of a personality that enables an individual to interpret a situation. In addition to individual attitudes, repertory grids enable the expression of how individuals interpret a task procedure or technical system feature, or perceive others' behavior in a certain situation. Recording an individual's point of view concerning an object or phenomenon as well as the individual's contrasting viewpoint constitutes an individual's value system. Hereby, a phenomenon is likely to have a certain meaning for each individual as expressing it, enabling even varying experiences of situations to be represented. Yielding insights in such a way about an individual's value system may prompt a discussion regarding whether a certain behavior or fact can be accepted by members of an organization or not. The repertory grid technique therefore allows for the maximum bandwidth of responses.

Moreover, when utilizing repertory grid elicitation, responses are expressed in the stakeholder's own words. However, it encourages others to delve further into the responses to understand better and thus elicit richer information. This allows for a thorough understanding of stakeholders' perceptions. Finally, the rich data collected enables a thorough examination of the underlying meaning behind each proposed transaction and role characteristic, allowing for a specification of value exchanges to be derived. In this way, a stakeholder can also identify any new characteristics not evident in prior holomaps to become part of an improved organizational work practice.

Based on the findings and arguments brought up when discussing repertory grids, it was decided to test this method in the context of VNA, to identify mismatches and matches between individual value systems and VNA-created values influencing the flow between participants of a value network.

Externalizing Value Systems through Repertory Grids

The field test targeted the individual value systems involved in the already exemplified development case of the Sales and Presales groups. The application of the repertory grid technique comprises several steps (Fransella & Bannister, 1977; Fromm, 1995; Hemmecke & Stary, 2006; Honey, 1979; Jankowicz, 2004):

- Identification and specification of elements (objects of investigation), being part of the considered setting.
- Elicitation of constructs, that is, setting up a personal construct system.
- Rating of constructs, that is, in a sense a consolidation of constructs.
- Saturation of constructs, that is, identifying the "completeness" of a construct system.
- Analysis and synthesis of the result, that is, element-specific interpretation, such as an ideal personality or object.
- Testing plausibility, in order to ensure a sound data basis.

The result is a validated value system in a given context that can be applied to reflect on value creations, in the test case in the participants' context of sales activities.

Identification and specification of elements: The elements in the repertory grid technique are the objects of concern. Examples that can be given to the participants for element types are physical entities (e.g., products), living things (e.g., customers), events in a timeline (e.g., critical incidents in a customer relationship), social entities (e.g., teams in a firm), behavior and activities (e.g., techniques in performing a certain sales task), appraisals and abstractions (e.g., "positive" events in a customer relationship setting, criteria for sales reports), and emotions (e.g., emotions during professional socialization).

Furthermore, the participants need to be told that elements should have certain properties to support the elicitation of meaningful constructs. Ideally, they should be: (1) discrete—the element choice should contain elements on the same level of hierarchy, and should not contain sub-elements; (2) homogeneous—it should be possible to compare the elements, for example, things and activities should not be mixed within one grid; (3) comprehensible—the person from whom the constructs are elicited should know and understand the elements, otherwise the results will be meaningless; and (4) representative—the elicited construct system will reflect the individually perceived reality, once the element choice is representative for the domain of investigation.

The focus of the field test was work behavior and thus stakeholders acting in particular roles, particularly those involved in the setting addressed by the VNA holomapping. These are likely to form the repertory grid's elements. In the field test, the four participants were asked to identify between five and eight elements as entities they could refer to in the business case addressed by the holomap. These elements should allow naming certain properties of the situations relevant for meeting the objectives of that business case. Since these elements (e.g., roles, persons, objects, and task settings) stem from a running business operation involving actual stakeholders, the participants were briefed to use anonymous abbreviations, labels, or names when denoting the elements. The participants were also asked to include an element they consider to be ideal in the addressed business case, to gain an understanding of the anticipated improvements of the transactional behavior through value creation.

Elicitation of constructs: When it comes to identifying the constructs (i.e., properties of the identified elements), we need to be aware that they are generated through learning about the world when acting in the world. Hence, construct systems, and thus value systems, depend on the experiences individuals make during their lifetime and, moreover, depend on the common sociocultural construct systems. Since construct systems change over time, they need to be considered dynamic entities. As such, elicited knowledge through grids is only viable for a certain person in a certain physical and social context.

Elicitation of constructs in a repertory grid interview in the field test was concerned with understanding the participant's perception or personal construct system of work behavior. Constructs may be elicited by using one, two, or three elements at a time. We used the triad form, that is, three out of the specified five to eight elements during each elicitation iteration (see also Table 8.3, presenting a repertory grid for Sales participant X who identified A, B, C, D, and I (Ideal) as elements).

Our participants' sample reflects an average age of 29 years (median of 33), with the youngest being 24 and the oldest being 60. There were slightly more men than women (three vs one). According to the objectives, the participants have a tight range of employment, from Sales to Presales. All participants have lived in Austria for a substantial part of their life.

Table 8.3   Sample Repertory Grid of a Sales Person (Participant X)

Sales X Construct	A	B	C	D	I	Contrast
Openness	1	1	−3	−2	1	Information hiding
Openness	1	1	−2	3	−3	Formal reporting
Upfront information	−1	2	1	2	3	Unprepared service
Informed about product	2	−2	1	2	3	Outdated product education
Interested in method development	−1	−1	1	−1	1	Mainly interested in method application

In the triad elicitation, three elements are compared with each other according to their similarities and differences. The person is asked to specify "some important way in which two of them are alike and thereby different from the third" (Fransella & Bannister, 1977, p. 14). The elicited similarity between the two elements is recorded as the construct. Subsequently, the person has to detail the contrast with the following question: "In what way does the third element differ from the other two?" The construct (pair) elicitation continues as long as new constructs can be elicited. Table 8.3 shows a sample grid that was elicited from Sales participant X following the elicitation steps 1 to 3.

Step 1: Select triad: Each element selected by the interviewed participant was written on a single card. The front of each card showed the name of the respective element. In the case at hand, to ensure that the participants could remember the addressed elements when using pseudonyms, abbreviations, or symbols, they were asked to make a coding list before starting the construct elicitation procedure. After that, the interviewer turned the index cards upside-down and mixed them, and the participant was asked to pick three of the index cards, at random. This move prevented the participant from knowingly selecting a specific card.

Step 2: Elicitation of raw constructs: After the participant had selected three index cards, the participant was asked: "In the context of sales, how are two of these elements the same, but different from the third?" The participant expressed his or her construct system concerning the characteristic that differentiates the selected elements. The participant was then asked to provide a short label to describe the emergent pole (how two are the same). For instance, Sales participant X identified "openness" for the A and B elements in the first elicitation round (data lines 1 and 2, respectively, of Table 8.3), however these were contrasted by two constructs, namely "information hiding" and "formal reporting."

Step 3: Laddering: As the participant provided insights into his or her personal construct system, the interviewer applied the laddering technique by asking "how" and "why" questions to encourage the participant to provide more information in areas where certain ambiguities still existed ("openness"). A construct with two contrasting constructs was indicated in the first two lines of the table. The laddering technique incited the participants to pose contrasting constructs to the emergent pole. The following example and transcript illustrate the laddering technique.

The second participant provided the emergent pole for two role carriers. He explained how performing formal reporting in contrast to openness induces a feeling of not being open to each other. He declared that the feeling was present in regards to the third person (see line 2 in Table 8.3). The interviewer asked for an elaboration of the contrasting pole, and the second participant replied as follows:

If you get a report without any context and personal purpose, you feel like a mere fact checker without any background information on how a sales process went so far and you are lost if you are supposed to arrange the next step in customer relationship management. In particular, once the data indicates you could have known more about a sales process when studying the subsequent report from that person which indicates information you should have received before launching the latest intervention.

Rating the constructs for each element: The third phase of a repertory grid elicitation session is the rating of the elements according to the elicited constructs. The mutual relations of elements and constructs can be explored by using a rating procedure. The rating can range from 5 to 7 on a point scale. In the case at hand, the rating ranged from +3 to −3 (excluding 0), with +3, +2, +1 for rating the strength of property (+3 is the strongest) for left-side constructs (left row of Table 8.3), and −3, −2, −1 for rating the strength of property (−3 is the strongest) for right-side constructs, also termed contrasts (right row of Table 8.3). For instance, "openness" was rated for element D as very strong (+3 in line 2), and also "formal reporting" for element I (−3, since it refers to the contrast, rated from −3 to −1).

Construct saturation: The overall aim of construct saturation is to obtain as many unique constructs as required to externalize the value system of a participant. Since even people acting in the same role (e.g., Sales) in a certain organizational context are likely to have different backgrounds, there is no common rule regarding when to stop eliciting constructs. Once all possible combinations of elements have been selected, a second round of selections could make sense, in case the participant wants to add more constructs to the already elicited ones in the grid. Typically, a repertory grid with four to six elements has not more than 10 to 15 lines of entry. However, a dedicated check of whether construct saturation has been reached is strongly advised. In the field test this strategy was applied. We even asked the participants whether certain constructs were considered unique to ensure that a new construct was elicited when the participants gave a statement reflecting on a triad.

Analysis and synthesis of the result: The goal of the analysis of a repertory grid is to represent the construct system in a way that the participant gets explicit insights into his or her own view about the corresponding elements. Finally, other individuals should come to an understanding about the participant's way of thinking. The most straightforward consideration of a grid is to look at the ideal (I) element entries and interpret the pattern representing the ideal case, object, or person. In addition, bi-plots derived from principal component analysis and dendrograms derived from cluster analysis can be generated. In bi-plots the relations of elements and constructs can be displayed, whereas in dendrograms only the relations of elements or the relations of constructs can be visualized. However, it is not necessary to use computer programs for analysis, especially when the results of a single person are subject to analysis. Content analysis (e.g., as proposed by Fromm, 1995) allows not only determining the meaning of constructs, but also clusters them according to a category system.

Testing plausibility: A check of the plausibility of the named constructs should be conducted with independent third parties to ensure the reliability and stability of the results, as well as the accuracy of the data (Jankowicz, 2004). Initially, we verified the clarity of constructs by asking two domain experts not participating in the elicitation to read each construct carefully and explain its meaning in his or her own words in the context of the addressed setting (i.e., sales). The constructs were adjusted in cases where constructs were not self-explaining. Statements from the laddering step in the elicitation phase served as a valuable input to that task.

Developing Commitment for an Organizational Move

In this step the constructs were checked for operational improvement with respect to the value creation in the VNA, that is, to what extent the proposed value creations represent new insights about the addressed role-specific behavior. In this way, new patterns of exchange can be specified and verified, as the repertory grid insights are pertinent for the development of novel role-specific behavior.

As shown in Table 8.2 (value creation analysis), for Sales participant X an important value creation is to add value to product features received from the Product Development group, in particular by providing timely information for interviewers and the Customer Service group. When checking his or her repertory grid, such an improvement is in line (since positively rated for the ideal Sales participant) with the positively rated constructs "openness," "upfront information," and "interested in method development." The same holds for the participant's role in contributing to timely information in cooperation with the Product Development group. Becoming an active agent, and thus extending value through targeted requests and other efforts to reach the Product Development group, is consistent with all positively rated constructs of the Ideal element. In this way, Sales participant X can check whether the value creation is in line with his or her individual value system, before actively promoting this change in the ongoing organizational development process.

In receiving new input from others (e.g., the Presales group), the same approach can be applied. In our case, a Presales participant in the value creation analysis converted the intangible deliverable "incomplete information" to a tangible one. Hence, a possible value creation for this Presales participant is to add value to customer relations and product features received from the Product Development group by providing indicators for incomplete information, since it may affect the work of interviewers and the Customer Service group. Again, by checking his or her repertory grid, such an improvement is totally acceptable to Sales participant X, since it is in line (i.e., positively rated for the Ideal Sales participant) with the constructs of "openness" and "upfront information." Sales participant X can support this change proposal based on his or her elicited and rated constructs in the context of sales processes.

As these two examples reveal, the individually-valid repertory grid can be utilized to check incoming and outgoing transactions as part of the value creation analysis, and ascertain whether they fit to the individual value system. Based on this information, value creation can either be promoted or opposed to be implemented on the organizational level.

Conclusive Summary

Value congruence plays a crucial role in organizations once individual proposals are discussed for improving business operation. We have looked at the case of individual stakeholder transaction analyses, since VNA focuses on value creations while targeting stakeholder- and business-conform exchange patterns. As the approach considers both the exchange of tangible (or contracted) and intangible (or voluntarily) provided deliverables, value creations need to be aligned with individual value systems before becoming a collective practice.

Various methods, in particular the repertory grid technique, the critical incident technique, and storytelling, exist to elicit individual value systems. Several experts have been interviewed taking into account the listed candidates as well as their experiential knowledge, to identify the best choice for eliciting value systems in stakeholder-driven change management. They agreed on the consensus-building process supported by individually generated repertory grids. These grids allow individuals in certain roles to reflect on their value system when acting in that role.

In a field test, individual stakeholders were able to rethink organizational behavior changes offered by other stakeholders, as they could assess them according to their role-specific value system. Further studies will deepen this methodological insight, since the reported field test only revealed the feasibility of the approach. There are other points of interventions (e.g., the holomapping phase) when applying repertory grids in the context of VNA. They are currently studied in ongoing field tests in other service and production industries.

References

Allee, V. (1997). *The knowledge evolution: Expanding organizational intelligence.* Amsterdam, the Netherlands: Butterworth-Heinemann.

Allee, V. (2003). *The future of knowledge: Increasing prosperity through value networks.* Amsterdam, the Netherlands: Butterworth-Heinemann.

Allee, V. (2008). Value network analysis and value conversion of tangible and intangible assets. *Journal of Intellectual Capital, 9*(1), 5–24.

Augl, M., & Stary, C. (2015). Communication- and value-based organizational development at the university clinic for radiotherapy-radiation oncology. In *S-BPM in the wild* (pp. 35–53). Berlin, Germany: Springer International Publishing.

Augl, M., & Stary, C. (2017). Adjusting capabilities rather than deeds in computer-supported daily workforce planning. In M. S. Ackerman, S. P. Goggins, T. Herrmann, M. Prilla, & C. Stary (Eds.), *Designing healthcare that works. A sociotechnical approach* (pp. 175–188). Cambridge, MA: Academic Press/Elsevier.

Boje, D. M. (1991). The storytelling organization: A study of story performance in an office-supply firm. *Administrative Science Quarterly, 36*(1), 106–126.

Boyle, T. A. (2005). Improving team performance using repertory grids. *Team Performance Management: An International Journal, 11*(5/6), 179–187.

Butterfield, L. D., Borgen, W. A., Amundson, N. E., & Maglio, A. S. T. (2005). Fifty years of the critical incident technique: 1954–2004 and beyond. *Qualitative Research, 5*(4), 475–497.

Callan, R. J. (1998). The critical incident technique in hospitality research: An illustration from the UK lodge sector. *Tourism Management, 19*(1), 93–98.

Cavarero, A. (2014). *Relating narratives: Storytelling and selfhood.* New York, NY: Routledge.

Dalkir, K. (2011). *Knowledge management in theory and practice.* Cambridge, MA: MIT Press.

Denning, S. (2001). *The springboard: How storytelling ignites action in knowledge-era organizations.* New York, NY: Routledge.

Durand, M. (2016). Employing critical incident technique as one way to display the hidden aspects of post-merger integration. *International Business Review, 25*(1), 87–102.

Easterby-Smith, M. (1976). The repertory grid technique as a personnel tool. *Management Decision, 14*(5), 239–247.

Edwards, J. R., & Cable, D. M. (2009). The value of value congruence. *Journal of Applied Psychology, 94*(3), 654.

Eriksson, K., Wikström, L., Fridlund, B., Årestedt, K., & Broström, A. (2016). Patients' experiences and actions when describing pain after surgery—A critical incident technique analysis. *International Journal of Nursing Studies, 56*, 27–36.

Ferretti, V. (2016). From stakeholders analysis to cognitive mapping and Multi-Attribute Value Theory: An integrated approach for policy support. *European Journal of Operational Research, 253*(2), 524–541.

Firestone, J. M., & McElroy, M. W. (2003). *Key issues in the new knowledge management.* New York, NY: Routledge.

Flanagan, J. C. (1954). The critical incident technique. *Psychological Bulletin, 51*(4), 327.

Ford, K., Perty, F., Adams-Webber, J., & Chang, P. (1991). An approach to knowledge acquisition based on the structure of personal construct systems. *IEEE Transactions Knowledge and Data Engineering, 4*(1), 78–88.

Fransella, F., & Bannister, D. (1977). *A manual for repertory grid technique.* London, UK: Academic Press.

Fromm, M. (1995). *Repertory grid technique: A Tutorial (in German).* Weinheim, Germany: Deutscher Studienverlag.

Gaines, B. R., & Shaw, M. L. G. (1992). Knowledge acquisition tools based on personal construct psychology. *Knowledge Engineering Review—Special Issue on Automated Knowledge Acquisition Tools, 8,* 49–85.

Goffin, K. (2002). Repertory grid technique. In D. Partington (Ed.), *Essential skills for management research* (pp. 199–225). London, UK: Sage.

Goffin, K., Lemke, F., & Szwejczewski, M. (2006). An exploratory study of "close" supplier-manufacturer relationships. *Journal of Operations Management, 25,* 189–209.

Goldman, A. W., & Kane, M. (2014). Concept mapping and network analysis: An analytic approach to measure ties among constructs. *Evaluation and Program Planning, 47,* 9–17.

Gremler, D. D. (2004). The critical incident technique in service research. *Journal of Service Research, 7*(1), 65–89.

Hemmecke, J., & Stary, C. (2006). The tacit dimension of user tasks: Elicitation and contextual representation. In *Proceedings of the International Workshop on Task Models and Diagrams for User Interface Design* (pp. 308–323). Berlin, Germany: Springer.

Hewlin, P. F., Dumas, T. L., & Burnett, M. F. (2017). To thine own self be true? Facades of conformity, values incongruence, and the moderating impact of leader integrity. *Academy of Management Journal, 60*(1), 178–199.

Hodgkinson, G. P., Wright, R. P., & Paroutis, S. (2016). Putting words to numbers in the discernment of meaning: Applications of repertory grid in strategic management. In G. B. Dagnino & M. C. Cinici (Eds.), *Research methods for strategic management* (pp. 201–226). New York, NY: Routledge.

Honey, P. (1979). The repertory grid in action: How to use it to conduct an attitude survey. *Industrial and Commercial Training, 11*(11), 452–459.

Hrovat, A., & Luke, M. (2015). Is the personal theoretical? A critical incident analysis of student theory journals. *Journal of Counselor Preparation and Supervision, 8*(1), 5.

Jankowicz, D. (2001). Why does subjectivity make us nervous? Making the tacit explicit. *Journal of Intellectual Capital, 2*(1), 61–73.

Jankowicz, D. (2004). *The easy guide to repertory grids.* Chichester: Wiley.

Kawaf, F., & Tagg, S. (2017). The construction of online shopping experience: A repertory grid approach. *Computers in Human Behavior, 72,* 222–232.

Kelly, G. A. (1955). *The psychology of personal constructs: Volume one—A theory of personality.* New York, NY: Norton.

Lau, L., Yang-Turner, F., & Karacapilidis, N. (2014). Requirements for big data analytics supporting decision making: A sensemaking perspective. In N. Karacapilidis (Ed.),

Mastering data-intensive collaboration and decision making (pp. 49–70). Berlin, Germany: Springer International Publishing.

Livo, N. J., & Rietz, S. A. (1986). *Storytelling: Process and practice*. Littleton, CO: Libraries Unlimited.

Lundby, K. (2008). *Digital storytelling, mediatized stories: Self-representations in new media* (Vol. 52). Frankfurt/Main, Germany: Peter Lang.

Novak, J. D., & Cañas, A. J. (2006). The origins of the concept mapping tool and the continuing evolution of the tool. *Information Visualization, 5*(3), 175–184.

Payne, A. F., Storbacka, K., & Frow, P. (2008). Managing the co-creation of value. *Journal of the Academy of Marketing Science, 36*(1), 83–96.

Posner, B. Z., & Schmidt, W. H. (1993). Values congruence and differences between the interplay of personal and organizational value systems. *Journal of Business Ethics, 12*(5), 341–347.

Rich, B. L., Lepine, J. A., & Crawford, E. R. (2010). Job engagement: Antecedents and effects on job performance. *Academy of Management Journal, 53*(3), 617–635.

Selby, R. W. (2009). Analytics-driven dashboards enable leading indicators for requirements and designs of large-scale systems. *IEEE Software, 26*(1), 41–49.

Senior, B. (1996). Team performance: Using repertory grid technique to gain a view from the inside. *Journal of Managerial Psychology, 11*(3), 26–32.

Senior, B., & Swailes, S. (2004). The dimensions of management team performance: A repertory grid study. *International Journal of Productivity and Performance Management, 53*(4), 317–333.

Song, S. R., & Gale, A. (2008). Investigating project managers' work values by repertory grids interviews. *Journal of Management Development, 27*(6), 541–553.

Stary, C. (2007). Intelligibility catchers for self-managed knowledge transfer. In *Proceedings of the 7th international conference on advanced learning technologies, ICALT 2007, IEEE* (pp. 517–521).

Stary, C. (2014). Non-disruptive knowledge and business processing in knowledge life cycles—Aligning value network analysis to process management. *Journal of Knowledge Management, 18*(4), 651–686.

Stary, C., Maroscher, M., & Stary, E. (2012). *Knowledge management in praxis: Methods—tools—Examples (in German)*. Munich, Germany: Hanser.

Stewart, V., Stewart, A., & Fonda, N. (1981). *Business applications of repertory grid*. Maidenhead: McGraw-Hill.

Stoyanov, S., Jablokow, K., Rosas, S. R., Wopereis, I. G., & Kirschner, P. A. (2017). Concept mapping—An effective method for identifying diversity and congruity in cognitive style. *Evaluation and Program Planning, 60*, 238–244.

Strube, M. J. (2012). From "out there" to "in here": Implications of self-evaluation motives for self-knowledge. In S. Vazire & T. D. Wilson (Eds.), *Handbook of self-knowledge* (pp. 397–412). New York, NY: Guilford Press.

Swap, W., Leonard, D., Shields, M., & Abrams, L. (2001). Using mentoring and storytelling to transfer knowledge in the workplace. *Journal of Management Information Systems, 18*(1), 95–114.

Tantalo, C., & Priem, R. L. (2016). Value creation through stakeholder synergy. *Strategic Management Journal, 37*(2), 314–329.

Treem, J. W., & Leonardi, P. M. (2017). Recognizing expertise: Factors promoting congruity between individuals' perceptions of their own expertise and the perceptions of their coworkers. *Communication Research, 44*(2), 198–224.

Trochim, W. M., & McLinden, D. (2017). Introduction to a special issue on concept mapping. *Evaluation and Program Planning, 60,* 166–175.

Vargo, S. L., Maglio, P. P., & Akaka, M. A. (2008). On value and value co-creation: A service systems and service logic perspective. *European Management Journal, 26,* 145–152.

Vogel, R. M., Rodell, J. B., & Lynch, J. W. (2016). Engaged and productive misfits: How job crafting and leisure activity mitigate the negative effects of value incongruence. *Academy of Management Journal, 59*(5), 1561–1584.

Wang, M., Cheng, B., Chen, J., Mercer, N., & Kirschner, P. A. (2017). The use of web-based collaborative concept mapping to support group learning and interaction in an online environment. *The Internet and Higher Education, 34,* 28–40.

Chapter 9

Data Visualization Practices and Principles

Jeonghyun Kim and Eric R. Schuler

Contents

Introduction

The ever-increasing growth of Big Data has impacted every aspect of our modern society, including business, marketing, government agencies, health care, academic institutions, and research in almost every discipline. Due to its inherent properties of excessive volume, variety, and velocity, it is very difficult to store, process, and extract the intended information from it; that is, it becomes harder to grab a key message from this universe of data. In this direction, the demand for data analytics, which is the process of collecting, examining, organizing, and analyzing large data sets for scientific understanding and direct decision-making, has increased in recent years. Various forms of analytics, including predictive analytics, data mining, and statistical analysis, have been applied and implemented in practice; this list can be further extended to cover data visualization, artificial intelligence, natural language processing, and database management to support analytics.

Among those capabilities that support analytics, data visualization[1] has a critical role in the advancement of modern data analytics (Bendoly, 2016). It has become an active and vital area of research and development that aims to facilitate reasoning effectively about information, allowing us to formulate and test hypotheses, to find patterns and meaning in data, and to easily explore the contours of a data collection from different perspectives and at varying scales. As Keim et al. (2008) asserted, the goal of visualization is to make the way of processing data and information transparent for an analytic discourse. The visualization of these processes provides the means for communicating about them, instead of being left with the results. Visualization fosters the constructive evaluation, correction, and rapid improvement of our processes and models, and ultimately, the improvement of the management of knowledge on all levels, including personal, interpersonal, team, organizational, inter-organizational, and societal, as well as enhancement of our decision-making process and management capabilities.

To achieve this goal, various visualization techniques have been developed and proposed, evolving dramatically from simple bar charts to advanced and interactive 3D environments, such as virtual reality visualizations, for exploring terabytes of data. A rapidly evolving landscape in the business intelligence market led to interesting innovations in areas such as behavioral and predictive analytics; now visualization is increasingly being integrated with analysis methods for visual analytics. This integration offers powerful and immediate data exploration and understanding. The number of tools and software solutions that integrate visualization and analysis, including Power BI,[2] Tableau,[3] Qlik,[4] and Domo,[5] has grown rapidly during the past several decades while keeping pace with the growth of data and data types. Toolkits for creating high-performance, web-based visualizations, like Plotly[6] and Bokeh[7], are becoming quite mature. Further, various web applications supporting visualization features have emerged. For instance, web-based notebooks like Jupyter[8]

---

[1] Elsewhere, other terms are being used, such as scientific visualization, information visualization, and visual analytics in a more restricted sense (Keim et al., 2008; Tory & Möller, 2004). In this chapter, the term data visualization will be used in a broad sense, referring to the use of visual representations to explore, make sense of, and communicate data.

[2] https://powerbi.microsoft.com

[3] https://www.tableau.com

[4] https://www.qlik.com

[5] https://www.domo.com

[6] https://plot.ly

[7] http://bokeh.pydata.org

[8] http://jupyter.org

are interactive interfaces that can be used to ingest data, explore data, visualize data, and create small reports from the results.

In recent years, discussions on data visualization have arisen, including effective data visualization. Important questions have been raised regarding principles for data visualization: What constitutes a good visualization? How can we ensure that we select the best visualizations to expose the value in the data sets? These questions are critical and deserve empirical investigation, because a good visualization can make the process of understanding data effective, while a bad visualization may hinder the process or convey misleading information. Thus, the purpose of this chapter is to address these questions by providing a review of good data visualization practice techniques as articulated by the leading scholars and practitioners, and discussing leading data visualization components to ensure data visualizations are used effectively.

Data Visualization Practice

There are numerous methods to visualize data. It is critical to utilize the correct visualization based on the nature of the data and the point that the analyst is trying to present. Data visualizations can range from simple charts that portray frequencies and percentages to complicated visualizations that provide frequencies based on a geographical location. In this section, we provide a brief overview of two broad types of visualizations: (1) multidimensional visualization and (2) hierarchical and landscape visualization. Multidimensional data visualization enables users to visually compare data dimensions (column values) with other data dimensions using a spatial coordinate system (Bansal & Sood, 2011); they include bar charts, distributional graphs (i.e., histograms, boxplots, and violin plots), line graphs, scatter plots, and pie graphs. Hierarchical and landscape data visualization differs from normal multidimensional visualization in that they exploit or enhance the underlying structure of data set itself (Bansal & Sood, 2011); they include both tree visualizations and map visualizations.

To better understand how different data visualizations can be utilized, we will be referring to the following fictitious scenario: The Knowledge Management Corporation (KMC) is a company that helps other businesses collect and analyze various types of data to improve customer relations and profits. Recently, KMC has been working with More Fun Toys, a national toy company, to better understand the current trends in the market, customer feedback, and patterns of their product line across the country. We will be referring to this example in regard to the various data visualization tools and what types of questions each visualization tool can answer.

Multidimensional Visualization

Column charts, sometimes termed bar charts, are utilized to compare values from different groups (Ajibade & Adeiran, 2016). More specifically, if the groups are categorical, a bar chart can be utilized. The bars can be vertical or horizontal, depending on the number of groups and what is being presented. Bar charts help visualize differences in the frequencies among distinct categories. Specifically, bar charts allow individuals to accurately read numbers off the chart and view trends across the groups (Zoss, 2016). There are some weaknesses of bar charts that need to be noted. For example, if the categories have long titles, they can be difficult to read. Additionally, it is important to pay attention to the numerical values of the *y*-axis as any axis that does not start at zero could distort the visualization.

Another useful visualization is the line chart. Line charts are used to visualize the relationship among several items over time (Ajibade & Adediran, 2016). Additionally, various categories could be added to the visual so that each color line reflects a different group. This allows readers to look at changes over time and across groups. Using line graphs can be advantageous over stacked bar charts in that the exact values can be easily read and the exact dates do not need to necessarily match on the *x*-axis (Zoss, 2016). Line charts can be difficult to read if there are numerous groups and the lines often intersect.

For example, More Fun Toys wanted to know which of their recent toys was the most popular across the country in the last year. Since they are only interested in a visual of what toy was most frequently bought, we can use a bar chart (Figure 9.1). Based on Figure 9.1, we can see that Super Fast Racer was the most popular toy across the nation, closely followed by Happy Hamster and Shufflebot. Furthermore, More Fun Toys was curious how three of their new toys sold over the last four fiscal quarters; the three new toys were Happy Hamster,

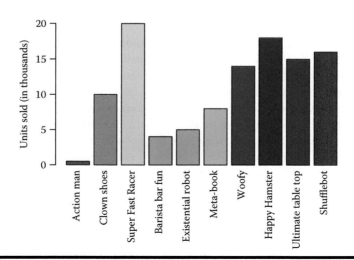

Figure 9.1 Bar chart that shows how many units of each toy were sold over the 2016 fiscal year.

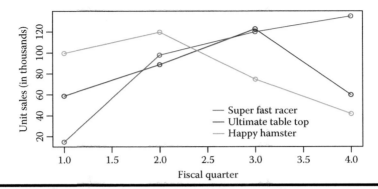

Figure 9.2   Line graph that presents the trends in sales for three toys over each 2016 fiscal quarter.

Ultimate Table Top, and Super Fast Racer. Figure 9.2 presents the number of units (in thousands) that each toy sold in the last four fiscal quarters. As can be seen in Figure 9.2, the product Happy Hamster had high initial sales that diminished over time, whereas Super Fast Racer had a slow start, but sales increased each quarter.

Usually, data analysts are interested in not only describing the data but also making inferences from the data, termed inferential statistics (Tabachnick & Fidell, 2007). Prior to running any sort of inferential statistics, it is important to understand the nature of the distributions. The most insightful methods to view the distributions are through basic data visualizations to assess for normality, outliers, and linearity of variables. Oftentimes, this starts with a histogram or distribution graph. A histogram is akin to a bar chart in that it has each value of a variable on the x-axis and frequencies or number of occurrences on the y-axis (Soukup & Davidson, 2002). A histogram can inform us of how data is spread out, whether the data is skewed positively or negatively, and whether there are potential extreme scores or outliers within the data that could affect measures of central tendency (i.e., mean). An additional visual for the distribution of values for a variable is a density plot. The density plot is a smoothed histogram that can have a normal curve superimposed over it to compare the actual distribution with a hypothesized normal distribution. This type of visualization can help researchers determine if the distribution of scores is normal or non-normal, as well as visually assess for skewness. For example, More Fun Toys was interested in the distribution of customer satisfaction scores, with scores ranging from 0 (unsatisfied) to 100 (satisfied). The distribution of scores (Figure 9.3) appears to be mostly above a score of 50 (e.g., neutral).

The boxplot is another visualization to graphically depict the distribution of values for a single quantitative variable. This allows researchers to view aspects of the measure's central tendency (i.e., mean) and the distribution of values around the

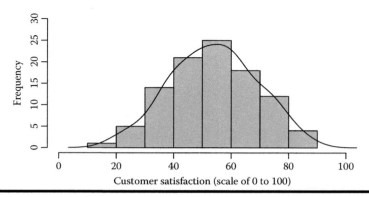

Figure 9.3   Histogram that shows the frequencies of scores on a customer satisfaction survey.

mean (i.e., standard deviation or variance) (Soukup & Davidson, 2002). A boxplot is designed so that the data from the first quartile and third quartile (50% of the data) are visualized in a box with the median dividing the box. The remaining upper and lower 25% of the data create "whiskers" to reflect how spread out the data are. This can be helpful to determine if there are potential floor effects (i.e., if many of the cases have a score of zero on the variable), which would be noted in the visual as there being no lower whisker. Alternatively, ceiling effects, or cases that are clustered at the highest value, could be depicted by the lack of an upper whisker. In our scenario, More Fun Toys was interested in looking at the marketing budget for the various products across each of the three divisions (i.e., Atlantic, Midwest, and West Coast). The amounts spent to advertise each product were collapsed across the division. As can be seen in Figure 9.4, the West Coast

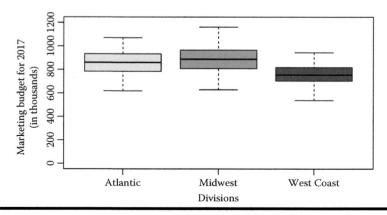

Figure 9.4   Boxplot that represents the median, quartiles, and potential outliers for the aggregated project marketing budgets for the three divisions of More Fun Toys.

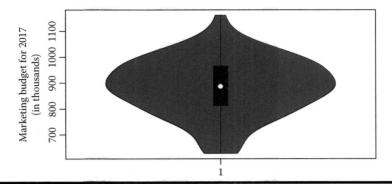

Figure 9.5 **Violin plot that shows the quartiles and distribution of the marketing budget for the Midwest division. The magenta color is the distribution plot that has been mirrored on both sides. The black rectangle represents the data within the first and third quartile. The white circle is the median and the lines extending from the first and third quartiles are the whiskers of the boxplot. The data was from the simulated data for the Midwest division (Figure 9.4).**

division spent far less on the various marketing programs, whereas the Midwest division far exceeded the average budget figures.

Related to boxplots are violin plots (Hintze & Nelson, 1998), which combine a boxplot and a doubled kernel density plot. As presented in Figure 9.5, violin plots combine the features of the density plot with the information of a boxplot in one easy-to-interpret visualization (Hintze & Nelson, 1998). The information about the quartiles is denoted by lines and a thin box within the mirrored density plot. This provides more information than either plot alone, however, it is not available within some of the commercial software packages, with the exceptions of R and SAS.

A scatterplot can be used when there is an interest in determining if there is a linear or non-linear relationship between two or more quantitative variables (Tabachnick & Fidell, 2007). Simply put, the scores on one quantitative variable (X) are plotted on the *x*-axis with the corresponding second variable scores (Y) on the *y*-axis. If there is an ellipse or a line, then there could be a linear relationship between the variables, whereas a U-shaped pattern could be indicative of a curvilinear relationship (Howell, 2013; Tabachnick & Fidell, 2007). These visuals are often used to assess for statistical assumptions of linearity and as a first step to determine the relationship of two variables prior to running additional analyses (Howell, 2013). An additional benefit to a scatterplot is that it can be used to detect multivariate outliers, where a case has normal scores on each measure separately, but when assessed together, the combination of both scores makes it further removed from the other cases (Khan & Khan, 2011; Tabachnick & Fidell, 2007).

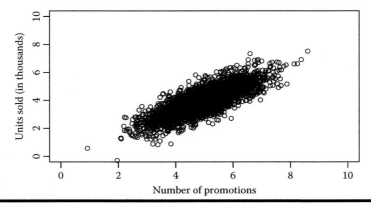

Figure 9.6   Scatterplot that visualizes the relationship between units sold and the number of marketing promotions in the Midwest division.

If there are more than two variables of interest, a three-dimensional scatterplot can be utilized. The three-dimensional scatterplot provides an *x*, *y*, and *z*-axis, and the corresponding three scores for each case are placed in the visual space as an orb in the space. As shown in Figure 9.6, the Midwest division of More Fun Toys was interested in seeing if there was a relationship between the number of units sold and the number of promotional deals that were done in the last year. Additionally, the Midwest division was interested in whether the units sold and promotions had a relationship with the customer satisfaction survey that was completed by some customers after their purchase. This was done to get an idea of whether there was a relationship with promotions and the other two variables, and if more time and money should be utilized on promotional offers, as presented in Figure 9.7.

When there are three variables, an alternative to the scatterplot, called a bubble chart, can be used. Bubble charts are typically used to represent the relationships among three variables (Ajibade & Adediran, 2016). Specifically, the relationship between the outcome (Y) and the first variable (X1) are graphed on the *x* and *y*-axis, like a scatter plot. Then the third variable's relationship with the other two variables is reflected by the size of the bubble. Like a scatterplot, each bubble represents a single observation. Figure 9.8 shows the relationship among the number of units sold, the number of promotions from the Midwest division, and the customer feedback survey that was completed by individuals after making a purchase.

Although line graphs are useful to look for trends in a continuous outcome variable over time and across groups, they do not visualize the relative amounts of each group to the whole over time. An area chart can be utilized to show changes over time and view the groups and the total values simultaneously (Soukup & Davidson, 2002). More specifically, area graphs are used to visually compare categorical values or groups over a continuous variable in the *y*-axis and how each group relates to the

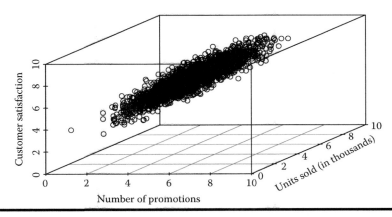

Figure 9.7 **Three-dimensional scatterplot that shows the relationship among units sold, number of marketing promotions, and customer satisfaction for the Midwest division.**

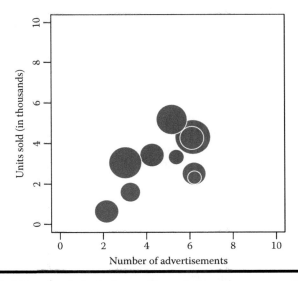

Figure 9.8 **Bubble chart that shows the relationship among the units sold, amount of marketing promotions, and customer satisfaction for the Midwest division.**

whole. In Figure 9.9, More Fun Toys wanted to determine whether it would be beneficial to update their online store. The company was interested in whether the new layout would appear more welcoming to potential customers. Furthermore, they were curious if there were gender differences between the number of pages that were viewed and the overall time spent on the site. To determine this, KMC conducted

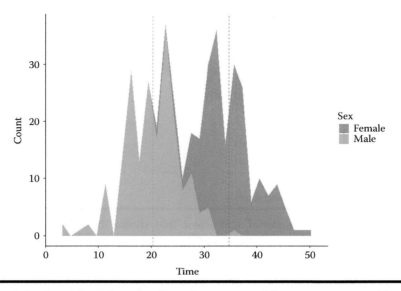

Figure 9.9 Area chart that visualizes the amount of time spent on searching the retail website and the number of page views by gender.

a small pilot study and monitored how long individuals spent on the More Fun Toys website and recorded the number of pages viewed. The amount of time and pages clicked are shown by gender in the area chart in Figure 9.9. As can be seen in Figure 9.9, men tended to spend less time on the website compared to women; however, the number of pages viewed appeared to be consistent across gender.

The last frequently used multidimensional visualization technique is a pie chart that contains wedges of data that correspond to categories or groups. Typically, the wedges of data represent percentages of the data, with all wedges adding up to 100% (Khan & Khan, 2011). There are three subtypes of pie charts: the standard pie chart, an exploding pie chart, and a multilevel pie chart. The exploding pie chart separates the wedges from the rest of the wedges to visualize interactions (Khan and Khan, 2011). Multi-level pie charts are useful when the data are hierarchical in nature; the inner circle represents the parent level (i.e., upper level of the nested data), and the outer circle is the child level (i.e., lower level of the nested data). Pie charts create a visual to compare parts to the whole of one variable across multiple categories. However, it is important to note that pie charts can make it difficult to infer the precise value of each wedge, especially if there are numerous categories that are then turned into very small wedges (Zoss, 2016).

Hierarchical Data Visualization

Oftentimes, individuals are clustered in a region or departments are clustered by the type of business, making it nested or hierarchical data. When the data are hierarchical in nature, it is important to preserve the nested relationships since the child group is dependent on the parent group (Bickel, 2007). One method to visualize the hierarchical structure of the data is through tree visualizations, which include tree maps and tree diagrams. Tree diagrams are useful in displaying hierarchical data as the levels of the tree diagram visually depict the location of each piece of data in a way that quickly and intuitively conveys to a user both the location of the data within the hierarchy and the relationship of that data to other data in the hierarchy. To represent hierarchical data, tree maps partition the display space into a collection of rectangular bounding boxes representing the tree structure; they use a space-filling technique to reflect the quantitative data (Yeh, 2010). The color and size of the boxes reflect the amounts for each of the categories in the data (Morabito, 2016). In our scenario, the executive board of More Fun Toys was interested to see if there are any patterns among the sales of the Super Fast Racer toy and the stores that were nested within each state that were also nested within the three divisions of More Fun Toys. Figure 9.10 depicts the tree map to highlight sales by the nested locations within the three divisions.

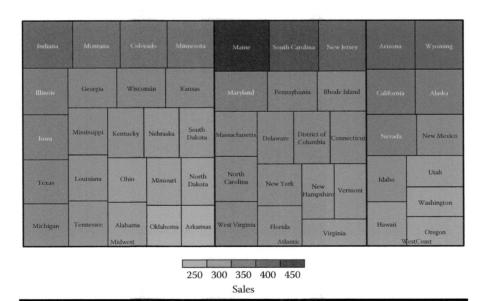

Figure 9.10 **Tree map that shows the number of unit sales for each state within the three regional divisions during the 2016 fiscal year. The data is hierarchical in that states are grouped within the three divisions (Atlantic, Midwest, and West Coast divisions). The green shading depicts the amount of sales and the size of each box is based on the number of stores that carry Super Fast Racer in each state.**

Within business information, we may be interested in not only the quantity of purchases, but also the geographical regions in which there are a high degree of purchases and which areas have a low market. Map visualizations provide a way to visualize the number of purchases or other quantitative variables based on the location of the data points (Morabito, 2016). The map visualizations can provide information on where certain markets are meeting their quotas, whereas others may need to use a different approach to reach the targeted audience. Map visualizations require geographic information (e.g., ZIP codes or GPS coordinates), and certain programs allow additional variables to be overlaid on the map (e.g., gender, income levels, etc.) to provide an interactive visualization. Interactive maps allow individuals to sort, filter, and visualize patterns within the data. More Fun Toys was interested in viewing trends in the purchases of Super Fast Racer by state (Figure 9.11) and seeing if certain stores sold more units than others (Figure 9.12). An additional interest was overlaying the 2017 per capita for each state on the map visualization. Specifically, was the toy selling better in states with higher income?

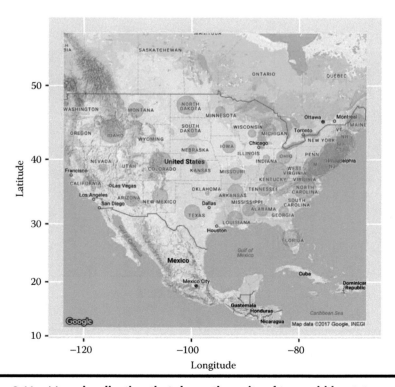

Figure 9.11 Map visualization that shows the units of toys sold by state.

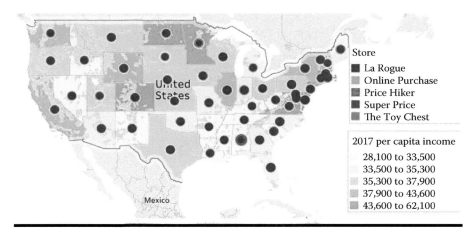

Figure 9.12 Interactive map visualization that shows the relationship among units of the toys sold by store, geographical location, and estimated 2017 per capita income. The filters include the five stores that carry the Super Fast Racer (i.e., La Rogue, Online Purchase, Price Hiker, Super Price, and The Toy Chest). An additional data layer of 2017 per capita income was added to explore for patterns in the data.

Tree graphs can aid researchers in visualizing hierarchical data by focusing on the center of the hierarchy. At the center of the hierarchy is a person, place, or thing, termed a node, and the relationship of that node with all the other nodes in the data are graphically represented. There are two specific types of tree graphs: radial trees and hyperbolic trees. The radial tree focuses on splitting the branches from the chosen nodes equally (Morabito, 2016). It allows the connections among the central node of the hierarchy and its relationship to the various branches (Morabito, 2016). Furthermore, the center node can be relocated to view the relationship of the specified node with the others. On the other hand, hyperbolic trees utilize a nonlinear procedure to show the connections among one node and the others (Morabito, 2016). Figure 9.13 represents a radial tree graph that visualizes the hierarchical structure of More Fun Toys' products with a parent node of product age category (i.e., toddler, young, and teenage), toy type (i.e., educational, action, and doll), and genre of the toy (i.e., horror, real life, wild west, fantasy, and space). Based on Figure 9.13, we can see that under the toddler age group, there are two types of toys, educational and doll. Common to both educational and doll type toys under the toddler age group are the genres of space and real life, however, wild west-themed toys for

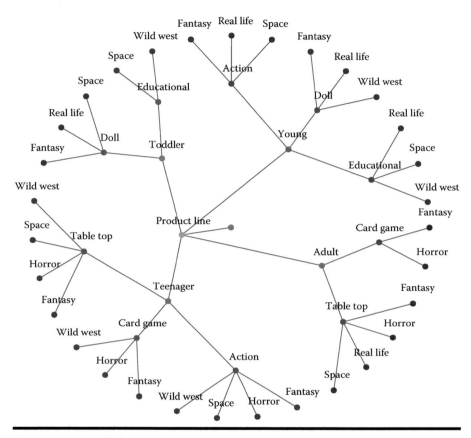

Figure 9.13 Radial tree graph that shows the hierarchical relationships of the product age group, toy type, and toy genre classifications.

toddlers are only educational type toys and there are only fantasy dolls for toddlers. Furthermore, only the young age group had all three types of toys. Based on these findings, More Fun Toys could look into a future line of educational space-themed toys for toddlers, since there are currently no toys that fall within that classification. Additionally, More Fun Toys could add additional genres to their adult card game selections, since they only carry two genres: a fantasy card game and a horror card game.

Parallel coordinates plots allow the visualization of multidimensional relationships of several variables rather than finite points (Inselberg & Dimsdale, 1990). More specifically, data elements are plotted across several dimensions, with each dimension related to a single *y*-axis (Khan & Khan, 2011). If a line chart or scatterplot is utilized with multi-dimensional data, the relationships across multiple

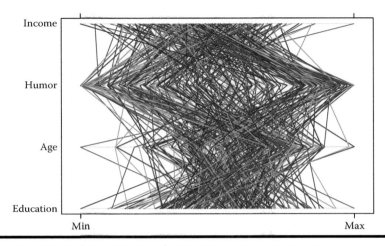

Figure 9.14 **Parallel coordinates plot that shows the relationships among educational level, age, humor level, and income.**

variables can be obscured (Khan & Khan, 2011). Within a parallel coordinates plot, individual cases can be viewed across numerous dimensions to detect patterns in a large data set (Khan & Khan, 2011). Parallel coordinates plots can be used to identify groups of individuals based on several variables that share the same patterns (Johansson, Forsell, Lind, & Cooper, 2008). However, it is important to know that if the parallel lines overlap, it can be difficult to identify characteristics and subsequent patterns (Ajibade & Adediran, 2016). More Fun Toys was interested in classifying customers into distinct groups to better market their products based on levels of education, age, humor levels, and income. Figure 9.14 presents the relationships among these variables in a sample of 250 customers. There was a wide spread of education levels among the customers, with most being older based on the lines from education to age. There appeared to be three potential groups based on humor levels denoted by the clusters of lines from age to humor. Interestingly, from humor scores to income, it appeared that individuals who had lower levels of humor had an associated higher income and individuals who had higher humor scores had lower incomes.

Data Visualization Principles

As the field of data analytics and visualization has matured, the principle of data visualization has evolved from a single universe set into a more complex, but still loosely defined structure. In fact, researchers in many fields have considered the problem of how to construct effective data displays for a long time. Educational researchers who work with graphical displays are mainly concerned with the effectiveness of charts and diagrams as instructional tools. Human factors researchers are principally concerned with psychophysical properties of displays, especially dynamic displays, rather than the interpretation of static displays of quantitative data (Sanders & McCormick, 1987). Designers, statisticians, and psychologists have developed rules for constructing good graphs and tables.

Yau (2009) asserted "Data has to be presented in a way that is relate-able; it has to be humanized. Oftentimes we get caught up in statistical charts and graphs, which are extremely useful, but at the same time we want to engage users so that they stay interested, continue collecting data, and keep coming back to the site to gauge their progress in whatever they are tracking. Users should understand that the data is about them and reflect the choices they make in their daily lives (p. 7)."

In this context, good data visualizations should engage the audience as well as improve both the accuracy and the depth of their understanding; both are critical to making smarter decisions and improving productivity. Poorly created visualizations, on the other hand, can be misleading and undermine your credibility with your audience. They make it more difficult for them to overcome the daily data onslaught.

The most important principle of visualizing data is to address the following two questions: What is the message you would like to convey? and Who is your audience? You should construct data visualization so it conveys the message in the most efficient way, and the message one wants it to convey. You also need to work closely with your intended audience to understand what actions they want to be able to take based on the data they consume; thus, you need to identify the needs of your audience. If the visualization doesn't portray what the audience needs, it does not work. Then the next step is to choose the right visualization type.

Different types of data require different visualizations. As discussed in the previous section, line charts, for example, are most useful for showing trends over time or a potential correlation between two variables. When there are many data points in your dataset, it may be easier to visualize the data using a scatterplot instead. Histograms, on the other hand, show the distribution of the data, and the shape of the histogram can change depending on the size of the bin width.

In the next section, a general vision of graphing data, which is often considered as philosophical guidelines, is presented. The following section describes some detailed rules and best practices to ensure data visualization is efficient. It should be noted that the set of principles and rules for designing data visualizations presented in this chapter is neither complete nor addresses all the issues relevant in visualization design; but it is expected that this set would provide some guidelines for visualization designers in selecting various visualization tools and primitives to effectively display the given data.

General Principles

A number of scholars have published lists of principles governing the definition or construction of a "good" visualization. Tufte (1983), who is known as a pioneer in the field of data visualization, described graphs as "instruments for reasoning about quantitative information (p. 10)" and argued that graphs are usually the most effective method to describe, examine, and summarize even the largest data-sets, and if properly designed, displays of statistical information can communicate clearly even complex information and are one of the most powerful data analytic tools. In his classic book, "*The Visual Display of Quantitative Information*," Tufte provided principles for graphical excellence, which is defined as follows:

- Graphical excellence is the well-designed presentation of interesting data of substance, of statistics, and of design.
- Graphical excellence consists of complex ideas communicated with clarity, precision, and efficiency.
- Graphical excellence is that which gives to the viewer the greatest number of ideas in the shortest time with the least ink in the smallest space.
- Graphical excellence is nearly always multivariate.
- Graphical excellence requires telling the truth about the data.

Building on Tufte's standards, Wainer (1984) published his principles for displaying data badly, which are so clearly and humorously illustrated with real-life graphs and tables. He defined that "the aim of good data graphics is to display data accurately and clearly" (p. 137) and further breaks this definition into three parts: showing data, showing data accurately, and showing data clearly, as presented in Table 9.1. His principles provide some guidance for the design of graphs in the form of a checklist of mistakes to be avoided. It is noteworthy that his principles address the question of how to construct a graph, but do not address the question of why things should be done this way.

Cleveland (1985) provides a more scientific treatment of the display of science data that summarizes much of the research on the visual perception of graphic rep-resentations. Unlike other scholars who took a more theorist's intuitive approach, Cleveland has actually tested some of his rules for making "good" graphs. He and his colleagues have conducted many psychophysical investigations of the elementary constituents of graphical displays by abstracting and then ordering a list of physical dimensions. The results of those empirical works have influenced Cleveland's (1985) listing of the principles of graph construction. He has organized his principles under headings that give the reader clues about how he thinks each of his principles contributes to better data displays: clear vision, clear understanding, scales, and general strategy, as shown in Table 9.2.

Many other authors have also reported studies attempting to figure out the best methods of displaying data (Day, 1994; Wilkinson, 1999). Such works have been

Table 9.1   Wainer's Twelve Principles for Displaying Data Badly

Principles	Description
Showing data	• Show as few data as possible. • Hide what data you do show.

Showing data accurately	• Ignore the visual metaphor altogether.
	• If the data is ordered and if the visual metaphor has a natural order, a bad display will surely emerge if you shuffle the relationship.
	• Only order matters.
	  • For example, use length as the visual metaphor when area is what is perceived.
	• Graph data out of context.
	  • For example, Fidell with the scale or the interval displayed in order to misrepresent the facts.

Showing data clearly	• Change scales in mid-axis. • Emphasize the trivial. (Ignore the important). • Jiggle the baseline. • Make the graph worse by using different baseline scales for variables on the same graph. • Austria First! • Order graphs and tables alphabetically to obscure structure in the data that would have been obvious had the data been ordered by some aspect of the data. • Label: (a) Illegibly, (b) Incompletely, (c) Incorrectly, and (d) Ambiguously. • More is murkier: (a) More decimal places and (b) More dimensions. • If it has been done well in the past, think of another way to do it.

Table 9.2 Cleveland's Listing of the Principles of Graph Construction

Principles	Description
Clear vision	• Make the data stand out. Avoid superfluity. • Use visually prominent graphical elements to show the data. • Use a pair of scale lines for each variable. Make the data region the interior of the rectangle formed by the scale lines. Put tick marks outside of the data region. • Do not clutter the data region. • Do not overdo the number of tick marks. • Use a reference line when there is an important value that must be seen across the entire graph, but do not let the line interfere with the data. • Do not allow data labels in the data region to interfere with the quantitative data or to clutter the graph. • Avoid putting notes, keys, and markers in the data region. Put keys and markers just outside the data region and put notes in the legend or in the text. • Overlapping plotting symbols must be visually distinguishable. • Superimposed data sets must be readily visually discriminated. • Visual clarity must be preserved under reduction and reproduction.
Clear understanding	• Put major conclusions into graphical form. Make legend comprehensive and informative. • Error bars should be clearly explained. • When logarithms of a variable are graphed, the scale label should correspond to the tick mark labels. • Proofread graphs. • Strive for clarity.

(Continued)

Table 9.2 (*Continued*) Cleveland's Listing of the Principles of Graph Construction

Principles	Description
Scales	• Choose the range of the tick marks to include or nearly include the range of data. • Subject to constraints that scales have, choose the scales so that the data fills up as much of the data region as possible. • It is sometimes helpful to use the pair of scale lines for a variable to show two different scales. • Choose appropriate scales when graphs are compared. • Do not insist that zero always be included on a scale showing magnitude. • Use a logarithm scale when it is important to understand percent change or multiplicative factors. • Showing data on a logarithmic scale can improve resolution. • Use a scale break only when necessary. If a break cannot be avoided, use a full-scale break. Do not connect numerical values on two sides of a break.
General strategy	• A large amount of quantitative information can be packed into a small region. • Graphing data should be an interactive, experimental process. • Graph data two or more times when it is needed. • Many useful graphs require careful, detailed study.

described in literature reviews; for instance, Cleveland (1987) provided a bibliography of graphical statistics; Dibble (1997) provided a number of sources contrasting tables and graphs; and Lewandowsky and Spence (1989) focused on the perception of graphs.

Specific Principles: Text

Simplicity is a key principle for data visualization. To support this principle, you should remove unnecessary text in your visualization. However, nonintrusive text elements can help increase understanding without detracting from a visual's meaning. Such text elements include a descriptive title, subtitle, and annotation. The title should state clearly what is being graphed or represented; it has to be short and concise. The title should be recognizable as such because it is clearly set off from the rest of the visualization; it should not be close enough to any line to be grouped perceptually with it. A large font size will also prevent the title from being perceptually grouped with other labels or parts (Kosslyn, Pinker, Simcox, & Parkin, 1983). Additionally, subtitles need to be formatted to be subordinated to the main title, and annotations are normally located alongside or below a specific visualization. Both subtitles and annotations can add explanatory and interpretive power to a visualization.

A data label identifies the name of the variable or series plotted, whereas a data value is a number indicating a specific value in the data. Data labels should be employed to improve the readability of data values; they need to be strategically positioned, sized, and worded so that the text complements the visualization without overshadowing it. Data labels are often displayed on or next to the data components (i.e., bars, areas, lines) to facilitate their identification and understanding. For data values, the best practice to is put scale values on both scales, such as the *x*-axis and *y*-axis, and to make sure that they are closer to the correct tick mark than to anything else (Kosslyn et al., 1983).

A font needs to be selected so that text is easy to read. A good rule of thumb is that any font that looks fancy rather than simple should be examined closely for readability. Multiple fonts can sometimes make sense as a tool for engaging a viewer or emphasizing a point, but too many different fonts tend to distract from a message. Some experts suggest three as a reasonable limit (Evergreen, 2014).

Specific Principles: Color

In data visualization, color should not be used for decorative or non-informational purposes. Color needs to be used to show your audience what questions you are answering as it provides information that makes for improved assessment and decision making (Campbell & Maglio, 1999); thus, the best practice is to use action colors (e.g., bright or dark colors) to guide your audience to key parts of the display and soft colors for less important or supporting data (Few, 2012). However, you should limit the number of colors shown in a visualization because use of multiple colors can be distracting. For nominal and ordinal data types, discrete color palettes must be used. For the bar chart, for instance, research suggests the best option would be choosing the same color for all bars except for the bar that needs the most attention, such as those failing below the cut score of those representing your particular program (Evergreen & Merzner, 2013). For interval and ratio data, one can use both discrete and continuous color palettes, according to the data structure and to the visualization aims.

Legibility is another very important criteria that needs to be considered. The best practice is to use solid blocks of color and avoid fill patterns, which can create disorienting visual effects. Further, background colors should generally be white or have very reduced colors (Tufte, 1990) but graph text should be black or dark gray (Wheildon, 2005). Color also needs to be legible for people with color-blindness; red–green and yellow–blue combinations need to be avoided as those colors touch one another on the color wheel.

It is also important to note that many colors depend on, and are constrained by, the cultural traditions or technical experience of the user (Bianco, Gasparini, & Schettini, 2014); in fact, there are many norms. Some are more universal. For instance, most people associate green with positive or above-goal measurements, while judicial use of red generally indicates peril or numbers that need improvement. Some are political or cultural, such as red state, blue state. Some are dependent on specific domain expertise; in the black, in the red (finance).

Specific Principles: Layout

It is important to lay out the visualizations in an appealing and understandable arrangement. Improper arrangement of visualization elements can confuse and mislead readers. To avoid distortion in the data, the visualization should not be too long or too narrow. The aspect ratio of a data rectangle is its height divided by width; it is the aspect ratio that decides the shape of the visualization, so it should be chosen with care. According to *Banking to 45°*, a classic design guideline first proposed and studied by Cleveland (1994), aspect ratios that have center slopes around 45 degrees minimize errors in visual judgments of slope ratios. Kosslyn et al. (1983) suggested including a second vertical axis on the right of the outer frame of the visualization if the horizontal axis is more than twice as long as the vertical axis.

The axes should be uniform and continuous. The numbers on the visualization increase from left to right on the x-axis and from bottom to top along the y-axis. They increase in uniform increments, and the scales on each axis should be chosen carefully so that the entire range of values can be plotted on the visualization. For bar charts, histograms, line or plot charts where absolute magnitudes are important, zero should be included on the y-axis (Robbins, 2005). When the quantitative scale corresponds to the y-axis, it can be placed on the left side, right side, or on both sides of the visualization. When it corresponds to the *x*-axis, it can be placed on the top, bottom, or both, but it is usually sufficient to place the scale in one place (Few, 2012). Additionally, the numbers on the axis must follow a proper interval in the same unit and be evenly spaced; you should not skip values when you have numerical data.

Graphical objects inside the visualization must be sized to present ratios accurately as a visualization that displays data as objects with disproportionate sizes can be misleading. This is true for the map visualization, where a proportional symbol is used to display absolute values. It should be noted that it may be difficult to identify the unit that the symbol refers to when the size of the symbol is bigger than the size of corresponding spatial unit. In a plot, both length and position are better quantitatively perceived than size, line width, or color hue, meaning that the data values that they represent and how those values compare to other values are easily determined (Cleveland & McGill, 1984). The order of graphic objects is important, too. It is generally acknowledged that comparing numbers ordered alphabetically, sequentially, or by value is best; more importantly, data should be displayed in an order that makes logical sense to the viewer. When using bar or pie charts, you should sort data from smallest to largest values, so they are easier to compare.

Consistent text layout and designs are also recommended to avoid distortion. It is recommended that you avoid steep diagonal or vertical text type as it can be difficult to read; lay out text horizontally. This includes titles, subtitles, annotations, and data labels. Line labels and axis labels can deviate from this rule. Consistency in the use of other elements, including line widths and box sizes, is also critical unless those elements enable us to isolate all marks belonging to the same category.

Implications for Future Directions

It is hard to deny that data visualization is already ubiquitous. As the data rate has increased, visualization has become more common for us to be able to get information and insight out of the data. Advanced data visualization tools have already offered new ways to view data and this resulting increase in data visualization has improved our visualization literacy.

Now data visualization is entering into a new era and realm. Conceptual and theoretical developments, new and emerging sources of intelligence, and advances in multidimensional visualization are reshaping the potential value that analytics and insights can provide, with visualization playing a key role. In this context, the principles of effective data visualization will never change. But new technologies and evolving cognitive frameworks will be opening new horizons, moving data visualization from art to science (Towler, 2015).

However, despite the huge impact data visualization has had, it still faces considerable challenges in the future. Here we outline important challenges in this context.

Data: Big Data has introduced an enormous opportunity to utilize visualization tools to generate insights. Visualization applications need to accommodate and graphically represent high-resolution input data as well as continuous input data streams of high bandwidth. Additionally, data visualization should process data from multiple and heterogeneous sources and different data types. Further, it should be noted that more and more unstructured data are generated but current visualization practices are more appropriate for analyzing structured data. Thus, methods for modeling and visualizing unstructured data need to be developed.

Infrastructure: Managing large amounts of data for visualization requires special data structures, both in memory and on disks. Further, to fully support the analytical process, the history of the analysis should also be recorded and interactively edited and annotated. These requirements call for a novel software infrastructure, built upon well understood technologies such as databases, software components, and visualization but augmented with asynchronous processing, history managements, and annotations.

Evaluation: Although this chapter provides guidelines for effective data visualization, there are still clearly defined metrics or evaluation methods to measure effectiveness for tools and systems. A theoretically founded evaluation framework needs to be developed to assess the contribution of any visualization tools and systems toward the level of effectiveness and efficiency achieved regarding their requirements.

Note

All the visualizations presented in this chapter used fictitious data. They were simulated and created using R and Tableau.

References

Ajibade, S. S., & Adediran, A. (2016). An overview of big data visualization techniques in data mining. *International Journal of Computer Science and Information Technology Research, 4*(3), 105–113.

Bansal, K. L., & Sood, S. (2011). Data visualization: A tool of data mining. *International Journal of Computer Science and Technology, 2*(3). Retrieved from http://www.ijcst.com/vol23/1/kishorilal.pdf

Bendoly, E. (2016). Fit, bias, and enacted sense making in data visualization: Frameworks for continuous development in operations and supply chain management analytics. *Journal of Business Logistics, 37*(1), 6–17. doi:10.1111/jbl.12113.

Bianco, S., Gasparini, F., & Schettini, R. (2014). Color coding for data visualization. In M. Khosrow-Pour (Ed.), *Encyclopedia of information science and technology*. Hershey, PA: IGI Global.

Bickel, R. (2007). *Multilevel analysis for applied research: It's just regression.* New York, NY: Guilford Press.

Campbell, C. S., & Maglio, P. P. (1999). Facilitating navigation in information spaces: Road-signs on the World Wide Web. *International Journal of Human-Computer Studies, 50*(4), 309–327.

Cleveland, W. S. (1985). *The elements of graphing data.* Monterey, CA: Wadsworth Advanced Books and Software.

Cleveland, W. S. (1987). Research in statistical graphics. *Journal of American Statistical Association, 82*(398), 419–423.

Cleveland, W. S. (1994). *The elements of graphing data* (2nd ed.). Summit, NJ: Hobart Press.

Cleveland, W. S., & McGill, M. E. (1984). Graphical perception: Theory, experimentation, and application to the development of graphical methods. *Journal of American Statistical Association, 79*(387), 531–554.

Day, R. A. (1994). *How to write and publish a scientific paper.* Phoenix, AZ: Oryx Press.

Dibble, E. (1997). *The interpretation of tables and graphs.* Seattle, WA: University of Washington.

Evergreen, S. (2014). *Presenting data effectively: Communicating your findings for maximum impact.* Thousand Oaks, CA: Sage Publications.

Evergreen, S., & Merzner, C. (2013). Design principles for data visualization in evaluation. In T. Azzam & S. Evergreen (Eds.), *Data visualization, Part 2. New directions for evaluation* (pp. 5–20). New York, NY: John Wiley & Sons.

Few, S. (2012). *Show me the numbers* (2nd ed.). Oakland, CA: Analytics Press.

Hintze, J. L., & Nelson, R. D. (1998). Violin plots: A box plot-density trace synergism. *The American Statistician, 52*(2), 181–184. doi:10.1080/00031305.1998.10480559.

Howell, D. C. (2013). *Statistical methods for psychology* (8th ed.). Belmont, CA: Wadsworth Cengage Learning.

Inselberg, A., & Dimsdale, B. (1990). Parallel coordinates: A tool for visualizing multi-dimensional geometry. In *Proceedings of the 1st conference on visualization'90* (pp. 361–378). San Francisco, CA: IEEE Computer Society.

Johansson, J., Forsell, C., Lind, M., & Cooper, M. (2008). Perceiving patterns in parallel coordinates: Determining thresholds for identification of relationship. *Information Visualization, 7*(2), 152–162.

Keim, D., Andrienko, G., Fekete, J-D., Görg, C., Kohlhammer, J., & Melançon, G. (2008). Visual analytics: Definition, process, and challenges. In A. Kerren, J. Stasko, J-D. Fekete, & C. North (Eds.), *Information visualization* (pp. 154–175). Berlin, Germany: Springer.

Khan, M., & Khan, S. S. (2011). Data and information visualization methods and interactive mechanisms: A survey. *International Journal of Computer Applications, 34*(1), 1–14.

Kosslyn, S., Pinker, S., Smicox, W., & Parkin, L. (1983). *Understanding charts and graphs: A project in applied cognitive science* (NIE-40079-0066). Washington, DC: National Institute of Education.

Lewandowsky, S., & Spence, I. (1989). The perception of statistical graphs. *Sociological Methods and Research, 18*, 200–242.

Morabito, V. (2016). *The future of digital business innovation: Trends and practices.* Switzerland: Springer International Publishing. Heidelberg, Germany: Springer Verlag.

Robbins, N. (2005). *Creating more effective graphs.* Hoboken, NJ: Wiley-Interscience.

Sanders, M. S., & McCormick, E. J. (1987). *Human factors in engineering and design* (6th ed.). New York, NY: McGraw-Hill.

Soukup, T., & Davidson, I. (2002). *Visual data mining: Techniques and tools for data visualization and mining.* New York, NY: John Wiley & Sons.

Tabachnick, B. G., & Fidell, L. S. (2007). *Using multivariate statistics* (6th ed.). New York, NY: Pearson Education.

Tory, M., & Möller, T. (2004). Rethinking visualization: A high-level taxonomy. In *INFOVIS '04 proceedings of the IEEE symposium on information visualization* (pp. 151–158). Washington, DC: IEEE Computer Society.

Towler, W. (2015, January/February). Data visualization: The future of data visualization. *Analytics Magazine,* pp. 44–51.

Tufte, E. (1983). *The visual display of quantitative information* (1st ed.). Cheshire, CT: Graphic Press.

Tufte, E. (1990). *Envisioning information.* Cheshire, CT: Graphics Press.

Wainer, H. (1984). How to display data badly. *The American Statistician, 38*(2), 137–147.

Wheildon, C. (2005). *Type and layout: Are you communicating or just making pretty shapes?* Mentone, Australia: The Worsley Press.

Wilkinson, L. (1999). *The grammar or graphics.* New York, NY: Springer-Verlag.

Yau, N. (2009). Seeing your life in data. In T. Segaran & J. Hammerbacher (Eds.), *Beautiful data.* Farnham, CA: O'Reilly.

Yeh, R. K. (2010). *Visualization techniques for data mining in business context: A comparative analysis.* Retrieved from http://www.swdsi.org/swdsi06/proceedings06/papers/kms04.pdf

Zoss, A. M. (2016). Designing public visualizations of library data. In L. Magnuson (Ed.), *Data visualization: A guide to visual storytelling for libraries* (pp. 19–44). Lanham, MD: Rowan & Littlefield.

Chapter 10

Analytics Using Machine Learning- Guided Simulations with Application to Healthcare Scenarios

Mahmoud Elbattah and Owen Molloy

Contents

Introduction

Simulation modeling (SM) was considered a standalone discipline that encompassed designing a model of an actual or theoretical system, executing the model on a digital computer, and analyzing the execution output (Fishwick, 1995). Based on a virtual environment, simulation models provide extended capabilities to model real systems in a flexible build-and-test manner. In this respect, Newell and Simon (1959) asserted that the real power of the simulation approach is that it provides not only a means for stating a theory, but also a very sharp criterion for testing whether the statement is adequate. Similarly, Forrester (1968) emphasized in one of his principles of systems that simulation-based solutions present as the only feasible approach to represent the interdependence and nonlinearity of complex systems, whereas analytical solutions can be impossible.

The complexity of systems can be interpreted in terms of several dimensions. One possible dimension can be attributed to the data and metadata that represent the system knowledge. For instance, the data complexity can correspond to one, or more, of the four aspects that characterize the notion of Big Data including: volume, velocity, variety, or veracity. In such scenarios, further burdens can be unavoidably placed on the modeling process, which go beyond human capabilities.

Recently, the community of systems modeling and simulation has started to consider the potential opportunities and challenges facing the development of simulation models in an age marked by data-driven learning. For instance, a study (Taylor et al., 2013) introduced the term "big simulation" to describe one of the grand challenges for the simulation research community. Big simulation is intended to address issues of scale for big data input, very large sets of coupled simulation models, and the analysis of big data output from these simulations, all running on a highly-distributed computing platform. Another more recent position paper (Tolk, 2015) envisioned that the next generation of simulation models will be integrated with machine learning (ML), and deep learning in particular. The study argued that bringing modeling, simulation, Big Data, and deep learning all together can create a synergy delivering significantly improved services to other sciences.

In line with that direction, the chapter endeavored to spur a discussion on how the practice of modeling and simulation can be assisted by ML techniques. The initial discussion focused on why and how Big Data and ML can provide further support to simulations. Subsequently, a practical scenario is presented in relation to healthcare planning in Ireland to demonstrate the applicability of our ideas. First, unsupervised ML was utilized in a bid to discover potential patterns. The knowledge learned by ML was then used to build simulation models with a higher level of confidence. Second, simulation experiments were conducted with the guidance of ML models trained to make predictions on the system's behavior. The key idea was to realize ML-guided simulations during the phases of model design or experimentation.

This chapter can be viewed as structured into two main parts as follows. The first part initiates a discussion regarding the prospective integration of simulation models and ML in a broader sense. That discussion is intended to serve as an opening to the rationale underlying our approach. Starting in Section 5, the second part provides a more practical standpoint for integrating SM and ML through a realistic use case.

Motivation

Simulation Modeling and Machine Learning: Toward More Integration

The fields of SM and ML are long-established in the world of computing. However, both of them tended to be employed in separate territories with limited, if any, integration. This lack of integration might be attributed to a couple of issues.

First, the development of ML models is highly data-driven compared to SM. Simulation models have been largely developed with the aid of domain experts. Hence, the subjective expert-driven knowledge can stipulate to a great extent the behavior of a simulation model in terms of structure, assumptions, and parameters. On the other hand, ML models can be developed with slight involvement, if any, of experts.

Second, SM and ML were often considered to be addressing different types of analytical questions. From the perspective of data analytics, ML is largely concerned with predicting what is likely to happen. However, SM goes beyond that question and addresses further questions for more complex scenarios (e.g., "What if?" or "How to?"). Figure 10.1 illustrates the position of SM and ML within the

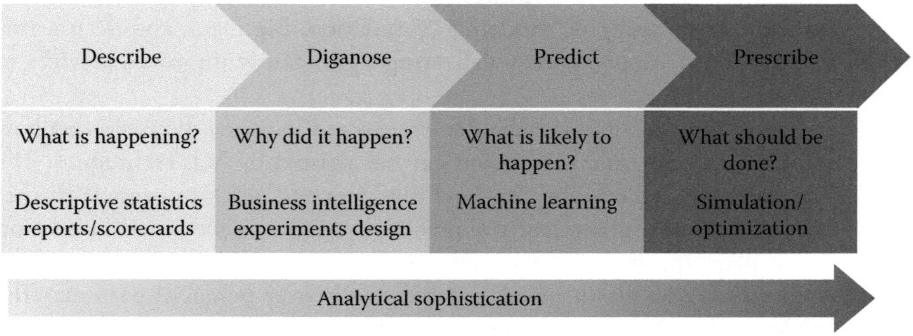

Describe	Diganose	Predict	Prescribe
What is happening?	Why did it happen?	What is likely to happen?	What should be done?
Descriptive statistics reports/scorecards	Business intelligence experiments design	Machine learning	Simulation/ optimization

Analytical sophistication

Figure 10.1 The spectrum of data analytics. (Adapted from Barga, R. et al., ***Predictive Analytics with Microsoft Azure Machine Learning,*** **Apress, 2015; Maoz, M., How IT should deepen big data analysis to support customer-centricity,** ***Gartner G00248980, 2013.)***

landscape of data analytics. The figure classifies data analytics into four categories with an increasing level of analytical sophistication.

The Prospective Role of Machine Learning in Simulation Modeling

"Even though the assumptions of a model may not literally be exact and complete representation of reality, if they are realistic enough for the purpose of our analysis, we may be able to draw conclusions which can be shown to apply to the world." (Cohen and Cyert, 1965)

The "realism" of simulation models has always been a controversial issue. It has been continuously argued (Shreckengost, 1985; Stanislaw, 1986) that all models, including simulation models, are considered invalid regarding complete reality. However, a simulation model does not need to be a complete representation to be useful for analytics purposes, as stated earlier. A simulation model should largely attempt to represent a sufficient level of reality, which can be acceptable regarding the questions of interest.

In this regard, the integration of SM with ML may help simulations attain a higher level of model realism. Our main view is that simulation experiments can be supported by predictive models trained to carefully make predictions, which reflect the real world with a relatively lower degree of uncertainty. To explain further, it is assumed that a new state of the system can generate new data, as illustrated in Figure 10.2. Based on new data (i.e., feedback), ML models can be trained to predict the future behavior of the system. The ML models can be continuously trained and refitted to echo feedback, and update the system knowledge. In this manner,

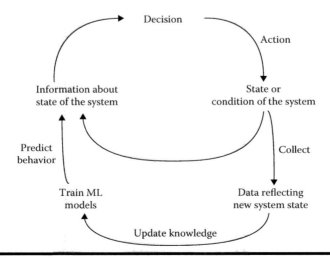

Figure 10.2 Data-driven feedback loop.

the new system states can be "learned" aided by ML models, which can in turn help simulation models become dynamic and more realistic.

From a more practical standpoint, ML can be utilized to predict the behavior of system variables that may not be feasible to express analytically. For example, Zhong et al. (2016) trained ML models within a use case for crowd modeling and simulation. The ML models were used to learn and predict the flow of crowds. Likewise, unsupervised ML techniques (e.g., clustering) can be used to learn about key structural characteristics of systems, especially if tackling big data scenarios.

In a broader sense, ML can be employed as an assistive tool to help reduce the epistemic uncertainty (Oberkampf et al., 2002) underlying simulation models. This kind of uncertainty can be attributed to the subjective interpretation of system knowledge by modelers, simulationists, or subject matter experts.

Related Work

Owing to the multifaceted nature of the presented work, we believe that the study can be viewed from different perspective. Therefore, we reviewed studies with relevance to the following: hybrid simulations, artificial intelligence (AI)-assisted simulations, and simulation-based healthcare planning.

Hybrid Simulations

This study can be viewed from the perspective of developing hybrid simulations. As suggested by Powell and Mustafee (2014), a hybrid modeling and simulation study refers to the application of methods and techniques from disciplines like operations research, systems engineering, or computer science to one or more stages of a simulation study. Likewise, the study here attempted to integrate simulation models with a method from the computer science discipline (i.e., ML). Viewed this way, we aimed to review examples of hybrid studies that incorporated simulation methods with ML techniques. To focus our search, two main sources were selected for review over the past 10 years (i.e., 2016–2007): Winter Simulation conference and ACM SIGSIM Conference on Principles of Advanced Discrete Simulation (PADS) conference. It is acknowledged that other relevant studies could have been published in other conferences or journals, but we believe that the selected venues provided excellent, if not the best, sources of representative studies in accordance with the target context.

One good example is Rabelo et al. (2014) who applied hybrid modeling, where SM and ML were used altogether in a use case related to the Panama Canal operations. A set of simulation models was developed to make predictions about the future expansion of the canal. This information was further used to develop ML models (e.g., neural networks and support vector machines) to help with the analysis of the simulation output. With a comparable hybrid approach, Elbattah and Molloy (2016) embraced an approach that integrated SM with ML with application to a healthcare case. The study claimed that the use of ML improved the predictive power of the simulation model. Another example is Zhong et al. (2016) who utilized ML to assist with crowd modeling and simulation. The ML models were used to learn and predict the flow of crowds.

Artificial Intelligence-Assisted Simulations

In an early insightful suggestion, Shannon (1975) envisioned that:

> "The progress being made in AI technology opens the door for a rethinking of the SM process for design and decision support."

Considering ML as one of the cornerstones of AI, the chapter should also be viewed from that standpoint. The promise of AI-assisted simulations was that the model behavior can be generated or adjusted aided by AI techniques. Examples of approaches were developed in this regard including knowledge-based simulations and qualitative simulations.

Fishwick and Modjeski (2012) defined knowledge-based simulation as the application of knowledge-based methods within AI to the field of computer simulation. As such, the knowledge about the model can be expressed as rule-based facts that characterize experimental conditions. Early endeavors of knowledge-based simulation started with building Prolog-based simulation languages such as TS-Prolog (Futó and Gergely, 1986), and T-CP (Cleary et al., 1985). A more recent example of knowledge-based simulations is Lattner et al. (2010) who developed an approach for knowledge-based adaptation of simulation models. The approach was used as a means to enable an automated generation of model variants, with a specific focus on automated structural changes in simulation models.

On the other hand, qualitative simulations followed a different path, which was introduced as an attempt to replicate the process of human reasoning (i.e., qualitative variables) into computer simulations. Cellier (1991a) defined qualitative simulations as evaluating the behavior of a system in qualitative terms. However, Fishwick and Modjeski (2012) described that the aim of qualitative simulations is not to adopt qualitative methods instead of quantitative methods, but rather to use qualitative methods as an augmentation to qualitative methods. The Qualitative Simulation (QSIM) algorithm (Kuipers, 1986) was an early qualitative simulation study.

Simulation-Based Healthcare Planning

From the context of healthcare decision-making, the study can also be related to similar simulation-based studies that endorsed care planning. The literature is actually rife with examples of such studies. For instance, Harper and Shahani (2002) used SM for the planning and management of bed capacities within an environment of uncertainty. The simulation model was utilized to help understand and quantify the consequences of planning and management policies. Similarly, Rashwan et al. (2013) developed a simulation model to map the dynamic flow of elderly patients in the Irish healthcare system. The model was claimed to be useful for inspecting the outcomes of proposed policies to overcome the delayed discharge of elderly patients. However, the literature generally laid little emphasis on endeavors toward incorporating simulation methods and ML techniques.

Background: Big Data, Analytics, and Simulation Modeling

Our initial view was that the practice of modeling and simulation can be supported by emerging technologies driven by big data analytics. For instance, ML can play a vital role for building more realistic simulation models. This section presents a brief background to Big Data, analytics, simulation, and the prospective opportunities for integration.

Definitions of Big Data

Although the term Big Data has been ubiquitously used over the past years, it has been difficult to establish a commonly accepted definition. For instance, the Oxford Dictionary (2017) defines Big Data as "Extremely large datasets that may be analyzed computationally to reveal patterns, trends, and associations, especially relating to human behavior and interactions." However, different interpretations of Big Data can be found based on context, domain, and computation standards. In this regard, many studies such as Ward and Barker (2013), Beyer and Laney (2012), Gandomi and Haider (2015), and De Mauro et al. (2015) aimed to survey the common interpretations of Big Data in a bid to articulate a broader definition.

In this section, we aim to review the various definitions of Big Data. In a timeline fashion, Figure 10.3 overviews the diversity of Big Data definitions over a timespan of more than 15 years since 1997. The early definitions of Big Data appeared to focus on the size of data in storage. The variety of definitions clearly signifies the multifaceted interpretations of Big Data. However, the definitions can be considered generally in terms of: volume, complexity, and technologies.

Characteristics of Big Data

Over the years, increasing dimensions have been continuously added to distinguish the case of Big Data. Below we present the main viewpoints that were developed for describing the characteristics of big data.

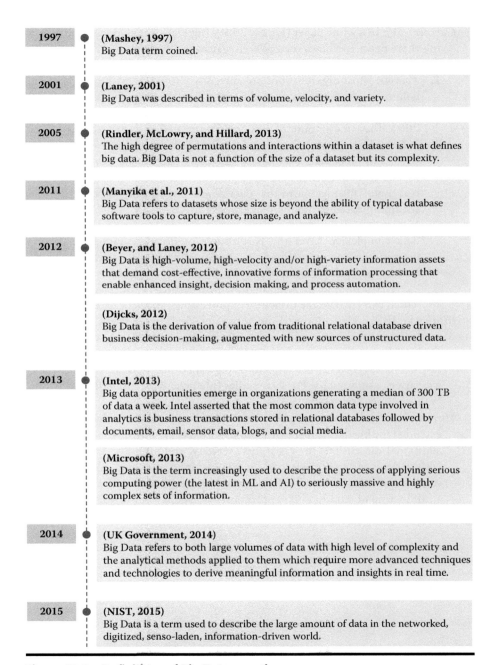

1997 ● (Mashey, 1997)
Big Data term coined.

2001 ● (Laney, 2001)
Big Data was described in terms of volume, velocity, and variety.

2005 ● (Rindler, McLowry, and Hillard, 2013)
The high degree of permutations and interactions within a dataset is what defines big data. Big Data is not a function of the size of a dataset but its complexity.

2011 ● (Manyika et al., 2011)
Big Data refers to datasets whose size is beyond the ability of typical database software tools to capture, store, manage, and analyze.

2012 ● (Beyer, and Laney, 2012)
Big Data is high-volume, high-velocity and/or high-variety information assets that demand cost-effective, innovative forms of information processing that enable enhanced insight, decision making, and process automation.

(Dijcks, 2012)
Big Data is the derivation of value from traditional relational database driven business decision-making, augmented with new sources of unstructured data.

2013 ● (Intel, 2013)
Big data opportunities emerge in organizations generating a median of 300 TB of data a week. Intel asserted that the most common data type involved in analytics is business transactions stored in relational databases followed by documents, email, sensor data, blogs, and social media.

(Microsoft, 2013)
Big Data is the term increasingly used to describe the process of applying serious computing power (the latest in ML and AI) to seriously massive and highly complex sets of information.

2014 ● (UK Government, 2014)
Big Data refers to both large volumes of data with high level of complexity and the analytical methods applied to them which require more advanced techniques and technologies to derive meaningful information and insights in real time.

2015 ● (NIST, 2015)
Big Data is a term used to describe the large amount of data in the networked, digitized, senso-laden, information-driven world.

Figure 10.3 Definitions of Big Data over the years.

Gartner's 3Vs: In a white paper (Laney, 2001), Big Data was characterized as having three main attributes (i.e., 3Vs). Regarded as the basic dimensions of Big Data, the 3Vs can be explained as follows:

1. *Volume*: Most organizations are currently struggling with the increasing volumes of their data. According to an estimate by *Fortune* magazine (Fortune, 2012), about five exabytes of digital data was created until 2003. The same amount of data could be created in just two days by 2011.

2. *Velocity*: Data velocity describes the speed at which data is created, accumulated, and processed. The rapidly increasing pace of the world has placed further demands on businesses to process information in real time or near real time. This may mean that data should be processed on the fly or in a streaming-based fashion to make quicker decisions (Minelli et al., 2012).

3. *Variety*: The assortment of data variety represents a critical factor of the data complexity. Over the past couple of decades, data have become increasingly unstructured as the sources of data have varied beyond the traditional operational applications. Therefore, large-scale datasets may likely exist in different structured, semistructured, or unstructured forms, which can escalate the difficulty of processing tasks to a greater extent.

IBM's 4Vs: IBM added another dimension, "veracity", to Gartner's 3Vs. The reason behind the additional dimension was justified as that IBM's clients started to face data-quality issues while dealing with big data problems (Zikopoulos, 2013). Hence, IBM (2017) defined the big data dimensions as: volume, velocity, variety, and veracity. Further studies (Demchenko et al., 2014) added the "value" dimension to IBM's 4Vs.

Microsoft's 6Vs: For the purpose of maximizing business value, Microsoft extended the big data dimensions into 6Vs (Wu et al., 2016). The 6Vs included additional dimensions for variability, veracity, and visibility. In comparison with variety, variability refers to the complexity of data (e.g., the number of variables in a dataset), while visibility emphasizes that there is a need to have a full picture of data to make informative decisions. Figure 10.4 below summarizes the common dimensions of Big Data as explained before.

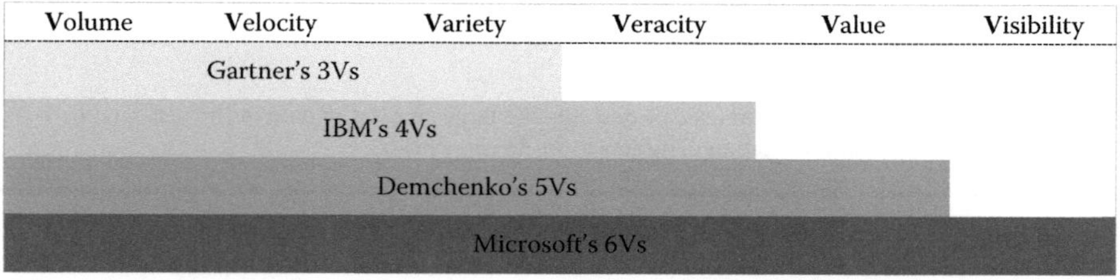

Volume	Velocity	Variety	Veracity	Value	Visibility
Gartner's 3Vs					
IBM's 4Vs					
Demchenko's 5Vs					
Microsoft's 6Vs					

Figure 10.4   The articulations of big data dimensions.

Analytics

The opportunities enabled by Big Data led to a significant interest in the practice of data analytics. Thus, data analytics has evolved into a vibrant and broad domain that incorporates a diversity of techniques, technologies, systems, practices, methodologies, and applications.

Similar to Big Data, various definitions were developed to describe analytics. Table 10.1 presents some common definitions used to describe analytics. In the same context, Figure 10.5 portrays the interdisciplinarity involved within

Table 10.1 Common Definitions of Analytics

Definition of Analytics	Reference
The extensive use of data, statistical, and quantitative analysis; explanatory and predictive models; and fact-based management to drive decisions and actions. The analytics may be input for human decisions or may drive fully automated decisions. Analytics are a subset of what has come to be called business intelligence.	Davenport and Harris (2007)
Delivering the right decision support to the right people at the right time.	Laursen and Thorlund (2016)
The scientific process of transforming data into insight for making better decisions.	Liberatore and Luo (2011)

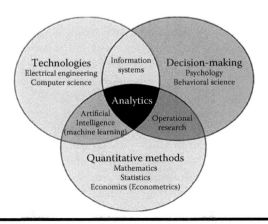

Figure 10.5 A taxonomy of disciplines pertaining to data analytics. (From Mortenson, M. J. et al., *Eur. J. Oper. Res.*, 241, 583–595, 2015.)

the practice of analytics. According to Mortenson et al. (2015), these disciplines can fit into one or more of the following categories:

■ *Technological*: Refers to the various tools used including hardware, software, and networks, which together support the efficient processing of large-scale datasets.
■ *Quantitative methods*: Refers to the applied quantitative approaches to analyzing data, such as statistics, ML, econometrics, and Operations Research (OR).
■ *Decision-making*: An inherently interdisciplinary area including tools, theories, and practices used to support and understand the decision-making process (e.g., human–computer interaction and visualization in information systems, or problem structuring methods in OR/MS).

Simulation Modeling

"When experimentation in the real system is infeasible, simulation becomes the main, and perhaps the only, way to discover how complex systems work." (Sterman, 1994)

Shannon (1975) described simulation as the process of designing a model of a real system and conducting experiments with this model for the purpose of understanding the behavior of the system and/or evaluating various strategies for the operation of the system. With a virtual build-and-test environment, simulation provides the feasibility to model real systems that exhibit adaptive, dynamic, goal-seeking, self-preserving, and sometimes evolutionary behavior (Meadows, 2008).

In view of the analytics spectrum as presented earlier in Figure 10.1, simulation models endeavor to answer more complex questions that fall under the category of prescriptive analytics. Therefore, SM can present as a vital component for data analytics. This section provides a brief background of the common simulation approaches, and the main distinctions between them.

A simulation model is developed largely based on the "world view" of a modeler. A world view reflects how a real world system is mapped to a simulation model. In this respect, there are three primary approaches including: system dynamics (SD), discrete event simulation (DES), and agent-based modeling.

The SD approach assumes a very high degree of abstraction, which can be considered adequately for strategic modeling. On the other hand, discrete-event models maintain medium and medium–low abstraction, whereas a model comprises a set of individual entities that have particular characteristics in common. Agent-based models are positioned in the middle, which can vary from very fine-grained agents to highly abstract models. Figure 10.6 portrays the three approaches with respect to the level of abstraction. Further, Table 10.2 makes a more detailed comparison based on Brailsford and Hilton (2001), Lane (2000), and Sweetser (1999).

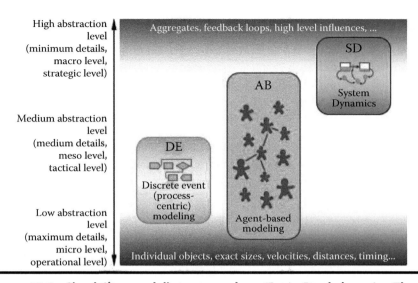

Figure 10.6 Simulation modeling approaches. (From Borshchev, A., *The Big Book of Simulation Modeling: Multimethod Modeling with AnyLogic 6***, AnyLogic North America, Chicago, IL, 2013.)**

Table 10.2   Comparison between the SD and DES Approaches

	System Dynamics Models	*Discrete Event Models*
Scope	Operational/Tactical	Strategic
Nature of models	Continuous/Mostly deterministic	Discrete/Mostly stochastic
World view	A series of stocks and flows, in which the state changes are continuous. Entities are viewed as a continuous quantity, or aggregates rather than individual entities (e.g., products and people).	Entities flowing through a network of queues and activities. Entities have attributes that describe specific features of an entity (e.g., entity type, dimensions, weight, and priority).
Building blocks	Stocks, inflows, outflows, tangible variables, and soft variables.	Individual entities, attributes, queues, and events.
Model outputs	A full picture of the system.	Estimates of system's performance.

What Can Big Data Add to Simulation Modeling?

As described by Cellier (1991b), a model for a system (S) and an experiment (E) is anything to which E can be applied to answer questions about S. In other words, every simulation model attempts to formulate a representation of system knowledge to answer questions about that system. We here review the common sources of knowledge utilized to develop simulation models, and the potential links to Big Data as well.

System knowledge can be learned from different sources. Huang (2013) mentioned four categories of knowledge that can be utilized to build a simulation model as below:

- *Formal knowledge*: This kind of knowledge can be explicitly presented and maintained. Examples of formal system knowledge can be theories, theorems, or mathematical models. The content of formal knowledge has the advantage of being readily accessible in its meaning and form.
- *Informal knowledge*: On the other hand, informal knowledge is implicitly represented in an intangible format in the heads of people who are familiar with the system, (e.g., domain experts or system users). Therefore, this kind of knowledge needs to be formulated using elicitation and formalization methods.
- *Empirical data*: Refers to the type of data obtained from recorded observations or measurements about the system. It potentially contains information about the behavior of a system and its subsystems.
- *System description*: Refers to data that can contain descriptive information about the system in terms of structure and relation information. Compared to empirical data, system description characterizes a system and its subsystems themselves. This type of knowledge is often produced by people such as domain experts or engineers involved with designing the system. Examples are documents, floorplans of factories, and manufacturing process maps. Figure 10.7 sketches the sources of system knowledge.

In the era of Big Data, it should be considered that system knowledge will increasingly become based on empirical data accumulated or generated autonomously. Specifically, more data will be increasingly utilized to learn about systems. Therefore, systems that deal with big data scenarios will inevitably place further burdens on the modeling process, which can be beyond the capabilities of humans in many situations. For instance, the knowledge of a system can be underlying huge amounts of data, or being accumulated with high velocity. In this regard, insightful studies such as Tolk (2015), and Tolk et al. (2015) emphasized the need for integrating SM with Big Data. In particular, it was stressed that big data techniques and technologies should be considered to avail of rapidly accumulating data that may be structured or unstructured.

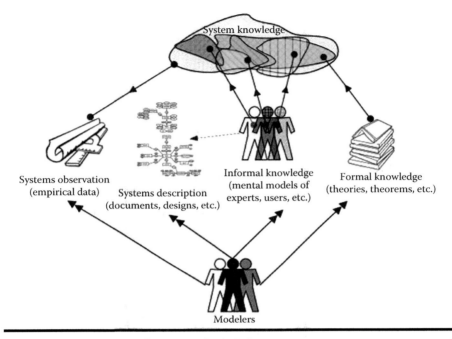

Systems observation (empirical data)

Systems description (documents, designs, etc.)

Informal knowledge (mental models of experts, users, etc.)

Formal knowledge (theories, theorems, etc.)

Figure 10.7 Sources of system knowledge. (From Huang, Y., *Automated Simulation Model Generation*, Doctoral dissertation, TU Delft, Delft University of Technology, 2013.)

Analytics Use Case: Elderly Discharge Planning

For the purpose of demonstrating the applicability of the key ideas presented, a case study was developed in relation to healthcare. The following sections elaborate the case setting, and the development of a set of simulation and ML models. The main goal here was to provide a practical scenario where simulation models can be designed or guided in concert with knowledge learned with the aid of ML.

Case Description

Approaching any analytics problem needs an in-depth understanding of the domain under study. Therefore, this section serves as a necessary background prior to building the simulation or ML models.

In Ireland, the population has been experiencing a pronounced demographic transition. The Health Service Executive (HSE) of Ireland reported in 2014 that the increase in the number of people over 65 is approaching 20,000 per year (HSE, 2014b). As a result, population aging is expected to have significant impacts on a broad range of economic and social areas, and on the demand for healthcare services.

Within the context of elderly care, the focus of the case was centralized around the care scheme of hip fracture. Hip fractures were considered for having two-fold significance. On one hand, hip fractures represent an appropriate example of elderly care schemes. An ample number of studies (Cooper et al., 1992; Melton, 1996) recognized that hip fractures increase exponentially with aging, though rates may vary from one country to another. Around 3,000 patients per year are assumed to sustain hip fractures in Ireland (Ellanti et al., 2014). Further, that figure may unavoidably increase due to the continuously aging population.

From an economic perspective, hip fractures can represent a major burden on the Irish healthcare system. According to the HSE, hip fractures were identified as one of the most serious injuries resulting in lengthy hospital admissions and high costs (HSE, 2014a). The median length of stay (LOS) was recorded as 13 days, and more than two-thirds of patients are discharged to long-stay residential care after surgery (NOCA, 2014). The cost of treating a typical hip fracture was estimated around €12,600 (HSE, 2014a), while a different study reported a higher cost of €14,300.

Questions of Interest

In relation to the elderly care of hip fracture, two complementary categories of questions were addressed by the use case including: population-level questions and patient-level questions. Table 10.3 poses the questions in detail.

Table 10.3   Questions of Interest

Scope	Questions
Patient-level	Q1. Given a patient's characteristics along with care–related factors, how to predict the inpatient length of stay in acute facilities?
	Q2. Given a patient's characteristics along with care–related factors, how to predict the discharge destination?
Population-level	Q3. On a population basis, what is the expected demand for long-stay care facilities such as nursing homes?
	Q4. On a population basis, what is the expected utilization of hospital resources in terms of length of stay?

Overview of Analytics Approach

The study embraced a multiple methodology approach that aimed to integrate SM with ML. In particular, the approach was conducted over four stages as follows: discovering clusters of patients, modeling clustered-flows of patients, modeling patient's care journey, and predicting care outcomes.

Reflecting on the questions of interest, the various stages were aimed to model the healthcare system at two interdependent levels (i.e., population level and patient level). Initially at the population level, we utilized unsupervised ML using clustering techniques in an endeavor to discover potential underlying structures. Based on patient records, clustering experiments were carried out to discover coherent patient clusters that have characteristics or care outcomes in common. In light of the discovered clusters, the flows of elderly patients were modeled. As such, each flow of elderly population represented a patient cluster of specific characteristics and care-related outcomes.

Subsequently at the patient level, DES modeling was used to mimic the patient's care journey at a finer-grained level of details. The DES model helped to deal with patient entities, rather than aggregate populations. Each patient entity could be treated individually in terms of characteristics (e.g., age, gender, and type of fracture), and care-related factors (e.g., time to admission [TTA] and time to surgery [TTS]). Eventually, ML models were trained to make predictions on care-related outcomes for the simulated patients. For every simulated patient, ML models were used to predict the LOS and the discharge destination. Figure 10.8 illustrates the four stages of our approach.

Various tools were used through the four stages. The R language was used intensively within building simulations and ML models as well. The R-package deSolve (Soetaert et al., 2010) was particularly used to build the SD model. The deSolve package facilitated solving the ordinary differential equations within the SD model. The DES model was built using the R language as well.

Furthermore, the Azure ML Studio was used to develop our ML models. The R language was used within the data preprocessing procedures. The visualizations were mostly produced using the R-package ggplot2 (Wickham, 2009). Figure 10.9 summarizes the methods and tools used by the study.

Data Description

The main source of data used by the study is the Irish hip fracture database (IHFD) (NOCA, 2017). The IHFD repository is the national clinical audit developed to capture care standards and outcomes for hip fracture patients in Ireland. Decisions to grant access to the IHFD data are made by the National Office of Clinical Audit (NOCA).

The IHFD contains ample information about the patient's journey from admission to discharge. Specifically, a typical patient record included 38 data fields such as gender, age, type of fracture, date of admission, and LOS. A thorough explanation of the data fields was available via the official data dictionary (HIPE, 2015).

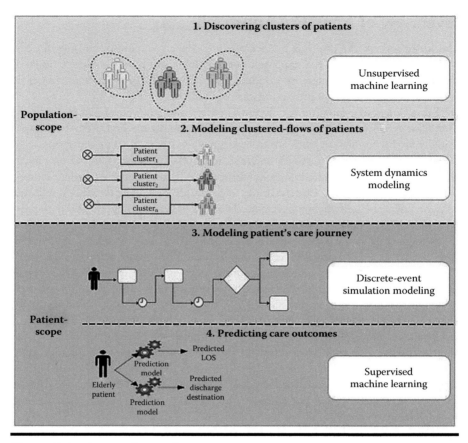

Figure 10.8   Approach overview.

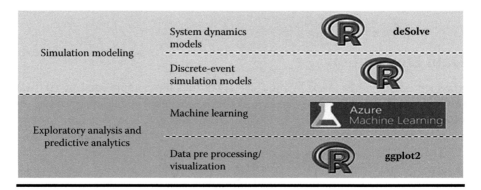

Figure 10.9   Tools used by the study.

The dataset included records about elderly patients aged 60 and over in particular. The data comprised about 8,000 records over three years, from January 2013 to December 2015. It is worth mentioning that one patient may be related to more than one record, in cases of recurrent fractures. However, we were unfortunately unable to determine the proportion of recurrent cases as patients had no unique identifiers, and records were completely anonymized for the purpose of privacy. Figure 10.10 plots a histogram of the age distribution within the dataset, while Figure 10.11 shows the percentages of male and female patients.

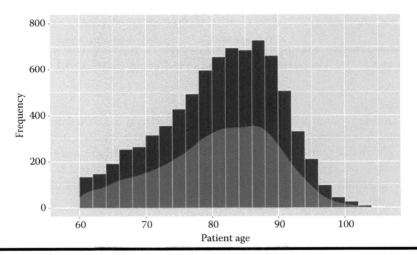

Figure 10.10   The distribution of age within the dataset.

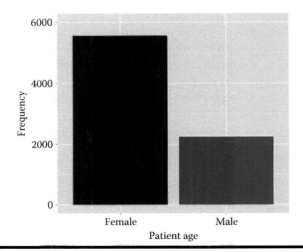

Figure 10.11   The distribution of gender within the dataset.

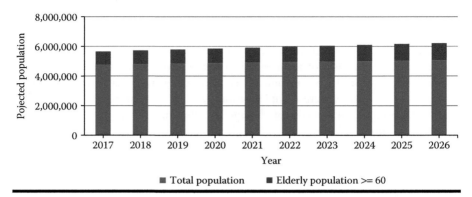

Figure 10.12　The projections of population in Ireland (2017–2026).

In addition, population projections were needed to address the population-level questions. In this regard, the study used population information prepared by the Central Statistics Office (CSO, 2017). The data contained comprehensive information about the population in specific geographic areas in terms of age and sex. However, we focused only on the elderly population aged 60 years and over, in line with the study scope. Figure 10.12 shows the population projections in Ireland from 2017 to 2026.

Unsupervised Machine Learning: Discovering Patient Clusters

As outlined in the approach overview, ML clustering was employed for learning about the system structure. In this regard, the study used k-means clustering to realize the segmentation of patients from a data-driven viewpoint. Specifically, elderly patients were grouped based on the similarity of patient's age, LOS, and elapsed TTS. The following sections explain data preprocessing procedures, and how clusters were computed.

Outliers Removal

As reported by NOCA (2014), the mean and median LOS for hip fracture patients were recorded as 19 and 12.5 days respectively. Therefore, we only considered the patients whose LOS were no longer than 60 days to avoid the odd influence of outliers. The excluded outliers represented approximately 5% of the overall dataset. Figure 10.13 plots a histogram of the LOS used to identify the outliers.

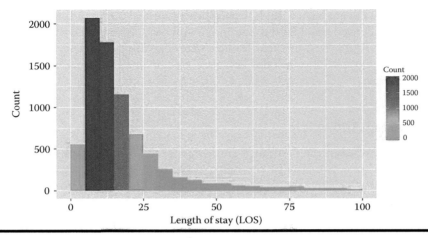

Figure 10.13 Histogram and probability density of the LOS variable. The density is visually expressed as a gradient ranging from green (low) to red (high). The outliers can be observed for LOS longer than 60 days.

Feature Scaling

Feature scaling is a necessary preprocessing step in ML in cases where the range of features values varies widely. Several studies such as Visalakshi and Thangavel (2009), and Patel and Mehta (2011) argued that large variations within the range of feature values can affect the quality of computed clusters. Therefore, the feature values were rescaled to a standard range.

The min-max normalization method was used, where every feature was linearly rescaled to the [0, 1] interval. The values were transformed using the formula below in Equation 10.1:

$$z = \frac{x - \min(x)}{\left[\max(x) - \min(x)\right]} \tag{10.1}$$

Features Extraction

In a report prepared by the British Orthopedic Association (2007), six quality standards for hip fracture care were emphasised. Those standards generally reflect good practice at key stages of hip fracture care including:

- All patients with a hip fracture should be admitted to an acute orthopedic ward within 4 hours of presentation.
- All patients with a hip fracture who are medically fit should have surgery within 48 hours of admission, and during normal working hours.

The raw data did not include fields that explicitly captured such standards. However, they can be derived based on the date and time values of patient arrival, admission, and surgery. In this way, two new features were added named as TTA and "TTS". Eventually, only TTS was included because TTA contained a significant amount of missing values.

Clustering Approach

The study embraced the partitional clustering approach using the k-means algorithm. The k-means is one of the most widely used clustering algorithms. The k-means clustering uses a simple iterative technique to group points in a dataset into clusters that contain similar characteristics. Initially, a number (*k*) is decided that represents centroids (center of a cluster). The algorithm iteratively places data points into clusters by minimizing the within-cluster sum of squares as the equation below (Jain, 2010). The algorithm converges on a solution when meeting one or more of these conditions: the cluster assignments no longer changes, or the specified number of iterations is completed. Equation 10.2:

$$J(C_k) = \sum_{X_i \in C_K} X_i - \mu_K^2 \qquad (10.2)$$

where:

μ_K is the mean of cluster C_k

$J(C_k)$ is the squared error between μ_K and the points in cluster C_k

Selected Features

The k-means algorithm was originally applicable to numeric features only, where a distance metric (e.g., Euclidean distance) can be used for measuring similarity between data points. Therefore, we considered the numeric features only. Specifically, the model was trained using the following features: LOS, age, and TTS. However, it is worth mentioning that there are some extensions of the k-means algorithm that attempted to incorporate categorical features, such as the k-modes algorithm (Huang, 1998).

Clustering Experiments

As usual, the unavoidable question while approaching a clustering task is how many clusters (*k*) exist? In our case, the quality of clusters was experimented using *k* ranging from 2 to 7. Table 10.4 presents the parameters used within the clustering experiments. We used the Azure ML Studio to train the clustering model.

Table 10.4   Parameters Used within the k-Means Algorithm

Parameter	Value
Number of clusters (k)	2–7
Centroid initialization	Random
Similarity metric	Euclidian distance
Number of iterations	100

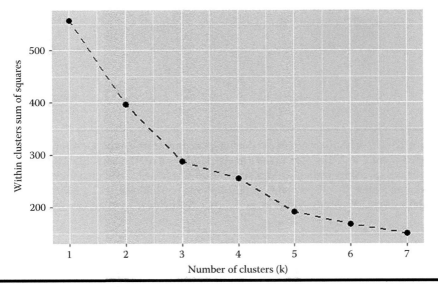

Figure 10.14   Plotting the sum of squared distances within clusters.

Initially, the quality of clusters was examined based on the within cluster sum of squares (WSS), as plotted in Figure 10.14. In view of that, it turned out that there may be three or four well-detached clusters of patients that can best separate the dataset. Furthermore, the suggested clusters were projected into two dimensions based on the principal component analysis (PCA) to determine the appropriate number of clusters, as in Figure 10.15. Each subfigure in Figure 10.16 represents the output of a single clustering experiment using a different number of clusters (k). Initially with $k = 2$, the output indicated a promising tendency of clusters, where the data space is obviously separated into two big clusters.

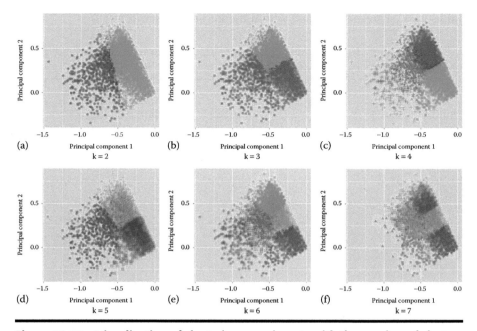

Figure 10.15 Visualization of clustering experiments with the number of clusters ranging from 2 to 7. The clusters are projected based on the principal components. A mixture of colors indicates less separation of clusters. The visualizations were produced using the R-package ggplot2. (From Wickham, H., *ggplot2: Elegant Graphics for Data Analysis,* Springer Science & Business Media, New York, 2009.)

Similarly for $k = 3$, the clusters are still well-separated. However, the quality of the clusters started to decline when $k = 4$ onwards. Thus, it turned out eventually that there were three clusters that divided the dataset into coherent cohorts of patients.

Exploring Clusters

In this section, we aim to explore the discovered clusters in a visual manner that can reveal potential correlations or insights, which can in turn assist with simulation model design. The clusters were particularly examined with respect to patient characteristics (e.g., age), care-related factors (e.g., TTS), and outcomes (e.g., discharge destination).

In Figure 10.16a, the inpatient LOS is plotted with respect to the three patient clusters. At first glance, it was obvious that the patients of cluster 3 experienced remarkably longer LOS periods compared to cluster 1 or cluster 2. In addition, cluster 1 and cluster 2 shared a very similar distribution of the LOS variable, apart from a few outliers in cluster 2.

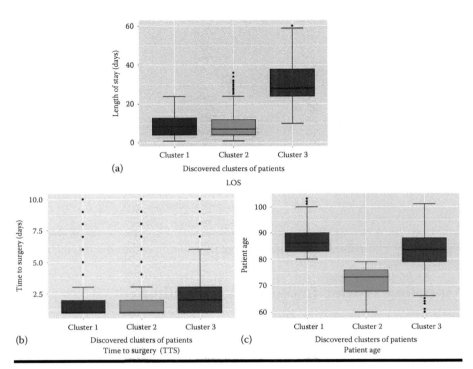

Figure 10.16 The variation of the LOS, TTS, and age variables within the three patient clusters.

Second, we examined the clusters with respect to the elapsed TTS. As mentioned in Section 8.3, the TTS has a particular significance for being one of the quality standards for hip fracture care. Once more, the patients of cluster 3 were observed for having a relatively longer TTS than those patients of cluster 1 and cluster 2. Likewise, the LOS, cluster 1, and cluster 2 had a very similar distribution of the TTS. Figure 10.16b plots the TTS variable against the three clusters of patients.

The patient age has a considerable relevance in elderly care schemes. In our context, the possibility of sustaining hip fractures can increase significantly with age. It turned out that cluster 2 and cluster 3 tended to have relatively older patients rather than cluster 1. Figure 10.16c plots the age distribution within the three clusters.

Furthermore, the clusters were inspected for possible gender-related patterns. Figure 10.17 shows the proportions of male and female patients within clusters. It can be clearly noticed that the number of female patients consistently exceeded males in all clusters.

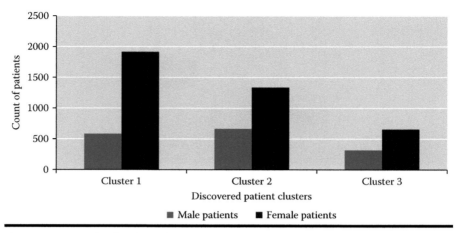

Figure 10.17   Distribution of male and female patients within the three clusters.

Modeling Cluster-Based Flows of Patients

Initial System Dynamics Model

The initial model was intended to provide a bird's eye view of the care scheme of hip fractures. The model mainly aimed at capturing the relationships among system entities in an SD manner. However, this preliminary version did not consider the different characteristics of patients learned by the ML clustering experiments.

Specifically, the model focused on describing the dynamic behavior pertaining to the continuous growth of population aging, and the consequent implications on the incidence of hip fractures among the elderly. The model defined the main system actors as follows: elderly patients, acute hospitals, and discharge destinations (e.g., home or long-stay care facilities). The model included a single reinforcing loop implied by patients with a fragility history, who are susceptible to resustain hip fractures or fall-related injuries. In this regard, the HSE (2014a) reported that one in three of elderly patients fall every year and two-thirds of them fall again within six months. Figure 10.18 illustrates the initial SD model.

Model Assumptions and Simplifications

A set of assumptions and simplifications was decided while maintaining the simulation model as a reasonably approximate representation of the actual healthcare system. Table 10.5 provides the assumptions and simplifications along with explanations. Table 10.6 lists the model variables and Table 10.7 presents the model equations.

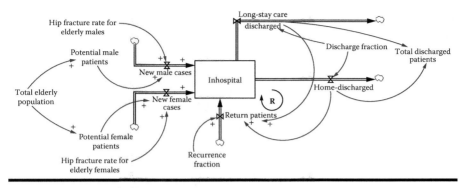

Figure 10.18   Initial SD model.

Table 10.5   Model Assumptions and Simplifications

Assumption/Simplification	Purpose/Reason
The rate of hip fracture in the total population aged 60 and over was set as 407 for females and 140 for males per 100,000.	The rate was defined by Dodds et al. (2009).
Elderly patients were assumed as aged 60 and over, though usually considered as aged 65 and older (Rosenberg and Everitt, 2001).	To conform to the defined hip fracture rate, which endorsed patients aged 60 and over.
The model did not consider the scenario of patient transfer from an acute hospital to another during treatment course.	Only for simplification, assuming that the treatment course was bounded within a single acute hospital.
The model used the same age distribution for both male and female elderly patients.	For the purpose of simplification, since both distributions were slightly different.

Cluster-Based System Dynamics Model

The SD model was redesigned in light of the clustering experiments. In particular, the model was disaggregated into three stocks representing the suggested clusters of patients. Furthermore, the auxiliary variables were decided based on the cluster analysis. For instance, the first and second patient clusters were set to undergo the same TTS delay (i.e., TimeToSurgery1), while the third cluster was assigned a different delay (i.e., TimeToSurgery2).

Table 10.6   Model Variables

Variable	Description
Total elderly population	The number of elderly population, aged 60 and over, in a particular year.
Potential male patients	Total male patients aged 60 and over.
Potential female patients	Total female patients aged 60 and over.
Hip fracture rate for elderly males	The rate of hip fracture in the total elderly male population = 140 cases per 100 K.
Hip fracture rate for elderly females	The rate of hip fracture in the total population aged 60 and over = 407 for females per 100 K.
InHospital	Stock variable representing the total number of elderly hip-fracture patients in acute hospitals nationwide.
Discharge fraction	Proportion of total elderly patients discharged to home or long-stay care.
Total discharged patients	The number of patients discharged to home and long-stay care.
Recurrence rate	The rate that defines the proportion of discharged patients who are susceptible to resustain a hip fracture and return to an acute hospital.

Equally important, the inflows of elderly patients were structured based on the age variation within clusters. In particular, both of the first and third patient clusters were modeled to include more elderly patients (i.e., ages 80–100), while the second cluster was associated with less elderly patients (i.e., ages 60–80). This reflected the age groups within the patient clusters. Figure 10.19 illustrates the cluster-based SD model.

In this manner, the clustering model was employed effectively for the purpose of understanding the system structure, where the SD model stocks actually represented the three discovered clusters of patients. Moreover, the variations within clusters in terms of patient characteristics (e.g., age), or care-related factors (e.g., TTS) assisted with shaping the model behavior. As such, it can be argued that the SD model was constructed with an established confidence predicated on the clustering model. The well-validated quality of clusters along with the compelling

Table 10.7 Model Equations

Equation	Type
1. Hip fracture rate for elderly males = 140 cases per 100,000.	Auxiliary
2. Hip fracture rate for elderly females = 407 cases per 100,000.	Auxiliary
3. New male cases = Hip fracture rate for elderly males * Potential male patients	Inflow
4. New female cases = Hip fracture rate for elderly females * Potential female patients	Inflow
5. Home-discharged = InHospital * Discharge fraction	Outflow
6. Long-stay care discharged = InHospital * (1-Discharge fraction)	Outflow
7. Recurrent patients = (Home-discharged * Recurrence rate) + (Long-stay care discharged * Recurrence rate)	Inflow
8. InHospital = Integ((New male cases + New female cases) – (Home-discharged + Long-stay care discharged) + Recurrent patients, initial value)	Stock

visualizations could support the rationale behind the SD model design in terms of structure and behavior as well.

Modeling Patient's Care Journey

Simulation Approach

The DES approach was used to model a fine-grained perspective of the patient's journey. The model was utilized in order to produce a realistic sequence of events corresponding to those within the care journey, as sketched in Figure 10.20. Further, the DES approach can facilitate the following:

- Representing the system's components in terms of entities (e.g., patients, hospitals etc.).
- The entity-based modeling allowed us to consider cluster-based aspects on the individual patient level. For instance, specific attributes of elderly patients (e.g., fracture type) were determined differently with respect to the cluster.
- With such entity-based structuring, ML models can subsequently be used to provide accurate predictions at the patient level.

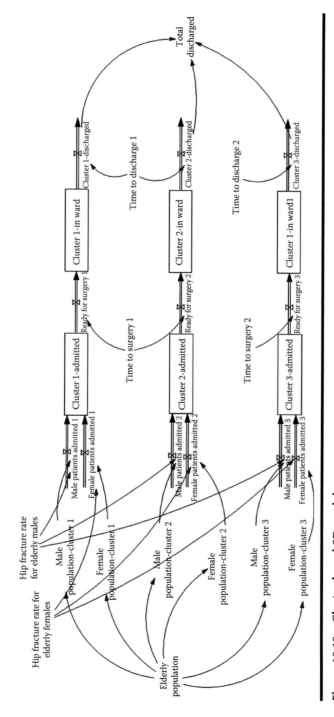

Figure 10.19 Cluster-based SD model.

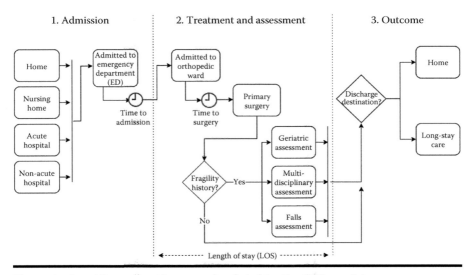

Figure 10.20   The patient journey simulated by the DES model.

Generation of Patients

The DES model made use of the projections produced by the SD model to generate individual patient entities. The generation process was implemented using the R language. The total number of generated patients reached around 30,000 for a simulated period of 10 years (i.e., 2017–2026). Table 10.8 presents the counts of elderly patients generated for every cluster.

Model Implementation

The DES model was mainly developed based on the empirical data acquired by the study. For instance, the probability distributions of patient attributes were set to mimic reality, as in the IHFD dataset. The simulation model was fully implemented using the R language. The source code can be accessed via our GitHub repository (Elbattah, 2017).

Table 10.8   Counts of Patients Generated by the DES Model

Patient Cluster	*No. of Simulation-Generated Patients*
Cluster 1	13,438
Cluster 2	10,782
Cluster 3	5,272

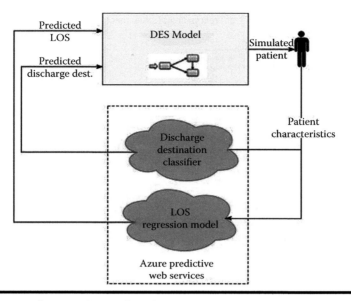

Figure 10.21 The experimental environment.

The main entity of the simulation model obviously represented the elderly patient. Each patient was assigned a set of attributes that characterized age, sex, area of residence, fracture type, fragility history and diagnosis type. The patients' characteristics varied based on the cluster they were assigned to. Further care-related factors (e.g., TTS) were considered on an individual basis as well.

The experimental environment consisted of two integrated parts. The DES model served as the core component. In tandem with the simulation model, ML models were then utilized to carefully predict the inpatient LOS and discharge destination for each elderly patient generated by the simulation model. The predictions were obtained from the ML models via web services enabled by the Microsoft Azure platform. Figure 10.21 illustrates the environment of simulation experiments where the DES was integrated with predictions from the ML models. In this manner, the ML models were employed to guide the simulation model with respect to the LOS and discharge destination of simulation-generated elderly patients.

Supervised Machine Learning: Predicting Care Outcomes

Overview of Predictive Models

As highlighted in Section 6, our approach endeavored to assist SM with ML. At the outset, the unsupervised data clustering helped to design the SD model structure and behavior to depict the real system in an accurate manner. In this section, we go through the development of supervised ML models (e.g., regression and classification), which were utilized to guide the DES model.

Based on the IHFD patient records, ML was employed to make predictions on important outcomes related to the patient's journey. At the micro level, the ML models were aimed at addressing patient entities generated by the simulation model. The ML models included a regression model for predicting the LOS, and a binary classifier for predicting discharge destinations. The predicted discharge destination included either home or a long-stay care facility (e.g., nursing home). The ML models were developed using the Azure ML Studio.

Significance of Predictive Models

The LOS and discharge destination are of vital importance within healthcare schemes. For instance, the LOS was considered as a significant measure of patient outcomes, as adopted by many studies such as (O'Keefe et al., 1999), and (Englert et al., 2001). From an operational perspective, the LOS was recognized as a valid proxy to also measure the consumption of hospital resources (Faddy and McClean, 1999; Marshall and McClean, 2003). Other studies (Brasel et al., 2007) argued that there was an evident correlation between the LOS and discharge destination. Further, it was reported by Johansen et al. (2013) that the LOS represents a major segment of the overall cost of hip fracture treatment.

Similarly, the early prediction of discharge destinations can be of significant benefit to care planning. On a population basis, the aggregated predictions can help estimate the desired capacity of long-stay care facilities (e.g., hospices and nursing homes) over the simulation period (i.e., 10 years).

Training Data

As mentioned previously, the study acquired a dataset of the IHFD repository. The IHFD dataset was used for training both the regression and classification models. Initially, we explored the variables that can serve as features for training the ML models. Based on our intuition, many irrelevant variables were simply excluded (e.g., admission time and discharge time). Table 10.9 lists the variables initially considered as candidate features.

Data Preprocessing

This section describes the data preprocessing phase conducted prior to training the ML models. The preprocessing procedures included: removing outliers, scaling features, tackling data imbalances, and extracting features. Removing outliers and scaling features were already conducted before building the clustering model, as

Table 10.9 Variables Explored as Candidate Features

Variables Explored		
Source hospital	Admission type	Discharge code
Residence area	Patient gender	Discharge status
Admission source	Hospital transferred from	Hospital transferred to
Age	LOS	ICD-10 diagnosis
Admission trauma type	Admission via ED	Fracture type
Pre-fracture mobility	Fragility history	Specialist falls assessment
Multi-rehabilitation assessment		

elaborated in Sections 8.1 and 8.2. Therefore, this section only explains the procedures performed for tackling imbalances and extracting features.

Feature Extraction

As mentioned in Section 8.3, the TTA and TTS represent important quality care-related factors for the hip fracture care scheme. The ML models utilized the TTA and TTS features, which were extracted earlier during the clustering model development.

Tackling Data Imbalances

Learning with imbalanced datasets was continuously recognized as one of the principal challenges for ML (Yang and Wu, 2006; Galar et al., 2012). The negative impacts of imbalanced data on the prediction accuracy were emphasized by numerous studies such as Japkowicz and Stephen (2002), and Sun et al. (2009).

In our case, the training data suffered from imbalanced class distributions, where a particular class predominated the dataset. In particular, imbalanced training samples were pronounced for patients who had LOS longer than 18 days. Likewise, imbalanced samples could be observed for patients who were discharged to home. Moreover, training samples for male patients, and particular age groups were obviously underrepresented.

Galar et al. (2012) identified two strategies to deal with the imbalance problem including algorithm-level approach and data-level approach, such as undersampling or oversampling. The oversampling technique (Chawla et al., 2002) was adopted in our case. The underrepresented samples were resampled at random until they contained sufficient examples. Figures 10.22 and 10.23 show the histograms of LOS and discharge destination respectively, where imbalanced distributions can be clearly observed.

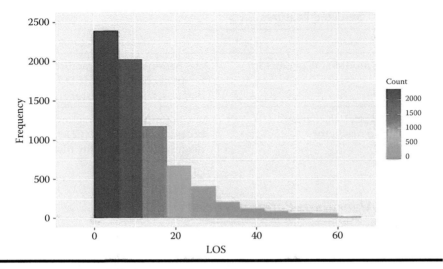

Figure 10.22   Data imbalance (LOS variable).

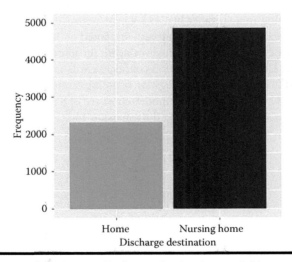

Figure 10.23   Data imbalance (discharge destination variable).

Feature Selection

The dataset initially contained 38 features; however, not all of them were relevant. Intuitively irrelevant features were excluded. In addition, the most important features were decided based on the technique of permutation feature importance (Altmann et al., 2010). Table 10.10 presents the set of features used by both models.

Table 10.10   Selected Features

Prediction Model	Selected Features
Regression model-LOS predictor	Age, patient gender, fracture type, hospital admitted to, ICD-10 diagnosis, fragility history, TTS, TTA
Classifier model-discharge destination predictor	Age, patient gender, fracture type, hospital admitted to, ICD-10 diagnosis, LOS, fragility history, TTS

Learning Algorithm: Random Forests

The random forests algorithm (Breiman, 2001) was used for regression and classification as well. A random forest consists of a collection of decision trees $\{h(x, \Theta_k), k = 1,...\}$, where the $\{\Theta_k\}$ are independent identically distributed random vectors and each tree casts a unit vote for the most popular prediction at input x. The main idea is that for the kth tree, a random vector Θ_k is generated, independently of the past random vectors $\Theta_1,..., \Theta_{k-1}$ but with the same distribution; and a tree is grown using the training set and Θ_k, resulting in a classifier $h(x, \Theta_k)$ where x is an input vector. Figure 10.24 portrays a simple example of a random forest composed of 3 decision trees.

Predictions are made through weighted voting for the most confident predicted class, and the trees that have a higher prediction confidence will have a greater weight in the final decision of the ensemble. The aggregation of voting can be done by a simple averaging operation as in Equation 10.3. Our models consisted of 8 decision trees. Table 10.11 presents the parameters used for training the regression and classifier models.

$$p(c \mid v) = \frac{1}{T} \sum_{t=1}^{T} p_t(c \mid v) \tag{10.3}$$

where $p_t(c \mid v)$ denotes the posterior distribution obtained by the t-th tree.

Predictors Evaluation

The predictive models were tested using a subset from the dataset described in Section 10.7. The randomly sampled test data represented approximately 40% of the dataset. The prediction error of each model was estimated by applying 10-fold cross-validation. Tables 10.12 and 10.13 present evaluation metrics of the regression and classifier models respectively. Further, Figure 10.25 shows the Area Under the Curve (AUC) of the classifier model.

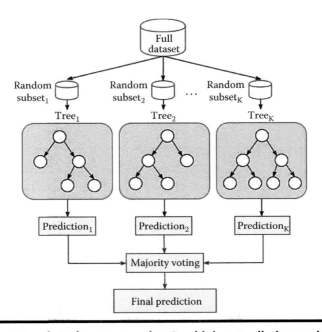

Figure 10.24 Random forest example: Combining predictions using majority voting.

Table 10.11 Parameters of Random Forests

No. of decision trees	8
Max. depth of decision trees	32
No. of random splits per node	128

Table 10.12 Average Accuracy Based on 10-Fold Cross-Validation (LOS Regression Model)

Relative absolute error	≈0.30
Relative squared error	≈0.17
Coefficient of determination	≈0.83

Table 10.13   Average Accuracy Based on 10-Fold Cross-Validation (Discharge Destination Classifier)

AUC	≈0.87
Accuracy	≈80%
Precision	≈81%
Recall	≈79%
F1 score	≈80%

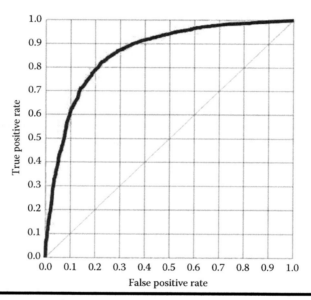

Figure 10.25   Classifier AUC = 0.865 (discharge destination classifier).

Results and Discussion

In line with the questions of interest, the simulation model output was interpreted in terms of the following: the expected number of discharged patients, the inpatient LOS, and the expected demand for long-stay care facilities as a discharge destination. Figure 10.26 plots the projections of elderly patients discharged from 2017 to 2027. It can be clearly observed that cluster 1 and cluster 2 steadily included the largest proportions of elderly patients. This adequately corresponded to the three clusters discovered by the ML clustering experiments, as shown earlier in Figure 10.16.

Figure 10.28a plots a histogram for the overall LOS expected over the simulated period for the three patient clusters. The histogram shows that the majority of

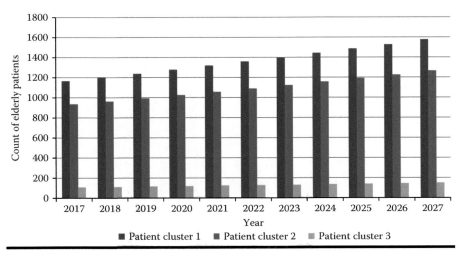

Figure 10.26   Cluster projections.

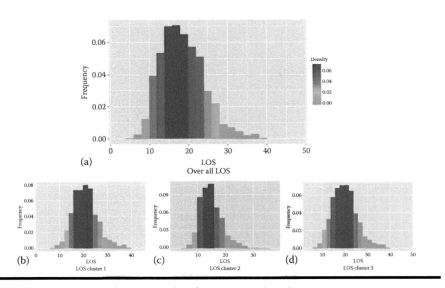

Figure 10.27   LOS experienced for the simulated patients.

patients experienced an LOS in the range of 10–30 days. In this regard, the model seemed to largely mimic reality, especially after excluding outliers. Moreover, Figures 10.27b, c, and d show the LOS with respect to every cluster individually. It turned out that cluster 1 and cluster 3 shared a similar distribution of the inpatient LOS, which to tended to be relatively longer compared to cluster 2 patients.

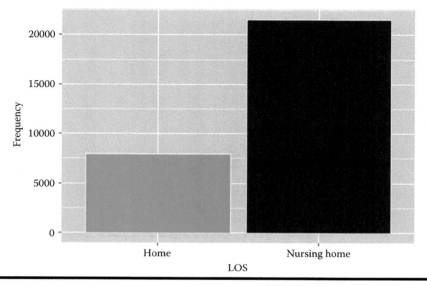

Figure 10.28 Expected demand for discharge destinations.

The similar LOS distribution might be due to including the same age group (i.e., aged 80–100). This can translate into the importance of considering early intervention schemes for the category of more elderly patients.

Furthermore, the simulation output anticipated a significant demand for long-stay care facilities (e.g. nursing homes) over the simulated period. As shown in Figure 10.28, the overall number of patients discharged to long-stay care is around 22,000 compared to only 7,000 of home-discharged patients. This pronounced difference raises an inevitable need for planning the capacity of nursing homes or similar residential care facilities.

Model Verification and Validation
Model Verification

To validate the logic and suitability of the simulation model, a set of verification tests (Martis, 2006) was conducted throughout the simulation model's development, as follows:

- *Structure-verification test*: The model structure was checked against the actual system. Specifically, it was verified that the model structure was a reasonable representation of reality in terms of the underlying patient clusters, and associated elderly populations.

■ *Extreme conditions test*: The equations of the simulation model were tested in extreme conditions. For example, the flows of elderly patients were set to exceptional cases (e.g., no elderly population aged 60 or over).

■ *Parameter-verification test*: The model parameters and their numerical values were inspected if they largely corresponded to reality. Specifically, the probability distributions of patient attributes (e.g. age, sex, and fracture types) were compared against those derived from the IHFD dataset.

Model Validation

According to Law (2008), the most definitive test of a simulation model's validity is comparing its outputs to the actual system. Similarly, we used the distribution of discharge destinations as a measure of the approximation between the simulation model and the actual healthcare system.

On one hand, Figure 10.29 provides a histogram-based comparison between the actual system and the simulation model regarding the discharge destination. The comparison showed that the distributions of the actual system and simulation output were largely similar. However, the comparison revealed that the model slightly underestimated and overestimated the proportion of patients discharged to homes and long-stay care facilities respectively.

Future Directions

A further direction is to consider more sophisticated ML techniques to extract the knowledge underlying further complex systems or scenarios. For example, it would be interesting to investigate how simulation models can be integrated with deep learning (DL). DL (LeCun et al., 2015) significantly helped approach hard ML problems such

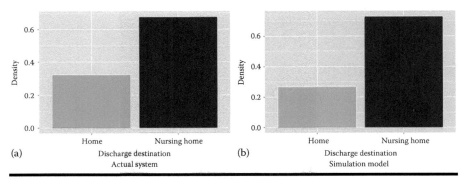

Figure 10.29 Histograms of the discharge destination output from the actual system and simulation model.

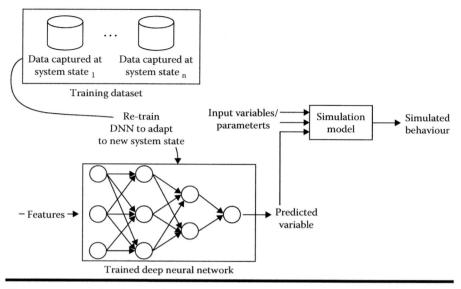

Figure 10.30 Adapting to the new system states via predictions from a trained DNN.

as speech recognition and visual object recognition for example. The capabilities of DL allow computational models that are composed of multiple processing layers to learn representations of data with multiple levels of abstraction. The multiple processing layers can effectively represent linear and nonlinear transformations.

We conceive that simulation models can be assisted by predictions from deep neural networks (DNN) trained to capture the system knowledge in a mostly automated manner. For instance, Figure 10.30 illustrates a simulation model where a variable (i.e., predicted variable) is input to the model. That variable can be predicted using a DNN, which was trained to capture the new system state. In this way, the DNN can be continuously trained in case of the arrival of new data that echo new system states or conditions.

Study Limitations

A set of limitations should be acknowledged as follows:

- More big data-oriented scenarios can better present the potentials of integrating SM and ML.
- The clustering of patients was based on a mere data-driven perspective. Adding a clinical viewpoint (e.g., diagnosis, procedures) may group patients differently.

■ All insights delivered by the study should be carefully considered as potential correlations limited by the dataset size and the number of variables included.
■ The IHFD dataset included records coming from public hospitals only.
■ The rate of hip fractures was assumed to be constant over the simulated interval; however, it might increase or decrease in reality.
■ Due to lack of information, the study could not distinguish between patients discharged to nursing homes or other residential care (e.g., rehabilitation institutions).

Conclusions

The integration of simulation modeling and ML can help address further complex questions and scenarios of analytics. We believe that the present work contributes in this direction. The study can serve as a useful example of how ML can assist the practice of modeling and simulation at different stages of model development.

From a practical standpoint, it was also attempted to deliver useful insights in relation to healthcare planning in Ireland, with a particular focus on hip fracture care. The insights were provided based on a set of simulation models along with ML predictions. At the population level, simulation models were used to mimic the flow of patients, and the care journey, while ML provided accurate predictions of care outcomes at the patient level.

References

Altmann, A., Toloşi, L., Sander, O., and Lengauer, T. (2010). Permutation importance: A corrected feature importance measure. *Bioinformatics*, 26(10), 1340–1347.

Barga, R., Fontama, V., Tok, W. H., and Cabrera-Cordon, L. (2015). *Predictive Analytics with Microsoft Azure Machine Learning*. New York: Apress.

Beyer, M. A., and Laney, D. (2012). *The Importance of "Big Data": A Definition*. Stamford, CT: Gartner, pp. 2014–2018.

Borshchev, A. (2013). *The Big Book of Simulation Modeling: Multimethod Modeling with AnyLogic 6*. Chicago, IL: AnyLogic North America.

Brailsford, S. C., and Hilton, N. A. (2001). A comparison of discrete event simulation and system dynamics for modelling health care systems. In *Proceedings of ORAHS*, Glasgow, Scotland, pp. 18–39.

Brasel, K. J., Lim, H. J., Nirula, R., and Weigelt, J. A. (2007). Length of stay: An appropriate quality measure? *Archives of Surgery*, 142(5), 461–466.

Breiman, L. (2001). Random forests. *Machine Learning*, 45(1), 5–32.

British Orthopaedic Association. (2007). *The Care of Patients with Fragility Fracture*. London, UK: British Orthopaedic Association, pp. 8–11.

Cellier, F. E. (1991a). Qualitative modeling and simulation: Promise or illusion. In *Proceedings of the 23rd Conference on Winter Simulation*. IEEE Computer Society, pp. 1086–1090.

Cellier, F. E. (1991b). *Continuous System Modeling*. New York: Springer-Verlag.

Chawla, N. V., Bowyer, K. W., Hall, L. O., and Kegelmeyer, W. P. (2002). SMOTE: Synthetic minority over-sampling technique. *Journal of Artificial Intelligence Research*, 16, 321–357.

Cleary, J., Goh, K. S., and Unger, B. W. (1985). Discrete event simulation in prolog. In *Proceedings of the Artificial Intelligence, Graphics and Simulation*. San Diego, CA

Cohen, K. J., and Cyert, R. M. (1965). *Simulation of Organizational Behavior. Handbook of Organizations*. Chicago, IL: Rand McNally, pp. 305–334.

Cooper, C., Campion, G., and Melton, L. 3. (1992). Hip fractures in the elderly: A world-wide projection. *Osteoporosis International*, 2(6), 285–289.

CSO. (2017). Retrieved from http://www.cso.ie/en/statistics/population/.

Davenport, T. H., and Harris, J. G. (2007). *Competing on Analytics: The New Science of Winning*. Boston, MA: Harvard Business Press.

De Mauro, A., Greco, M., and Grimaldi, M. (2015). What is big data? A consensual definition and a review of key research topics. In *AIP Conference Proceedings* (Vol. 1644, No. 1). AIP, pp. 97–104.

Demchenko, Y., De Laat, C., and Membrey, P. (2014). Defining architecture components of the big data ecosystem. In *Proceedings of the 2014 International Conference on Collaboration Technologies and Systems* (*CTS*). IEEE, pp. 104–112.

Dijcks, J. P. (2012). *Oracle: Big Data for the Enterprise*. Oracle White Paper, Oracle Corporation.

Dodds, M. K., Codd, M. B., Looney, A., and Mulhall, K. J. (2009). Incidence of hip fracture in the republic of Ireland and future projections: A population-based study. *Osteoporosis International*, 20(12), 2105–2110.

Elbattah, M. (2017). Retrieved from https://github.com/Mahmoud-Elbattah/DESModelImp.

Elbattah, M., and Molloy, O. (2016). Coupling simulation with machine learning: A hybrid approach for elderly discharge planning. In *Proceedings of the 2016 annual ACM Conference on SIGSIM Principles of Advanced Discrete Simulation*. New York: ACM, pp. 47–56.

Ellanti, P., Cushen, B., Galbraith, A., Brent, L., Hurson, C., and Ahern, E. (2014). Improving hip fracture care in Ireland: A preliminary report of the Irish hip fracture database. *Journal of Osteoporosis*, 2014, 7.

Englert, J., Davis, K. M., and Koch, K. E. (2001). Using clinical practice analysis to improve care. *The Joint Commission Journal on Quality Improvement*, 27(6), 291–301.

Faddy, M. J., and McClean, S. I. (1999). Analysing data on lengths of stay of hospital patients using phase-type distributions. *Applied Stochastic Models in Business and Industry*, 15(4), 311–317.

Fishwick, P. A. (1995). *Computer Simulation: The Art and Science of Digital World Construction*. Gainesville, FL: Department of Computer and Information Science and Engineering, University of Florida.

Fishwick, P. A., and Modjeski, R. B. (2012). *Knowledge-Based Simulation: Methodology and Application* (Vol. 4). Berlin, Germany: Springer Science & Business Media.

Forrester, J.W. (1968). *Principles of Systems*. Cambridge, MA: MIT Press.

Fortune. (2012). What data says about us. *Fortune*, p. 163.

Futó, I., and Gergely, T. (1986). TS-PROLOG, a logic simulation language. *Transactions of the Society for Computer Simulation International*, 3(4), 319–336.

Galar, M., Fernandez, A., Barrenechea, E., Bustince, H., and Herrera, F. (2012). A review on ensembles for the class imbalance problem: Bagging-, boosting-, and hybrid-based approaches. *IEEE Transactions on Systems, Man, and Cybernetics, Part C* (*Applications and Reviews*), 42(4), 463–484.

Gandomi, A., and Haider, M. (2015). Beyond the hype: Big data concepts, methods, and analytics. *International Journal of Information Management*, 35(2), 137–144.

Harper, P. R., and Shahani, A. K. (2002). Modelling for the planning and management of bed capacities in hospitals. *Journal of the Operational Research Society*, 53(1), 11–18.

HIPE. (2015). Retrieved from http://www.hpo.ie/hipe/hipe_data_dictionary/HIPE_Data_Dictionary_2015_V7.0.pdf.

HSE. (2014a). Retrieved from http://www.hse.ie/eng/services/publications/olderpeople/Executive_Summary_-_Strategy_to_Prevent_Falls_and_Fractures_in_Ireland%E2%80%99s_Ageing_Population.pdf.

HSE. (2014b). *Annual Report and Financial Statements 2014*, Dublin, Ireland: Health Service Executive (HSE).

Huang, Y. (2013). Automated simulation model generation. Doctoral dissertation, TU Delft, Delft University of Technology.

Huang, Z. (1998). Extensions to the K-means algorithm for clustering large data sets with categorical values. *Data Mining and Knowledge Discovery*, 2(3), 283–304.

IBM. (2017). Retrieved January 15, 2017, from http://www.ibmbigdatahub.com/infographic/four-vs-big-data.

Intel. (2013). Peer Research report: Big data analytics. Retrieved February 10, 2017, from http://www.intel.com/content/www/us/en/big.

Jain, A. K. (2010). Data clustering: 50 years beyond K-means. *Pattern Recognition Letters*, 31(8), 651–666.

Japkowicz, N., and Stephen, S. (2002). The class imbalance problem: A systematic study. *Intelligent Data Analysis*, 6(5), 429–449.

Johansen, A., Wakeman, R., Boulton, C., Plant, F., Roberts, J., and Williams, A. (2013). *National Hip Fracture Database: National Report 2013*. London, UK: Royal College of Physicians.

Kuipers, B. (1986). Qualitative simulation. *Artificial Intelligence*, 29(3), 289–338.

Lane, D. C. (2000). *You Just Don't Understand Me: Modes of Failure and Success in the Discourse Between System Dynamics and Discrete Event Simulation*. London, UK: LSE OR Department Working Paper LSEOR 00–34, London School of Economics and Political Science.

Laney, D. (2001). *3D Data Management: Controlling Data Volume, Velocity and Variety*. Stamford, CT: META Group Research Note, 6, 70.

Lattner, A. D., Bogon, T., Lorion, Y., and Timm, I. J. (2010). A knowledge-based approach to automated simulation model adaptation. In *Proceedings of the 2010 Spring Simulation Multiconference*. Orlando, FL: Society for Computer Simulation International, p. 153.

Laursen, G. H., and Thorlund, J. (2016). *Business Analytics for Managers: Taking Business Intelligence beyond Reporting*. Hoboken, NJ: John Wiley & Sons.

Law, A. M. (2008). How to build valid and credible simulation models. In *Proceedings of the 40th Conference on Winter Simulation*. Miami, FL: Winter Simulation Conference, pp. 39–47.

LeCun, Y., Bengio, Y., and Hinton, G. (2015). Deep learning. *Nature*, 521(7553), 436–444.

Liberatore, M., and Luo, W. (2011). INFORMS and the analytics movement: The view of the membership. *Interfaces*, 41(6), 578–589.

Manyika, J., Chui, M., Brown, B., Bughin, J., Dobbs, R., Roxburgh, C., and Byers, A. H. (2011). *Big Data: The Next Frontier for Innovation, Competition, and Productivity.* Washington, DC: McKinsey Global Institute.

Maoz, M. (2013). How IT should deepen big data analysis to support customer-centricity. *Gartner G00248980.*

Marshall, A. H., and McClean, S. I. (2003). Conditional phase-type distributions for modelling patient length of stay in hospital. *International Transactions in Operational Research,* 10(6), 565–576.

Martis, M. S. (2006). Validation of simulation based models: A theoretical outlook. *The Electronic Journal of Business Research Methods,* 4(1), 39–46.

Mashey, J. R. (1997). Big data and the next wave of infraS-tress. In *Computer Science Division Seminar.* Berkeley, CA: University of California.

Meadows, D. H. (2008). *Thinking in Systems: A Primer.* White River Junction, VT: Chelsea Green Publishing.

Melton, L. J. (1996). Epidemiology of hip fractures: Implications of the exponential increase with age. *Bone,* 18(3), S121–S125.

Microsoft. (2013). The big bang: How the big data wxplosion is changing the world—Microsoft UK enterprise insights blog-site home-MSDN blogs. Retrieved February 2, 2017, from http://blogs.msdn.com/b/microsoftenterpriseinsight/archive/2013/04/15/big-bang-how-the-big-data-explosion-is-changing-theworld.aspx.

Minelli, M., Chambers, M., and Dhiraj, A. (2012). *Big Data, Big Analytics: Emerging Business Intelligence and Analytic Trends for Today's Businesses.* New York: John Wiley & Sons.

Mortenson, M. J., Doherty, N. F., and Robinson, S. (2015). Operational research from taylorism to terabytes: A research agenda for the analytics age. *European Journal of Operational Research,* 241(3), 583–595.

Newell, A., and Simon, H. A. (1959). *The Simulation of Human Thought. Current Trends in Psychological Theory.* Pittsburgh, PA: University of Pittsburgh Press.

NIST. (2015). *S. 1500–1 NIST Big Data Interoperability Framework (NBDIF): Volume 1: Definitions.* Gaithersburg, MD.

NOCA. (2014). Irish hip fracture database national report 2014. National Office of Clinical Audit (NOCA), Dublin, Ireland.

NOCA. (2017). Irish hip fracture database. Retrieved from https://www.noca.ie/irish-hip-fracture-database.

O'Keefe, G. E., Jurkovich, G. J., and Maier, R. V. (1999). Defining excess resource utilization and identifying associated factors for trauma victims. *Journal of Trauma and Acute Care Surgery,* 46(3), 473–478.

Oberkampf, W. L., DeLand, S. M., Rutherford, B. M., Diegert, K. V., and Alvin, K. F. (2002). Error and uncertainty in modeling and simulation. *Reliability Engineering & System Safety,* 75(3), 333–357.

Oxford Dictionary. (2017). Retrieved from https://en.oxforddictionaries.com/definition/big_data.

Patel, V. R., and Mehta, R. G. (2011). Impact of outlier removal and normalization approach in modified K-Means clustering algorithm. *IJCSI International Journal of Computer Science Issues,* 8(5), 331–336.

Powell, J., and Mustafee, N. (2014). Soft OR Approaches in problem formulation stage of a hybrid M&S study. In *Proceedings of the 2014 Winter Simulation Conference.* IEEE Press, pp. 1664–1675.

Rabelo, L., Cruz, L., Bhide, S., Joledo, O., Pastrana, J., and Xanthopoulos, P. (2014). Analysis of the expansion of the panama canal using simulation modeling and artificial intelligence. In *Proceedings of the 2014 Winter Simulation Conference*. IEEE Press, pp. 910–921.

Rashwan, W., Ragab, M., Abo-Hamad, W., and Arisha, A. (2013). Evaluating policy interventions for delayed discharge: A system dynamics approach. In *Proceedings of the 2013 Winter Simulation Conference*. IEEE Press, pp. 2463–2474.

Rindler, A., McLowry, S., and Hillard, R. (2013). *Big Data Definition. MIKE2. 0, The Open Source Methodology for Information Development*. Retrieved from http://mike2. openmethodology.org/wiki/Big_Data_Definition.

Rosenberg, M., and Everitt, J. (2001). Planning for aging populations: Inside or outside the walls. *Progress in Planning*, 56(3), 119–168.

Shannon, R. E. (1975). *Systems Simulation: The Art and Science*. Englewood Cliffs, NJ: Prentice-Hall.

Shreckengost, R. C. (1985). Dynamic simulation models: How valid are they? In Beatrice, A. R., Nicholas, J. K., and Louise, G. R. (Eds.), *Self-Report Methods of Estimating Drug Use: Current Challenges to Validity*. Rockville, MA: National Institute on Drug Abuse Research Monograph, 57, pp. 63–70.

Soetaert, K. E. R., Petzoldt, T., and Setzer, R. W. (2010). Solving differential equations in R: Package deSolve. *Journal of Statistical Software*, 33.

Stanislaw, H. (1986). Tests of computer simulation validity: What do they measure? *Simulation & Games*, 17(2), 173–191.

Sterman, J. D. (1994). Learning in and about Complex Systems. *System Dynamics Review*, 10(2–3), 291–330.

Sun, Y., Wong, A. K., and Kamel, M. S. (2009). Classification of imbalanced data: A review. *International Journal of Pattern Recognition and Artificial Intelligence*, 23(4), 687–719.

Sweetser, A. (1999). A comparison of system dynamics (SD) and discrete event simulation (DES). In *Proceedings of the 17th International Conference of the System Dynamics Society* Wellington, New Zealand., pp. 20–23.

Taylor, S. J., Khan, A., Morse, K. L., Tolk, A., Yilmaz, L., and Zander, J. (2013). Grand challenges on the theory of modeling and simulation. In *Proceedings of the Symposium on Theory of Modeling & Simulation-DEVS Integrative M&S Symposium*. San Diego, CA: International Society for Computer Simulation.

Tolk, A. (2015). The next generation of modeling & simulation: Integrating big data and deep learning. In *Proceedings of the Conference on Summer Computer Simulation*. Chicago, IL: International Society for Computer Simulation.

Tolk, A., Balci, O., Combs, C. D., Fujimoto, R., Macal, C. M., Nelson, B. L., and Zimmerman, P. (2015). Do we need a national research agenda for modeling and simulation?. In *Proceedings of the 2015 Winter Simulation Conference*. IEEE Press, pp. 2571–2585.

UK Government. (2014). Retrieved January 4, 2017, from https://www.gov.uk/government/uploads/system/uploads/attachment_data/file/389095/Horizon_Scanning_-_Emerging_Technologies_Big_Data_report_1.pdf.

Visalakshi, N. K., and Thangavel, K. (2009). Impact of normalization in distributed K-means clustering. *International Journal of Soft Computing*, 4(4), 168–172.

Ward, J. S., and Barker, A. (2013). Undefined by data: A survey of big data definitions. *arXiv preprint arXiv:1309.5821*.

Wickham, H. (2009). *ggplot2: Elegant Graphics for Data Analysis*. New York: Springer Science & Business Media.

Wu, C., Buyya, R., and Ramamohanarao, K. (2016). Big data analytics= machine learning+ cloud computing. *arXiv preprint arXiv:1601.03115.*

Yang, Q., and Wu, X. (2006). 10 Challenging problems in data mining research. *International Journal of Information Technology & Decision Making*, 5(4), 597–604.

Zhong, J., Cai, W., Luo, L., and Zhao, M. (2016). Learning behavior patterns from video for agent-based crowd modeling and simulation. *Autonomous Agents and Multi-Agent Systems*, 30(5), 990–1019.

Zikopoulos, P. C. (2013). *Dirk deRoos, Krishnan Parasuraman, Thomas Deutsch, David Corrigan, and James Giles. Harness the Power of Big Data.* New York: The IBM Big Data Platform.

Chapter 11

Intangible Dynamics: Knowledge Assets in the Context of Big Data and Business Intelligence

G. Scott Erickson and Helen N. Rothberg

Contents

Introduction

The explosive growth of interest in Big Data and analytics has caught the attention of knowledge management (KM) researchers and practitioners. While some overlaps are clear between the fields, previous interest in data or information from the KM field was rather limited. Over the development of the KM discipline, highly valued knowledge was explicitly differentiated from purportedly uninteresting data. That perspective is changing.

This chapter will take an updated look at the nature of intangible assets, not just knowledge but related intangibles including data, information, and wisdom or intelligence. In clarifying the similarities and differences, we believe both practitioners and scholars can more effectively understand these potentially valuable assets, providing a greater opportunity to exploit them for competitive advantage. A structured understanding of all intangible assets, and their inter-relationships, can lead to better applications and results.

After a review of existing theory concerning types of intangibles and approaches for exploiting or managing them, we will turn to some data-driven examples of how a fuller understanding of the combined disciplines can help with strategic decision-making. In particular, we will consider how a firm's intangibles competencies translate into success in its industry sector as well as other directions for potential growth.

A Wider View of Intangibles

Even before the advent of KM and intellectual capital (IC) as well-established disciplines, an understanding existed of the differences between data and knowledge. In particular, Ackoff's (1989) data-information-knowledge-wisdom (DIKW) hierarchy has often been referred to as the basis for much of our discussion concerning all intangible assets, defining these as progressing from data and information to knowledge and wisdom. As KM developed in the 1990s, early scholarship often referred to these definitions for context. The field used the hierarchy to firmly entrench exactly what is a valuable intangible (knowledge, gained from learned know-how, placing information in context) versus what is not (Zack, 1999). The latter includes data (unorganized observations), information (categorized data), and wisdom (generally too personal to manage as knowledge). IC writings made some similar distinctions, often referring to knowledge as the key to IC but sometimes including some of the other intangibles in wider descriptions (Stewart, 1997).

As a result, much of the attention focused on knowledge had to do with defining its characteristics and exploiting it to its best advantage. Interestingly, the creation of knowledge, though acknowledged, did not receive as much attention as sharing the knowledge once it exists and is recognized. Part of this is due to the original insight from Nonaka and Takeuchi (1995) that individuals learn, not organizations. Knowledge comes from individuals learning from experience, gaining hands-on know-how. That can be hard for the organization to manage, though it can create the right conditions for learning, support learning, and provide opportunities to employ new learning. Even so, the study of innovation and from where new processes and new products come is generally seen as a separate field. While acknowledging that the knowledge needs to be created, almost exogenously, KM has focused heavily on what to do with it afterwards. The emphasis is on exchange and sharing, and most of what we study is centered on how to do that effectively.

If one considers the main streams of the literature over the years, scholars and practitioners have tended to try to understand the nature of knowledge assets, their different types, and then what to do with them based on that nature. On the KM side, the attention usually starts with the tacit or explicit distinction (Nonaka and Takeuchi, 1995; Polanyi, 1967); how explicit knowledge can be more easily codified, captured, and distributed by the organization while tacit knowledge is more personal and harder to communicate. From this difference, the field has developed approaches to better share the knowledge, from IT-based tools for more explicit knowledge to person-to-person tools for more tacit knowledge (and variations in between) (Brown and Duguid, 1991; Matson et al., 2003; Thomas et al., 2001). Other variables concerning the knowledge itself, from complexity to stickiness can also matter in how it is managed (Kogut and Zander, 1992; McEvily and Chakravarthy, 2002; Zander and Kogut, 1995). Similarly, variations across organizations such as social capital, social networks, and so forth can also be a factor in how firms plan for and use knowledge (Liebowitz, 2005; Nahapiet and Ghoshal, 1998).

On the IC side, we have seen similar things. Here, although tacit or explicit is also recognized, the distinction between human (job-related knowledge), structural (organization-related knowledge), and relational (knowledge concerning external parties such as customers) capital is also important (Bontis, 1999; Edvinsson and Malone, 1997; Stewart, 1997). Similar to KM, the idea is that when organizations better understand the nature of their assets, they can better manage them. Consequently, knowledge audits can be conducted (Marr et al., 2004) and deeper dives into the details of the IC pieces (e.g., human capital) can be done to better assess and then better exploit these knowledge assets. Increasingly specific differences between human capital have become a subject of some extensive interest in some scholarly circles, especially as regards unique individuals or unique talents ("stars") that might require individual attention (Crocker, 2014; Groysberg et al., 2010).

All of this can be important if one buys into the notion of intangibles having an impact on the competitive potential of organizations. Flowing from the resource-based view of the firm positing that an organization must possess some unique, sustainable differentiator to be successful (Wernerfelt, 1984), researchers in the field believe intangible assets such as knowledge or IC can be that unique resource (Grant, 1996; Teece, 1998). There is a school of thought that intangibles are the only really, truly unique resource to be found given the ubiquity of basic labor, capital, information technology (IT), and so on, in today's world. The only unique thing most firms having going for them is to be found in the heads of their key employees.

Consequently, the fields of KM and IC are fairly established in terms of defining the nature of knowledge assets, how they vary, and what to do with them. But over the past few years, the rapid growth of interest in big data applications has led to questions of how such matters relate to knowledge, particularly KM that has been dismissive of the value of "less-developed" data and information. Is there a means to reconcile the disciplines and include the full range of intangible assets in our discussions?

Big Data and Business Analytics/Intelligence

One of the important things to understand about Big Data and business analytics is the lack of understanding among much of the business community about what the terms actually mean. For those coming from a KM or IC perspective, the definitions are actually straightforward even if there is a general lack of appreciation for the potential of the fields. Big Data refers to the stream of bytes generated by information systems observing behavior, communications, or other activities. When organized, it can become information. The phenomenon of Big Data has to do with the exploding amounts of data suddenly available, characterized in the field by the "3 Vs" of volume, variety, and velocity (Laney, 2001). Ever increasing amounts of data (volume) in an expanding number of forms, including unstructured data such as text, images, and video (variety), are being generated, transmitted, and received ever more quickly (velocity) (Chen et al., 2012; McAfee and Brynjolfsson, 2012). The plunging costs of storing and processing such data are feeding these trends and are the reason so many participants are excited about the cloud—the manifestation of cheap data handling.

But this first stage of Big Data generally limits itself to collection, organization, and transmission of the data. There is exchange, and there can be action, but the data don't require additional analysis to be useful. Big Data, in and of itself, simply tracks observations of interest. It is exchanged across a company or across its extended network, but often just reported on dashboards or similar devices. Key performance indicators (KPIs) might be tracked and given special attention. The KPIs may even have tolerances attached to them, much the same as quality management systems track performance data to ensure adherence to standards. Something out of tolerance or with an abnormal pattern to the data might be further investigated or acted upon. Algorithms may be established so that the system automatically makes a decision to intervene. The types of data collected are often operational, transactional, or communications (especially digital media), but anything generating large amounts of data can be included in these big data platforms.

What is important to understand, however, is that these systems generally only organize and transfer the data. Someone or something may monitor the results but the data are not fundamentally changed in any way. Further analysis on the available data is not really a part of these systems. That capability is the interesting part of the second generation of Big Data, big data analytics. Also referred to as business intelligence, the analytics function subjects Big Data to analysis leading to new insights, effectively learning something new from studying the data. Specialized techniques, enabled by cutting-edge analysis software, are available for this type of work characterized by data mining, predictive analytics, clustering, and other approaches. Typically, analysts are doing a deep dive into the data looking for nonobvious correlations or differences. The new insights or learning require highly skilled individuals using advanced data processing and statistical analysis tools, though there is also some potential in artificial intelligence to uncover and test new insights in such environments.

Big Data and business analytics are major and connected advances in the use of intangible assets to better steer and run organizations but, as just explained, they are two different things. Big data systems collect and transfer observations, chiefly monitoring them and acting on deviations. Big data analytics organizes and subjects the data to deep analysis with advanced methods, looking for new insights. And in the context of how we see intangibles, how we define them, and how we make recommendations for applying them, those differences can be important. Important enough to reconsider how we characterize the entire field of intangible assets even if those in KM want to continue to refer to them as knowledge or as IC.

Reimagining the Intangibles Hierarchy

Earlier in this chapter, we mentioned the traditional DIKW hierarchy, the progression from data through information, knowledge, and wisdom. While this has had its uses over the years, it has also fallen into some disuse in the KM and IC fields specifically because they focus so precisely on just the knowledge component. Moreover, the advances in the fields of Big Data and business analytics, while having some clear connections to parts of the hierarchy (data) are less well-explained by the entire structure. If we are to incorporate both traditional KM and IC assets and the new potential we see from Big Data and from business analytics and intelligence, a different framework might have a better fit.

One possibility flows from Kurtz and Snowden's (2003) Cynefin sense-making framework. Though not necessarily intended in the manner of an intangibles hierarchy, it can and has been applied in such a manner (Rothberg and Erickson, 2017; Simard, 2014). The driver of the framework is understanding and processing different environments, from ordered to chaotic, but the real contribution is in describing those situations in a context related to identifying and collecting intangible inputs, making sense of them, and acting appropriately. In that sense, the best place to start might be with the dimensions of Cynefin.

Kurtz and Snowden divide the world into sectors based on the dimensions of centralization and hierarchy versus connections and networks ("meshwork" in one extension). In this conceptualization, do intangibles flow into the center (potentially redistributable), across individuals without going to the center, both, or neither? The different scenarios both help to explain how intangibles might be best understood and used most effectively. The basic ideas, including their potential contributions and applications, can be explained as follows:

■ *Known*: This domain is defined by centralization and boundary network connections. It is ordered, with observable, notable patterns. The intangibles are easily shared across all levels and locations and, if captured by the central core, easily dispersed back out into the network. This pattern fits what organizations do with data and information. It is only bytes, easily transferred from individual to individual or from the data-gathering point to the database at the center of the firm. Patterns are monitored, divergence from expected trends is noted and acted upon, and decisions are generally based on quick action, pattern recognition, and operational outcomes more than long-term strategic planning.

Kurtz (2009) characterizes the known domain similar to a clock mechanism: complicated but with observable, understandable repeated patterns. Verdun (2008) suggests that a traditional control hierarchy is appropriate for managing such circumstances as the operation is orchestrated through central direction and is focused on individual jobs and places. That being said, if the same environment recurs throughout the organization (multiple plants, multiple sales offices), data or information about the processes is certainly useful to centralize or just redistribute across individuals or locations. Going back to Kurtz, the scenario would have both central connections (feeding the center) and constituent connections (across the periphery).

■ *Knowable*: The knowable domain shows centralization but fewer connections to other networks. It is still ordered, but the patterns are less obvious or yet unknown, and require some experience to recognize and learn from. Once knowledge is created, it can be codified and shared with the center, again available for dispersion back out across the wider network. When appropriate, other individuals may see opportunities to apply the knowledge to their own situations. Patterns are noticed and individuals learn from them, though the decisions largely center on operations and customer relationships, day-to-day effectiveness, more than long-term insights. This scenario better fits what organizations do with explicit knowledge.

Verdon describes this as a blend of orchestrated and adaptive, with individuals given structure by the center but also learning things on their own as their circumstances dictate. Any new knowledge is fed back to the center from the periphery, then distributed back out to others in similar environments who may be able to use it. But it is not a center-driven knowledge exchange;

it depends on contributions from outside. Hence, a "modular heterarchical" structure connecting the periphery with the center and with communications running in both directions makes more sense. Kurtz characterizes knowable as having prominent central connections but less distinct constituent ones.

■ *Complex*: Complexity is a domain without centralization but with peripheral network connections. Order is reduced, fewer patterns are apparent. Knowledge is once again created by individuals but is more difficult to communicate. As a result, centralization or capture by the core organization is also difficult. But person-to-person knowledge exchange throughout the extended network is quite possible, so one sees the application of more tacit knowledge-related techniques such as communities of practice, mentoring, apprenticeships, and so on, being applied. Decisions largely remain about operational matters but now depend more on the level of the individuals doing the exchange. If at the senior manager level, they may well be about how to do strategic planning better, so operational decisions are still pre-eminent but strategic and tactical decisions now begin to enter the conversation. As noted earlier, tacit knowledge fits this domain.

Kurtz describes the complex domain as being unordered and with patterns but patterns that can unpredictably change. These may be simple patterns, discernible but based on large numbers of interactions or observations. Once individuals develop some knowledge, it is difficult to communicate that understanding to someone else without some context as contingencies, ability to anticipate pattern changes, and so forth make the knowledge harder to transfer. Verdun recommends a "customized—responsible autonomy" structure for management, providing individuals with the freedom to act given their own circumstances and the resulting ability to learn from their actions. Connections can be made wherever they are useful, across individual participants anywhere in the organization, so the context is person-centric and agile and adaptable. Kurtz' representation would be more constituent-connected and less central-connected. It is in this environment that Brown and Boudes (2005), commenting on the Cynefin approach, note the beginnings of its emphasis on a diversity of opinions, action outside the bureaucracy, and a willingness to accept complexity rather than standardized solutions. Problem solving in these environments starts to come from non-linear thinking and different perspectives.

■ *Chaos*: Chaos represents a lack of centralization and a lack of networking. Order and patterns are gone. Insights are created by unique individuals but they may find it difficult or even impossible to share their creative processes with others. If one is familiar with Snowden's work, this is a scenario where subconscious communication might be used (storytelling) but even that can be difficult. The learning is unique to the individual and often stays with them. Since such rare individuals are often valuable to the entire enterprise, now one sees the level of decision-making rise, and longer-term strategic, tactical, or innovative insights are more common. This environment not only fits C-suite executives but also the big data analysts discovering unexpected insights buried in the data lakes available to data-heavy organizations. Much like R&D innovators or charismatic leaders, what they do may be difficult to understand, let alone teach someone else to do. The results of their efforts might be perfectly understandable but how they get to that is opaque and potentially unlearnable. This domain characterizes intelligence, the ability to draw new insights by analyzing a variety of intangible assets.

This domain comes full circle to a non-bureaucratic approach to learning and creating intangibles of value to the organization. Kurtz now suggests intricate patterns (still liable to unexpectedly change) based on simple operations. She uses the example of a government project looking to identify weak signals in public information and news reports to predict a future environment, specifically noting how hard it was to teach participants to think outside their normal frames of reference. In chaos, it is the unexpected insight or spark of creativity that is critical, the different perspective, and while that solution might be communicable or teachable, the process to get to it might not be. Verdun uses the term improvisation and does not really recommend any structure, suggesting any and all connections may be useful. Kurtz would note the lack of both central connections and constituent connections. As just noted, this is where Snowden's emphasis on stories instead of rational linguistic explanation make sense as the learning, if possible, might take place subconsciously rather than consciously. And, once again, Brown and Boudes' emphasis on diverse perspectives, outside-the-network individual thinking, and non-conformity would resonate for organizations trying to operate and create or manage intangibles in such environments.

Following the logic, a new hierarchy evolves (Rothberg and Erickson, 2017; Simard, 2014). This time, the flow starts with data and information, moves through explicit knowledge, tacit knowledge, and on to insight and intelligence. Per Kurtz' (2009) own commentary on the framework, this hierarchy does not imply differences in value, the higher levels of the hierarchy are not necessarily better or worse than lower levels. But there are differences in environments, and the hierarchy helps with understanding how to manage under the different circumstances. Given the new circumstances engendered by Big Data and business analytics and intelligence, included now at opposite ends of this intangible asset hierarchy, we believe this structure helps to both explain the nature of the different environments with different intangibles present while also suggesting guidelines for handling them.

If the system is only designed to manage Big Data ("known" scenario), IT handling the exchange of data and information is enough. As noted, this may include dashboards to track metrics of high interest or established algorithms to react to results not at expected levels. But other than the way they are organized, these intangibles do not really require further analysis or more learning to have an impact.

If explicit knowledge is critical to the firm or within its industry ("knowable" scenario), then job-specific individual learning takes place that improves internal processes, external relationships, or other aspects of the organization. The appropriate system for managing such knowledge assets can be IT-based so as to leverage them through further distribution though decision makers. These are the KM systems that have been discussed in the literature for decades and are likely familiar to those with experience in the field.

If tacit knowledge is prominent ("complex" scenarios), individuals again learn and develop new knowledge but it is of a more personal nature and more difficult to communicate to others, especially in digital form. It can be done (e.g., case histories or best practices stored in IT systems) but is often more effective if transmitted on a person-to-person basis as with communities of practice, mentoring and apprenticeships, storytelling, and so forth. Again, these tacit KM approaches are well-known in the discipline; established tools for transferring more complex knowledge requiring context or some other deeper explanation.

Finally, if intelligence is evident in the firm or its industry ("chaos" scenarios), then a system should be available to more effectively manage the process. As noted in the discussion earlier, the path to individual insight and creativity may be unique and extremely hard, if not impossible to communicate to others. The intelligence just comes from the unique individual who can analyze intangible inputs of all types and find something new in the chaos. On the other hand, the literature does provide some guidance in how firms can structure themselves to encourage such individual brilliance, specifically learning organizations designed to feed inputs to individuals or diverse teams charged with finding insight by applying their different perspectives (Argyris, 1992, 1993, 1977; Senge, 1990). In another setting, we have called such structures intelligent learning organizations (Rothberg and Erickson, 2017), and they provide a targeted analytics process intended precisely at generating intelligence in a way Big Data or KM systems, focused only on exchange, do not.

Assessment of the Intelligence Hierarchy in Organizations

Much of this thought and the methodology behind it has been developed much more extensively in other research (Erickson and Rothberg, 2012, 2016; Rothberg and Erickson, 2017) and the intent is not to repeat that here. What is useful to note is that different sorts of intangibles exist, they are successfully applied in different types of circumstances, and so a strategic approach to identifying and developing the right intangibles to enhance a firm's competitiveness is required (Rothberg and Erickson, 2005). And we have a structured way to analyze those circumstances as well as appropriate metrics for identifying and then understanding them.

■ High levels of Big Data, knowledge, and intelligence suggest that all of the intangibles are present and important. Industries with these sorts of metrics include pharmaceuticals, software, and similar environments that not only require substantial human capital but also efficient and high-quality processes (structural capital) and effective, data- and relationship-driven marketing (relational capital). Pharmaceuticals, for example, employs Big Data in its operations, distribution, and marketing and sales; explicit knowledge in its operational processes and sales; but more tacit knowledge and intelligence in its labs and in analytics explorations to improve more strategic and tactical approaches throughout the firm (R&D, production, distribution channels, marketing, and consumer experience).

■ High levels of Big Data and evidence of intelligence but low knowledge metrics suggest that Big Data and analytics and insight (perhaps some tacit knowledge bleeding in as well) are important, but explicit knowledge is not. Big Data is available and value can come from conducting deeper analysis (data mining, predictive analytics) on it. But the ability to discover such insights is rare and difficult to share with others. Operations, transactions, and marketing are routine (structural capital and relational capital) and human capital is found in the rare analyst or team who can find something new in the data. Financial services are a typical industry, awash in data and aggressive competitive intelligence but very low knowledge metrics compared to other industries.

■ Moderate to high levels of big data and high knowledge metrics but little intelligence activity suggest that data and knowledge are involved in scale and efficiency, making processes of all kinds run better but that there is little in the way of new insights or innovation. Particularly notable is that we believe these types of industries have considerable explicit knowledge but probably less tacit. What is known can be and is widely shared. Human capital (supply chain, operations, and transactions), structural capital (in the same areas), and relational capital (customer relationships) are all important but generally operational rather than strategic. Characteristic industries are consumer goods and retail. Big brands (high relational capital) and operational size and efficiency (human and structural capital) are key competitive elements.

■ Moderate to low levels of big data, low knowledge, and low intelligence (very little of anything of value). What data and knowledge (explicit) there is should be supported, but operational, transactional, and marketing activities are well-understood and have little to be discovered that might improve them. Much of the IC is probably structural or relational, investing in high human capital employees never hurts but does not necessarily generate a great return either, so cost and benefit should be considered. These are usually mature manufacturing industries or regulated industries like utilities.

Again, the main idea in this chapter is not to recreate the existing logic and evidence behind these distinctions but to introduce them as evidence of the wide range of circumstances facing decision makers thinking about how to manage all these intangibles. Intangible dynamics are situational and call for different approaches. Circumstances may even call for different metrics and some of the all-encompassing measurement systems discussed in the literature may need to be broken back down to lower levels to really help. But, as a first pass, we already have some established metrics that allow an evaluation of different industries and individual firms' competitiveness within them.

Big Data can be measured directly by looking at relative data holdings in different industries. One often-cited report was done by McKinsey Global Services (Manyika et al., 2011). The industry categories are fairly broad but the data provide a good idea of where Big Data is prevalent (financial services, communications and media, discrete and process manufacturing, utilities) and where it is not (construction, consumer and recreation services, professional services, healthcare providers). Moreover, the study adds some details about the nature of the data and how it might be used in these industries.

KM metrics, of course, remains an area of considerable discourse and lack of consensus (Firer and Williams, 2003; Tan et al., 2007). Sveiby (2010) noted over 40 different approaches that have been applied by researchers or practitioners, specifically categorizing them by level of measurement (entire organization or individuals) and method (financial statements to survey instruments). In this chapter, as we try to evaluate multiple firms in multiple industries; the low-level, individual employee approaches do not really work. What has been used and what works in cross-industry comparisons are objective financial results, even if there can be issues with trying to apply them too precisely. Specifically, a modified Tobin's q can be used (market capitalization to book value or market capitalization to assets) to make broad distinctions between industries and even individual firms (Erickson and Rothberg, 2012; Tobin and Brainard, 1977). As noted above, however, we've come to the conclusion that a high Tobin's q, all by itself, probably only identifies explicit knowledge. Tacit knowledge does not scale enough to show up reliably in such metrics though it may also be present. But we can employ another metric that might flag it when used in combination with Tobin's q.

Intelligence, as defined in this chapter, signifies organizations seeking unique, creative insights from analytical techniques. Again, this may be highly individual and impossible to teach. Or it may be from a form of intelligent learning organization established to funnel inputs and provide room for learning and interchange between dedicated analysts with different perspectives. This capability can be particularly hard to identify, but we've found that organizations with an intelligence capacity in one area (e.g., R&D laboratories, competitive intelligence) know something about how to do it and so can extend the capability to other areas. So, again, an established metric exists of identifying competitive intelligence (CI) operations in organizations as a proxy for the ability to establish an analytics capability, drawing new insights or creativity from intangibles inputs (Erickson and Rothberg, 2012).

Measuring Intangible Asset Scenarios

The specific empirical data included in this chapter follows these general descriptions and, as a whole, can be used to identify, quantify, and explain the intangibles capabilities and related domains in different industry sectors. As characterized by the hierarchy and the prominent intangibles characteristics present, we have a foundation for assessing the competitive conditions in each sector. Once we possess a broad idea of what intangibles are typically present and necessary for success, we can also form an idea of the intangibles management capabilities are required to compete (especially for new entrants). And with deeper analysis, we can start to understand the actual activities, along the value chain, for example, that are related to the intangibles. We'll cover these shortly.

Regarding intangible assets such as data and information, the amount of data storage per industry is a good start to identifying the amount of useful data generated and kept. As noted, the McKinsey study (Manyika et al., 2011) which reports such numbers, has been very useful for a number of studies specifically on this question. Even though the study is starting to show its age, there's no reason to believe that the numbers, at least on a relative basis, have changed appreciably since publication. For our purposes, and even more interesting for assessing what firms need in particular sectors, we can employ the report's breakdown of data storage in each industry by firm and discover the average big data holding on a company-by-company basis. Here, we employ McKinsey's metrics on data holdings (terabytes) per industry per firm.

What large holdings of data indicate are systems capable of generating big data and some sense of value seen in processing it and holding it. As indicated in Table 11.1, the generation is often by machine, observing an activity or behavior, then feeding the observations into the data system. So typical areas generating such data would include supply chain logistics, operations (goods production, service delivery), transactions (retail, financial), communications (direct to customer, social media monitoring), and similar high-volume environments. As also noted, the systems used only to manage Big Data typically do no more than collect, process, and report the results on an ongoing basis. Given KPIs, established tolerances, performance levels, or other "triggers" will demand attention from a decision maker or an embedded algorithm will initiate a planned reaction. The presence of high big data holdings (high data storage per firm) and only high big data holdings indicates that only the big data system is needed and employed (a "known" sense-making environment). If high big data holdings are present, along with other intangibles, Big Data contributes to those, and one of the other sense-making environments exists.

As indicated earlier, to assess intangible assets beyond data, such as knowledge, we can employ a modified Tobin's q. The value of this metric is that it can be easily calculated and compared across firms, really the only way to do a cross-industry study such as this with large numbers of per firm observations. Reported here are two variations on the Tobin's q, market capitalization to book value and market

Table 11.1   Intangibles, Metrics, and Sense-Making Environments

	Data/Information	*Explicit Knowledge*	*Tacit Knowledge*	*Insight/ Intelligence*
Domain	Known	Knowable	Complex	Chaos
Definition	Collected observations, bytes	Learned know-how, codifiable, sharable	Learned know-how, personalized, hard to codify or share	Nonobvious insights from individual brilliance or applied analysis
Source	Data collectors, often machines	Individuals, learning over time	Individuals, learning over time	Individual or team, insightful discovery, often from analysis
Exchange	IT systems	IT systems or personal interaction	Personal interaction or IT systems	Personal interaction, if at all
Range	Extended organizational network	Organization	Groups	Individual
Management	Big data systems	Knowledge management systems (IT-centered)	Knowledge management systems (person-to-person centered)	Intelligence systems, learning systems
Metrics	Data storage per firm	Modified Tobin's q	Modified Tobin's q and CI activity	CI activity
Indicates	If only data storage per firm, only Big Data is present.	If only a high Tobin's q, only explicit knowledge is present.	High Tobin's q and high CI indicate tacit knowledge is present (along with other intangibles).	High CI activity, if only with high data storage, indicates business analytics present. If by itself or with other variables, other intelligence is present.

capitalization to asset value. We've tested both variations in multiple circumstances (Erickson and Rothberg, 2012), including specifically to evaluate whether critical differences are apparent (Erickson and Rothberg, 2013b). Where differences occur, they are often in circumstances where liabilities are considerable. The key difference in the metrics is performance given ownership of the assets (cap/book, where book is assets less liabilities) versus performance given asset levels regardless of ownership (cap/assets). Generally, we've found the two metrics agree, and one can push ahead with analysis with some confidence when that occurs. It is of interest and worth discussion when they do not (generally cap/assets is the more reliable one), and researchers would want to look more deeply into the reasons behind the result.

Here, both are reported. We have created two databases, one including financial results from 2005 to 2009 for all firms reporting on North American exchanges, the other for the same conditions but covering 2010–2014. The first database includes almost 2,000 firms and over 7,000 observations. The second is still under construction, some targeted results were pulled for this chapter, but looks as if it will generate similar descriptors from similar numbers of firms and observations. From the full databases, only yearly observations with firms generating $1 billion in revenue were included, as were industry sectors with at least twenty observations (sectors were sometimes expanded if tangential sectors seemed complementary).

High ratings on the Tobin's q variations (benchmarked against the database means) indicate significant intangible assets, interpreted here and elsewhere as knowledge, and an ability to turn a given amount of tangible assets into value for the firm. More value relative to assets suggests a better job in creating and managing intangibles. But it is relative to an industry or industry sector. The level of tangible assets and what intangibles can add to them varies dramatically by circumstance. Further, as discussed earlier, the skill in managing the knowledge-related intangible assets must be impactful enough to drive market capitalization, the perceived value of the firm. Tacit knowledge in a few heads isn't going to register in billion dollar corporations except in very unique circumstances. But explicit knowledge, scaled up and applied usefully throughout the firm might be. Specifically, skillfully-applied explicit knowledge can result in more efficient supply chains, operations, or marketing relationships (including brand equity). That shows up in the Tobin's q metric. So a high value on this measure is indicative of substantial explicit knowledge, all other things being equal, especially if paired with high Big Data stocks. The two together can drive efficiency and indicate productive operations and effective marketing, even without a high intelligence value associated with the firm.

That intangible asset characterized as intelligence can be observed in a metric used to assess competitive intelligence in firms. Here, we can apply data harvested from continuous surveys done by CI consultancy Fuld & Company across the years 2005–2009 (matching the terms of the McKinsey data and the first Tobin's q dataset). A later dataset is again in the process of construction but no concrete results are yet available. Early indications are that relative levels of CI activity have remained consistent over time.

This particular dataset is based on two pieces of data used to construct an index. The number of participants from firms within an industry sectors is one input. An industry sector with eighteen survey respondents is a quite different environment than one with one or none. Further, each survey asks respondents to report on the professionalism of their CI operation, from just starting to highly proficient. Again, this indicator shows considerable differences between inexperienced, perhaps one-person initiatives versus large, seasoned teams, even if only one respondent was included from each firm. Combined, the index presented gives a sense of both quantity and quality of CI efforts in a given industry sector. As discussed earlier, if a firm is capable of intelligence activity in one area (competitive intelligence), it has the ability to practice it in others (marketing intelligence and analytics, business intelligence and analytics).

Interpretation of the indicator, as with others, depends on circumstances and what other intangibles seem to be present. Just the intelligence metric, with no other intangibles (very rarely seen) or with Big Data (common in identifiable sectors), indicates an insight or analytic capability. Firms in these sectors are able to create new insights, particularly when analyzing big data inputs. Such capabilities are often unique to a particularly gifted individual or team within the firm and, unlike knowledge assets, can be extremely difficult if not impossible to teach to others. The results of the insight are replicable but the process to get there is associated only with the individual or team generating it. Such circumstances are seen when firms have gifted leaders or creative talent, the situations when "stars" are uniquely valuable and need to be managed as a rare asset in and of themselves.

Alternatively, when the intelligence indicator is present with a high Tobin's q or knowledge metric, tacit knowledge is also present. Again, unique insight comes from individuals, as indicated by the intelligence result. But at least some of these learnings occur more often and are more scalable—they occur in circumstances that make them scalable. That doesn't mean intelligence itself isn't available as well, just that tacit and explicit knowledge are apparent, too. Pharmaceutical companies, for example, typically have all the intangibles: Big Data, both types of knowledge, and intelligence. They have huge amounts of data from operations, their distribution channels, and their customers (providers, retailers, insurers, and end consumers). They also have explicit and tacit knowledge on how to make their processes better, their marketing relationships better, and their labs run better. But they still need individual insight as well, in the creativity and brilliance of their researchers as well as for occasional strategic concepts or new marketing directions.

In the end, we can identify the different circumstances where different intangibles are more or less prominent in an industry sector. We can spot industries using only Big Data, transferring and monitoring it to support operational and marketing efficiency. We can spot industries where explicit knowledge is critical, where employees learn to make improvements in logistics, operations, marketing, and any number of other functions of the firm. These learnings can then be scaled up for greater impact through appropriate KM systems. We can spot industries where tacit knowledge and intelligence are also present, either by themselves (more intelligence-oriented) or with explicit knowledge (indicating both tacit knowledge and intelligence).

As a consequence, decision makers gain a better appreciation of what intangibles contribute to competitive advantage in such sectors and how they might stack up against rival firms. The indicators also provide guidance as to where to focus attention in order to take a deeper look. So explicit knowledge might be indicated, for example, but what is the nature of it? Is it in manufacturing goods or service delivery? Is it in distribution efficiency? Is it in close customer relationships or deep customer knowledge? By knowing where to look and having a rough idea what to look for, strategists stand a better chance of truly understanding what intangibles are necessary and how they need to be managed.

Intangible Assets and Metrics: Illustrative Applications

To demonstrate, we'll look at a few applications, including pertinent data indicating different intangibles scenarios. In some ways, the conclusions might seem obvious, but we picked distinctive examples allowing clear discussion even at a fairly general level. The purpose here is demonstration of process, an actual strategic application might show less differences between sector values and would require considerably more depth in the follow-on data and information collection. A deeper dive could lead to a better understanding of what the difference in intangible assets mean to competitiveness.

The scenarios all revolve around insurance firms. In part, this is because we have profit pools already established in past studies by Bain (Eliades et al., 2012; Schwedel and Rodrigues, 2009) and don't need to construct new ones just for this chapter. The healthcare profit pool is included as is the financial services profit pool, both of which include insurance sectors. To this, we add a more general discussion on the free-for-all in the automobile industry as demonstrated by reported interest in self-driving cars. Although not a profit pool yet, this new direction will include multiple players from diverse sectors, and, to our eyes, it can provide another perspective on how an understanding of intangible dynamics can inform strategic decision-making.

Healthcare

Previous work drew on the metrics of the entire healthcare sector. As just noted, a report from Bain (Eliades et al., 2012) already defined sectors that might be included in a healthcare profit pool. And an earlier study (Erickson and Rothberg, 2013a) already gathered and presented some of the intangibles metrics we have been discussing. These are shown in Table 11.2, along with updated knowledge data from a new database.

Table 11.2 Intangible Dynamics from Healthcare Sectors

| Sector (SIC #) | Cap/Book | | Cap/Assets | | Intelligence Index | Data Storage/Firm (terabytes) |
	10–14	05–09	10–14	05–09		
Pharmaceuticals (2834)	4.49	4.39	2.06	1.94	64	967 (discrete mftg) 831 (process mftg)
Diagnostic/ Biological (2835/6)	7.88	4.37	2.73	2.41	14	
Surgical, medical, dental instruments (384)	3.09	3.43	1.73	2.02	19	
Lab, optical, measuring instruments (3826/9)	3.90	3.54	1.79	1.52	7	
Drugs, wholesale (512)	3.78	3.44	1.09	1.16	6	536
Medical, dental, hospital equip/ supplies (5047)	2.83	3.86	1.44	1.43	2	
Hospitals (806)	1.79	3.23	0.53	0.61	3	370
Offices, clinics, various health (801/5)	1.60	1.79	0.72	0.66	1	
Drugs, retail (5912)	6.94	2.38	1.04	1.23	1	697
Accident/health insurance (632)	1.22	2.29	0.34	0.80	36	870
Global means	3.59	2.68	1.06	1.02		

Source: Updated from Erickson, G.S., and Rothberg, H.N., *J. Inform. Knowl. Manag.*, 12, 1350031, 2013a.

One of the interesting things about healthcare is representation from just about all the different combinations of intangible assets and the domains they represent, starkly showing how this methodology is capable of distinguishing differences between the sectors. Further, because the sectors are so different in what they do, the activities behind the metrics and the competitive considerations they pose are fairly clear, even at the very general levels we'll be discussing in this chapter. Finally, insurance has such unique readings compared to other sectors that its place in the industry framework is also a straightforward matter to discuss.

As noted in the table, several industry sectors have relatively high levels of all the intangibles. These include pharmaceuticals, instruments, and diagnostics. They have highly complex, even chaotic environments demanding accumulation of big data on research processes, including clinical trials, on highly scrutinized operational processes, and on marketing relationships (both complicated distribution chains, third-party payors, and end consumers). Their processes and marketing need to be highly efficient and often meet rigorous quality standards explaining the presence of substantial explicit knowledge as well as some tacit knowledge related to more dramatic improvements. And they still require a steady stream of creative insights, especially in terms of new products but also including new strategic directions concerning targeted segments (organizational and consumer), distribution, marketing, and responses to competitor initiatives. Consequently, ample evidence exists of tacit knowledge and intelligence.

Alternatively, some sectors show very little of anything at all. Hospitals and other providers have low levels of big data holdings (though these will likely rise as the mandate to digitize records takes hold) and there is little evidence of their doing anything with them. Tobin's q levels are well under average, so not much is going on with knowledge development, particularly explicit, and intelligence activity appears to be virtually nonexistent as well. As pressure builds for U.S. providers to become more efficient, the data may feed into improvements in explicit knowledge levels and efficiency, but that is also hard to do when operations are not necessarily repeatable or systematized (each patient may have their own unique situation and personal attention is hard to develop knowledge around). Tacit knowledge is likely present with so many highly skilled workers but, again, it may be hard to share or to exploit. At least to any degree that would show up in the metrics.

Some sectors show considerable explicit knowledge (and Big Data) but no real evidence of intelligence activity. As discussed, that likely also means limited tacit knowledge in spite of the high knowledge metric. The prominent example of this is the retailers. Along with wholesalers, they generate considerable amounts of data from their supply chains, their in-store transactions, and their marketing relationships, especially those through loyalty programs and close connections to consumers. Further, those that have added pharmacy benefits management (PBM) to their activities are developing considerable databases that can draw consumers even closer. Much of their attention is on continuous improvements in efficiency, running ever tighter supply chains and retail operations and ever better relationships with consumers, physicians, and insurers paying the bills (explicit knowledge). While everyone likes a new, insightful idea, their businesses aren't predicated on creativity or the types of big steps forward promised by intelligence or even tacit knowledge.

And, finally, we come to the insurers. Their metrics are indicative of sectors with a lot of data, evidence of considerable intelligence work, but very low knowledge metrics. This is one of those applications where it is useful to apply both versions of Tobin's q as the financial sector has considerable tangible assets, far beyond what might be seen in other sectors, and so might show artificial differences in the metric not reflected in actual practice. That would be true of the cap/asset metric but the cap/book also includes the liabilities associated with the massive tangible assets, much of which is borrowed in one form or another. So the fact that both metrics show very low KM activity or success suggests that the sector really is low when compared to other sectors.

The basic scenario with insurers, and it's something we've seen in just about all financial services firms, includes massive amounts of data on transactions, individual client conditions and activities, and marketing relationships. Explicit knowledge is hard to develop as processes are well-understood and hard to improve while marketing relationships are also well-developed but brand equity is troublesome (very few consumers are enthusiastic about their insurance companies). Small improvements in process or efficiency aren't important to the sector. What is sought are big new ideas, new insights for targeting consumer groups, understanding risk profiles of different groups, identifying irregularities in the data (fraud, changes in prevalent symptoms, changes in prescribing patterns, etc.). Intelligence is pronounced, including competitive intelligence—when new knowledge or intelligence is so rare, competitors are going to be interested in uncovering it as quickly as possible.

The core question we're going to be asking in this section is what opportunities for growth might be open to insurers, in this case health insurers. The nature of in-house intangibles indicates considerable holdings of big data of a very specific nature. There are also indications of experience and competency in managing Big Data as well as in analyzing it (the analytics and intelligence ratings). The insurance firms show no particular holdings of or competency in knowledge assets, either explicit or tacit. Consequently, there would likely be limited potential in venturing into sectors requiring explicit knowledge or the related efficiencies regarding operations, supply chains, or external relationships (supply chains, providers, retailers, or consumers). In this case, insurers would not have good prospects in taking on such sectors themselves, whether in entry into new sectors, acquisition, or other partnerships based on such success requirements.

What healthcare insurers may have to offer are the skills developed in not only managing Big Data but also in analyzing it for deeper insights (intelligence). This could be of considerable value to many of the players in sectors with which the insurers interact, including the manufacturers (pharmaceuticals, medical devices) and the retailers/PBM. Although some data would be duplicative, the insurers have a different focus on their data and on how they analyze it, potentially adding value as a collaborator. Further, the insurers may be of particular help in increasing the capabilities of partners in sectors not so good at managing the intangibles, particularly Big Data. Although there are obvious relationship issues to overcome with hospitals and with other providers (as insurers are the payors in most cases), sharing data, data analysis capabilities and techniques, and acquired insights could benefit all and seems likely to be one of the strategic avenues with the greatest potential payoff.

Financial Services

The financial services profit pool looks much different from that in healthcare. There is no dedicated supply chain or manufacturing, nor is there much in the way of distribution or marketing channels not taken of by the providers themselves. Consequently, the profit pool described by Bain consultants (Schwedel and Rodrigues, 2009) is much more horizontal, encompassing banking, investment houses, and insurers but little else. This description still includes a number of different industry sectors, as illustrated in Table 11.3 but not much of an industry value chain. A few distinctions could be made (e.g., insurance providers versus insurance brokers) but the effective point is made of how different this profit pool looks compared to the healthcare industry and how that impacts strategic choices facing participants, especially once one looks more deeply at the intangible assets.

As noted, the results show a very different pattern from what we saw in healthcare industry sectors. The two highlighted sectors are outliers but also somewhat different in what services they provide. The rest are all mainstream financial services, either providing retail banking (commercial banks and savings institutions), investment services (security brokers and Real Estate Investment Trusts (REITs) REITs), or insurance (life, health, property). All of these involve very large databases. These databases include customer knowledge such as personal characteristics, transaction records, and relationship tracking (including all communications). All also track data on economic indicators, market financial results, and other world, country, or industry-wide information. In the McKinsey report, financial services take up most of the top spots in big data holdings and by a wide margin.

The knowledge data are a different story. Again ignoring the shaded rows for now, the rest of the sectors are uniformly well under the average cap/book and cap/assets ratios seen across the databases. This is true when looking at both metrics. Here, cap/assets may be artificially low just because of the huge level of tangible financial assets held by these types of institutions—inflating the denominator of the ratio. But the same pattern is also seen in the cap/book ratio which would level out the bias by taking into account any financial assets borrowed or employed on behalf of a customer who actually owns them. If there is some artificial deflation of the results, it is minimal and certainly doesn't change the overall conclusions. In financial services sectors, there is very little evidence of knowledge, particularly explicit knowledge.

Table 11.3   Intangible Dynamics for Financial Services

Sector (SIC #)	Cap/Book		Cap/Assets		Intelligence Index	Data Storage/ Firm (terabytes)
	10–14	*05–09*	*10–14*	*05–09*		
Commercial banks (6020)	1.13	1.61	0.12	0.14	45	1931
Savings institutions (6035/6)	1.16	1.14	0.16	0.11	18	
Security brokers and dealers (6211)	1.63	1.52	0.36	0.19	13	3866
Investment advice (6282)	2.54	4.73	1.18	1.57	3	
Life insurance (6311)	0.80	1.12	0.10	0.11	22	870
Accident/ health, hospital and, medical services (6321/24)	1.22	1.49	0.34	0.48	54	
Fire, marine and casualty (6331)	1.00	1.22	0.28	0.34	29	
Insurance agents and brokers (6411)	1.91	3.94	1.28	1.21	4	
Real estate investment trusts (6798)	2.36	1.45	0.91	0.47	16	3866
Global means	3.59	2.68	1.06	1.02		

There is, however, considerable evidence of intelligence. Given the low knowledge metric, one would conclude that much of this rating is attributable to insight or analytical abilities present in these sectors but not tacit knowledge. Assuredly, new tacit knowledge or intelligence insights are rare, valuable, and come by means of a process difficult or impossible to teach others. Organizations may set up structures to feed intangible inputs, especially data, to analysts and encourage insight and creativity, but success depends on hiring and retaining the right people more than on how they are taught their responsibilities.

This squares with what we know about these industry sectors. As noted, they manage huge amounts of tangible financial assets. Doing so requires little knowledge about efficiency of supply chains, operations, or customer relationships (indeed, consumers often actively dislike their banks and insurance companies, so brand equity is minimal). They do know how to execute transactions. They do know how to report to customers, regulators, and shareholders. But much of that is established and well-known, there is little new to discover in the nature of explicit (or tacit) knowledge improving activities or relationships.

What can be new are creative insights, crafting new investment strategies or new insurance risk profiles and products. When new knowledge and insights are so rare, they take on additional value, so the rewards for successful analytics are considerable. This is seen in the level of intelligence metric. A new, unique strategy is highly valued so other firms are extremely interested in uncovering competitor discoveries as quickly as possible. The competitive intelligence levels are quite high, reflecting this reality as well as the considerable weight all of these financial services sectors place on new solutions discovered through an effective analytics or intelligence process.

The exceptions validate the rule. The shaded rows have both higher knowledge and lower intelligence metrics. Both of these sectors, investment advisor and insurance brokers, depend more on individuals with knowledge that can be applied to variable conditions (advising consumers according to their specific situations and needs). For insurance brokers, this is fairly obvious as agents actually selling the insurance tend to be smaller concerns with close, personal client relationships and a need to match the client with the right insurance. They learn what works for what clients in what circumstances may have access to Big Data but may not know what to do with it, and rely more on their own assessment and conclusions than on repeatable patterns of action.

The investment advisors are more complex and, in some ways, more interesting. The firms making up this sector are quite a mixed bag. They range from high-powered investment firms (Blackstone, Carlyle, Invesco) to retail brokerages (Charles Schwab, T. Rowe Price). Generally, the former group has low knowledge metrics, similar to those seen with security brokers. The emphasis is again on Big Data and unique insights more than repeatable explicit or tacit knowledge. The latter group has the higher knowledge metrics as they do provide trade executions to retail buyers, a repeatable pattern that can yield knowledge based on segmented groups of customers. These firms also depend on close relationships with clients, something that we know requires good levels of customer knowledge and is again usually apparent in high explicit knowledge results.

To return to the overall point of this section, what do these results tell us about the strategic potential of insurers in this industry? Unlike the healthcare industries, in which insurers had little in common with any of the other industry sectors, the results are quite similar here across the categories. One could see substantial movement across sectors here as the different players have similar intangibles competencies. All have some substantial level of Big Data and apparently know what to do with it. All also have some substantial level of analytical or intelligence ability and, again, apparently know what to do with it. While partnerships to share data and insights are possible, as in healthcare, vertical movements across sectors in a more aggressive manner are also possible.

One indicator of this potential comes from recent news reports about insurance firms moving into pension management. Based on the intangibles, this makes a great deal of sense. As repeatedly noted, Big Data is present and important throughout all these sectors, so insurers' capabilities in managing large databases concerning client descriptions, activities, and more macro trends would be applicable across sectors. Perhaps even more importantly, insurers deal routinely with both individual retirement plans and actuarial data. Their experience with such data and their ability to find new insights in it after conducting deep analysis would be well-suited to pension management. Most also have experience with finding suitable investment opportunities for held funds as well as efficiency in client relationships (even if not overly warm and fuzzy). The intangibles results show a strong fit for this sort of move.

Automated Driving

Although a profit pool study for this burgeoning industry hasn't yet been compiled, work has been done on active and announced participants so we do have some sense of who might be involved. Further, Navigant Research (Abuelsamid et al., 2017) has released a report assessing the prospects of known aspirants according to a checklist of success criteria. Other observers might adjust the criteria or provide different ratings for individual firms, but the structure is there for us to consider the field with respect to intangibles.

The Navigant report, based on public announcements of the represented firms, includes the industry sectors listed in Table 11.4. To these, we have added both Apple (no announcement but substantial evidence of intentions) and Intel, which recently announced its own interest and a partnership with Mobileye. Apple's financial filings listed it as Computers (SIC 357) during the earlier reporting period but switched to Phones (SIC 3663) during the latter. The company, of course, competes in both sectors though its emphasis and revenue and profit streams have changed. Both sectors are listed here.

Table 11.4 Intangible Dynamics for Automated Driving Sectors

Industry Sector (SIC)	Cap/Book		Cap/Assets		Intelligence Index	Data Storage/ Firm (terabytes)
	10–14	05–09	10–14	05–09		
Auto manufacturing (3711)	8.45	0.80	0.82	0.47	9	967
Auto components (Delphi, ZF) (3714)	11.40	2.10	1.04	0.39	10	
Computers (Apple) (357)	5.63	4.48	0.74	1.58	22	
Phones (Apple) (3663)	2.57	2.71	1.18	1.56	17	
Semiconductors (Intel) (3674)	4.00	4.80	1.66	1.97	11	
Software/Web (Alphabet/ Waymo, Baidu, nuTonomy) (7370)	5.02	3.48	2.30	2.08	19	1792
Auto rental (Uber) (7510)	3.28	1.64	0.43	0.30	3	801
Fire, marine, casualty insurance (6331)	1.22	1.00	0.34	0.28	29	870
Database average	3.59	2.68	1.06	1.02		

For the record, the Navigant "leaders," those best placed to compete in the new area, are mainly traditional auto manufacturers (Ford, GM, Toyota, BMW, Tesla, etc.). The reasons behind their conclusions are probably best seen in the criteria used for assessment, including both strategy variables (vision, go-to-market strategy, partners, production strategy, and technology) and execution variables (sales, marketing, and distribution; product capability; product quality and reliability; product portfolio; and staying power). The details don't match up exactly, but one could easily associate much of the strategy component with tacit knowledge and intelligence. Alternatively, execution will often have more to do with explicit (and some tacit) knowledge.

What does the data tell us? The automakers have relatively low knowledge levels across the board, except for the very high 2010–2014 cap/book ratio. In looking into the data, this result is chiefly due to the presence of Tesla and is not mirrored in the cap/assets ratio. This is a case where using both metrics is useful. Tesla is highly leveraged, so the book value in the first calculation is artificially low. Cap/asset is probably the better and seemingly more consistent metric to pay attention to in this case though Tesla's considerable store of intangible assets should also be kept in mind. The auto industry has improved markedly from the earlier to the later time period in the knowledge metrics, recall that the earlier period included the 2008 financial downturn. The intelligence variable is average (anything in double digits is getting into more aggressive intelligence activity). Big Data is present, in substantial quantities but at a level that is also about average across all industries. All in all, the auto manufacturers have some capability in explicit knowledge (improving but still below average), evidence of some tacit knowledge and intelligence, and some Big Data. But nothing is outstanding, especially relative to the other industries. The manufacturers should be competent at supply chains, manufacturing, distribution, and customer relationships. They should also have some abilities in more creative insights such as new product development. But there is also nothing here that would scare anyone else off, perhaps why the field is so full of potential competitors from different sectors and interesting combinations of old-line manufacturers and new-line software and other firms.

Tesla is a bit of a special case. As this was being written, the electric vehicle manufacturer's market cap passed GM's for the first time, making it the most valuable U.S. auto company. As noted, the firm's knowledge metric is considerably higher than other competitors, particularly in the cap/book version but also in cap/asset. Even though we can't measure it directly, the firm likely also has higher tacit knowledge and, perhaps, intelligence levels given its formidable research and development efforts and presence of key players such as the founder. Moreover, the firm has specialist knowledge in certain areas (batteries) that may or may not be a feature of automated driving. This methodology gives a nice snapshot of the full industry sector but does require a deeper dive into the individual characteristics of firms to fully understand what might be happening, and Tesla might be explored in more detail.

The next four sectors, highlighted in the table, are a mix of manufacturing and services (software). All have relatively high knowledge metrics though the phone sector is showing signs of decline. Knowledge in the efficiency of design, execution, and delivery of computers, phones, semiconductors, and software is important in all these fields, even when some parts of that chain are outsourced (e.g., Apple). A number of high-powered brands are also present in these sectors (Apple, Google, Intel, Microsoft), and so relational capital is also good and likely pushes up the knowledge metric. Intelligence scores are also high in all the sectors and big data is present. Essentially, firms in these sectors possess relatively high levels of all the intangible assets, potentially adaptable to different circumstances and making dangerous competitors of key players in each.

Uber, slotted in the auto rental sector, is another special case. To some extent, it is a special player in the rental sector with a different approach than traditional agencies (who usually rent the entire car rather than just the single ride). Here, its financials are not included in the results of the full sector as it was not publicly traded when the data was reported. On the other hand, most of the major players in the sector are moving to new service models, whether car sharing, ride sharing, shorter-term rentals (e.g., hourly), or some other variation. So Uber (and Lyft's) impact on the sector should start to be visible.

What the data in auto rentals show are low but increasing knowledge levels. Once again, given the large fleet of tangible assets that may be debt-financed, the cap/asset ratio may be the more accurate metric though both cap/asset and cap/book generally agree. Intelligence activity is very limited, and big data is not overly pronounced either. One area where Uber may have an impact is in Big Data as its network thrives on collecting rider and driver data, creating greater efficiency by matching supply and demand. And this may drive up the knowledge metrics over time as managers get a better handle on logistics and competitors are forced to catch up. But, overall, the intangibles capabilities of this sector just don't suggest a serious player. Ride-sharing firms may have interest in autonomous vehicles but seem more likely to be customers or partners for the technology rather than developers.

Which brings us to the insurance sector once again. Here we have included the Fire, Marine, Casualty sector that would include auto insurers but, again, recall that just about all insurers have similar intangible profiles. Once again, lots of data. Huge amounts of data compared to some other prospective participants in the field. And, again, not much in the way of explicit or tacit knowledge, according to the knowledge metric. But what is there is some intelligence ability combined with the Big Data, suggesting a competency in accumulating, processing, and finding insights in databases.

There is no indication of the interest of any insurance firm in direct participation in the autonomous driving field. That might be a good thing. There is also no indication in the metrics of any capability to participate in development of key pieces of the technology such as sensors, software, vehicle manufacturing, or so forth. The car insurance firms, however, do have a vested interest in the progress of the technology, especially in terms of how it impacts accident rates, their severity, and where liability might be placed. As such, the auto insurance firms might be very valuable collaborators in this field, contributing their big databases on drivers, driving activity, and, particularly, accidents. The software and artificial intelligence systems that will guide autonomous driving cars are being developed in a trial-and-error manner by putting the vehicles on the road and teaching them to recognize circumstances and make the right decisions. While this makes sense, the process could likely be enhanced by incorporating the data the insurers already have on accident circumstances and correct or incorrect choices made by drivers. The insurers have potentially valuable data. They also have the analytical abilities to properly study the data and find the insights that more direct participants in the autonomous driving field could find useful. Considerable press has suggested that safe, mistake-free, driverless cars could kill the insurance sector. But since their intangibles are something not necessarily matched in-house by many of the would-be competitors, the insurers may instead take a role in the development process and find a way to be useful partners heading into the future.

Conclusions

Even though relatively young disciplines, KM, and IC have already faced a couple of waves of growth and decline in interest over the past few decades. The advent of interest in Big Data and business analytics and intelligence presents another possible threat or opportunity for the field. In some ways, the burgeoning investment in big data systems, including the cloud, and associated business analytics efforts have left KM installations behind. The possibilities of gathering massive amounts of data and applying it to make better and more informed decisions at all levels are intriguing. And the promise may be more compelling than some of the results seen from KM installations.

But opportunities are also present. The central question in KM and IC has always been about the systems used to identify, measure, and manage intangible assets. With knowledge, those assets originated in individuals and the systems had to encourage the human resources of organization to contribute their knowledge while being open to employing that gained from others. The complexities of humans interacting with IT systems or other tools have been at the core of much of the discussion over the past 30 years.

From that perspective, both big data systems and business analytics efforts can be seen as extensions, with different kinds of intangibles. As such, KM learnings may have something important to contribute, not only in what the systems look like for managing the intangibles but also how to execute the human element that makes the systems actually work. Further, KM and IC have been employed to help understand competitive environments, how intangibles are assessed and exploited in different circumstances in different industries. These sorts of applications can also be extended to the wider range of intangibles, not just knowledge varieties but also data, or information, and intelligence.

This chapter has brought together work from a variety of disciplines to explore this strategic perspective in more detail. In particular, we have presented an updated conceptualization of the intangibles hierarchy, a step that helps to explain the similarities and differences between data and information, explicit knowledge, tacit knowledge, and insight and intelligence. Traditional KM and IC covers explicit and tacit knowledge but the extensions to data and information, with connections to Big Data, and to insight and intelligence with its connections to business analytics, are useful in framing the full picture of contemporary intangible asset assessment and usage. This perspective adds even more complexity to our understanding of competitive environments and what intangibles are most effective in them but it also brings us closer to the reality of today's business battlegrounds.

With a firm grasp of the full range of intangibles, we can also look to accurately measure them. In this chapter, we have demonstrated how to do so in specific industry sectors and, in some cases, by individual firms. One can get a sense of the competitive capabilities of those individual firms but here we have looked primarily at what it takes in terms of intangible assets to be effective in the different sectors. From a strategic planning perspective, this approach provides guidance as to what a firm needs in its own sectors as well as what it might add in others, including whether it possesses the intangibles levels and management skills to be competitive on its own in a new environment.

With that in mind, we have provided analysis of three industries, guided by profit pool structures identifying the key sectors in two of those industries. Given the presence or potential presence of a single industry sector, insurance, we can more easily see how the methodology provides guidance to decision makers. In healthcare and automated driving, the insurers do not have the intangibles to be competitive players but might have a role to play in providing Big Data and business analytics capabilities to other industry challengers from other sectors (who may not have such intangibles or skill in managing them). In financial services, the intangibles capabilities of all types of insurers look similar and are also very much like those in other sectors (banking, investment services). In such a case, insurers may have the potential to be serious competitors themselves in tangential sectors.

Such analysis, in a short chapter like this, is necessarily only at the surface. But the approach illustrates what decision makers in these industry sectors, with considerably more knowledge of the competitive details behind the metrics, could do. The intangibles metrics provide guidance as to the state of affairs in the sector and alerts such decision makers where to focus their efforts. Their own, more specific knowledge can then lead to the deeper insights that help make better strategic choices.

References

Abuelsamid, S., Alexander, D., and Jerram, L. (2017). Navigant research leadership report: Automated driving (executive summary). Available at https://www.navigantresearch.com/wp-assets/brochures/LB-AV-17-Executive-Summary.pdf.

Ackoff, R. (1989). From data to wisdom, *Journal of Applied Systems Analysis*, 16, 3–9.

Argyris, C. (1977). Double loop learning in organizations, *Harvard Business Review*, September/October, pp. 115–125.

Argyris, C. (1992). *On Organizational Learning*, Blackwell, Cambridge, MA.

Argyris, C. (1993). *Knowledge for Action*, Jossey-Bass, San Francisco, CA.

Bontis, N. (1999). Managing organizational knowledge by diagnosing intellectual capital: Framing and advancing the state of the field, *International Journal of Technology Management,* 18(5–8), 433–462.

Brown, J.S., and Duguid, P. (1991). Organizational learning and communities-of-practice: Toward a unified view of working, learning, and innovation, *Organizational Science,* 2(1), 40–57.

Brown, L., and Boudes, T. (2005). The use of narrative to understand and respond to complexity, *E:CO,* 7(3–4), 32–39.

Chen, H., Chiang, R.H., and Storey, V.C. (2012). Business intelligence and analytics: From big data to big impact, *MIS Quarterly,* 36(4), 1165–1188.

Crocker, A. (2014). Combining strategic human capital resources: The case of star knowledge workers and firm specificity, *Academy of Management Proceedings,* January, p. 10705.

Edvinsson, L., and Malone, M. (1997). *Intellectual Capital,* Harper Business, New York.

Eliades, G., Retterath, M., Hueltenschmidt, N., and Singh, K. (2012). Healthcare 2020. Available at http://www.bain.com/Images/BAIN_BRIEF_Healthcare_2020.pdf.

Erickson, G.S., and Rothberg, H.N. (2012). *Intelligence in Action: Strategically Managing Knowledge Assets,* Palgrave Macmillan, London, UK.

Erickson, G.S., and Rothberg, H.N. (2013a). Alternative metrics for assessing knowledge assets, *Journal of Information & Knowledge Management,* 12(4), 1350031.

Erickson, G.S., and Rothberg, H.N. (2013b). A strategic approach to knowledge development and protection, *Service Industries Journal,* 33(13/14), 1402–1416.

Erickson, G.S., and Rothberg, H.N. (2016). Intangible dynamics in financial services, *Journal of Service Theory & Practice,* 26(5), 642–656.

Firer, S., and Williams, S.M. (2003). Intellectual capital and traditional measures of corporate performance, *Journal of Intellectual Capital,* 4(3), 348–360.

Grant, R.M. (1996). Toward a knowledge-based theory of the firm, *Strategic Management Journal,* 17(Winter), 109–122.

Groysberg, B., Lee, L.-E., and Abrahams, R. (2010). What it takes to make "star" hires pay off, *Sloan Management Review,* 51(2), 57–61.

Kogut, B., and Zander, U. (1992). Knowledge of the firm, combinative capabilities, and the replication of technology, *Organization Science,* 3(3), 383–397.

Kurtz, C.F. (2009). The wisdom of clouds, white paper. Available at http://www.cfkurtz.com.

Kurtz, C.F., and Snowden, D.J. (2003). The new dynamics of strategy: Sensemaking in a complex-complicated world, *IBM Systems Journal,* 42(3), 462–483.

Laney, D. (2001). 3D data management: Controlling data volume, velocity and variety. Available at November 1, 2013, http://blogs.gartner.com/doug-laney/files/2012/01/ad949-3D-Data-Management-Controlling-Data-Volume-Velocity-and-Variety.pdf.

Liebowitz, J. (2005). Linking social network analysis with the analytical hierarchy process for knowledge mapping in organizations, *Journal of Knowledge Management,* 9(1), 76–86.

Manyika, J., Chui, M., Brown, B., Bughin, J., Dobbs, R., Roxburgh, C., and Hung Byers, A. (2011). *Big Data: The Next Frontier for Innovation, Competition and Productivity,* McKinsey Global Institute, New York.

Marr, B., Schiuma, G., and Neely, A. (2004). The dynamics of value creation: Mapping your intellectual performance drivers, *Journal of Intellectual Capital,* 5(2), 312–325.

Matson, E., Patiath, P., and Shavers, T. (2003). Stimulating knowledge sharing: Strengthening your organizations' internal knowledge market, *Organizational Dynamics,* 32(3), 275–285.

McAfee, A., and Brynjolfsson, E. (2012). Big data: The management revolution, *Harvard Business Review*, 90(10), 60–66.

McEvily, S., and Chakravarthy, B. (2002). The persistence of knowledge-based advantage: An empirical test for product performance and technological knowledge, *Strategic Management Journal*, 23(4), 285–305.

Nahapiet, J., and Ghoshal, S. (1998). Social capital, intellectual capital, and the organizational advantage, *Academy of Management Review*, 23(2), 242–266.

Nonaka, I., and Takeuchi, H. (1995). *The Knowledge-Creating Company: How Japanese Companies Create the Dynamics of Innovation*, Oxford University Press, New York.

Polanyi, M. (1967). *The Tacit Dimension*, Doubleday, New York.

Rothberg, H.N., and Erickson, G.S. (2005). *From Knowledge to Intelligence: Creating Competitive Advantage in the Next Economy*, Elsevier Butterworth-Heinemann, Woburn, MA.

Rothberg, H.N., and Erickson, G.S. (2017). Big data systems: Knowledge transfer or intelligence insights, *Journal of Knowledge Management*, 21(1), 92–112.

Schwedel, A., and Rodrigues, A. (2009). Financial services' shifting profit pool. Available at http://www.bain.com/Images/BB_Financial_services_shifting_profit_pools.pdf.

Senge, P. (1990). *The Fifth Discipline*, Doubleday, New York.

Simard, A. (2014). Analytics in context: Modeling in a regulatory environment, in Rodriguez, E. and Richards, G. (Eds.), *Proceedings of the International Conference on Analytics Driven Solutions 2014*. Harvard University Press, Cambridge, MA, pp. 82–92.

Stewart, T.A. (1997). *Intellectual Capital: The New Wealth of Organizations*, Doubleday, New York.

Sveiby, K.E. (2010). Methods for measuring intangible assets. Available at http://www.sveiby.com/articles/IntangibleMethods.htm.

Tan, H.P., Plowman, D., and Hancock, P. (2007). Intellectual capital and the financial returns of companies, *Journal of Intellectual Capital*, 9(1), 76–95.

Teece, D.J. (1998). Capturing value from knowledge assets: The new economy, markets for know-how, and intangible assets, *California Management Review*, 40(3), 55–79.

Thomas, J.C., Kellogg, W.A., and Erickson, T. (2001). The knowledge management puzzle: Human and social factors in knowledge management, *IBM Systems Journal*, 40(4), 863–884.

Tobin, J., and Brainard, W. (1977). Asset markets and the cost of capital, in Nelson, R. and Balassa, B. (Eds.), *Economic Progress, Private Values, and Public Policy: Essays in Honor of William Fellner*. North Holland, Amsterdam, the Netherlands, pp. 235–262.

Verdun, J. (2008). The last mile of the market: How network technologies, architectures of participation and peer production transform the design of work and labour, *The Innovation Journal: The Public Sector Innovation Journal*, 13(3). Available at https://www.innovation.cc/.

Wernerfelt, B. (1984). The resource-based view of the firm, *Strategic Management Journal*, 5(2), 171–180.

Zack, M.H. (1999). Developing a knowledge strategy, *California Management Review*, 41(3), 125–145.

Zander, U., and Kogut, B. (1995). Knowledge and the speed of transfer and imitation of organizational capabilities: An empirical test, *Organization Science*, 6(1), 76–92.

Chapter 12

Analyzing Data and Words—Guiding Principles and Lessons Learned

---

Denise A. D. Bedford

Contents

Introduction

Business analytics and text analytics are not new. Work on the core analytics capabilities dates back to the 1950s and 1960s. What has changed in 2017 is the computing capacity we have—in the business and the research environment—to apply analytics. The increased capacity has created an expanded set of expectations for what is possible. It has also created opportunities to explore business and research questions we previously thought impossible due to scale, scope, cost, or reliability. The expanded capacity has allowed us to explore the ways in which humans leverage language to create and represent knowledge. Knowledge management methods and core concepts are critical to the intelligent use of analytics. The expanded foundation makes it possible to do much more than detect clusters and patterns in text. This new capacity holds great promise for advancing the discipline of knowledge management. Knowledge management has traditionally relied on qualitative methods that rely on human manual interpretation and analysis. The discipline of knowledge management, which has traditionally relied on qualitative methods, can be enhanced and expanded through the intelligent use of analytics. These expectations and opportunities are achievable, but only if we approach analytics intelligently. New linguistic and knowledge-based tools allow us to use machines to begin to truly understand text. Understanding, though, requires thoughtful design, investments, error and risk management, a fundamental understanding of qualitative methods and their machine-based transformation, a fundamental understanding of language and linguistics, and a willingness to navigate today's volatile "analytics" market. This chapter considers how we can leverage both of these disciplines to meet expectations and new demands. This chapter offers guidance for navigating these issues in the form of key questions and lessons learned.

Conceptual Framework

Over the past 35 years, four focus points have helped the author navigate the choice and use of analytical methods and tools to a range of projects. The focus points include: intent and focus of the project (why); the nature of analysis (how); sources we analyze and use for analysis (what); and tools and methods that support the kind of analysis we're doing. The remaining sections of the chapter walk through the key questions for each dimension. Each focal point is supported by a set of critical thinking questions (Table 12.1). The 27 questions are presented in sequential order. Lessons learned suggest that addressing each of these questions individually increases the probability of a successful effort.

In the sections that follow, we explore each of these key questions in the context of seven real world use cases. These use cases have all ended in success but could have resulted in significant failures had we not considered the key questions. The use cases are identified below.

- ■ *Use Case 1*: Causal Relationships between Arab Spring and Release of Wikileaked Cables (*manuscript in process*)
- ■ *Use Case 2*: Emotional and Social Tone of the Discourse of the 2012 U.S. Presidential Campaign (Bedford, 2012b)
- ■ *Use Case 3*: Language and Knowledge Structures of Mathematical Learning (Bedford & Platt, 2013)
- ■ *Use Case 4*: Knowledge Transfer Practices among Beekeepers (Bedford & Neville, 2016)
- ■ *Use Case 5*: Precision of Automated Geographical Categorization (Bedford, 2012)
- ■ *Use Case 6*: Analysis of Physician-Patient Communication (Bedford, Turner, Norton, Sabatiuk, & Nassery, 2017)
- ■ *Use Case 7*: Analysis and Detection of Trafficking in Persons (Bedford, Bekbalaeva, & Ballard, 2017)

Table 12.1   Key Questions by Dimensions of the Framework

Dimension	Key Questions
Research and business goals—why	1. What are you trying to achieve? 2. What good practice and theoretical models exist? 3. How will you measure the results? 4. What level of risk are you willing to assume? 5. What level of investment are you willing to make? 6. Is this a project or enterprise level of effort?
Analytical methods—how	7. What analytical method is best suited to the goal? 8. When is a quantitative analysis approach warranted? 9. When is a qualitative analysis approach warranted? 10. When should we choose a mixed methods approach? 11. Which of these methods are supported by tools? 12. Which of these methods are not supported by tools? 13. What opportunities are there for mixing tools?
Sources of evidence	14. What is the nature of language, information, and knowledge source evidence? 15. What meaning or understanding are we deriving from the language? 16. What linguistic registers are represented in the source evidence? 17. What knowledge structures are represented? 18. What types of semantic methods are available for us to work with? 19. What does workflow look like for each of these methods? 20. Is there a possibility of reducing or eliminating the subjective interpretation element? 21. What is the return on investment for the analysis? 22. What level of effort is involved in doing a rigorous analysis? 23. What types of competencies are needed to support the analysis? 24. What is the feasibility of the analysis without semantic methods?
Tools	25. Reviewing the tools used 26. Reviewing your successes 27. Reviewing your failures

Research and Business Goals (Why?)

Every organization will use analysis. We use analytics for many traditional reasons including making better decisions, achieving better research results, discovering ways in which our organizations can be more efficient or effective. We also use analytics because they allow us to leverage machines to achieve what is beyond the scale and scope of human manual analysis, and in some cases to remove the inherent subjective interpretation humans bring to any analysis. Twentieth century industrial economy organizations relied primarily on traditional analyses—quantitative methods and quantitative sources. Machine-based analytics offer the potential to go beyond what we have traditionally expected. An important new direction is that offered by knowledge management. Twenty-first century knowledge economy organizations require new and more robust analytical methods. Knowledge management uses all of the traditional methods and sources to "…. make the organization act as intelligently as possible and realize the best value from its knowledge assets, that is, to create a learning organization that is capable of measuring, storing, and capitalizing on the expertise of employees to create an organization that is more than the sum of its parts" (Bollinger & Smith, 2001).

The need for analysis to support knowledge management goes beyond what has traditionally been available. There are significant opportunities for leveraging business analytics to create a knowledge organization and for using knowledge management methods to enrich business analytics. Organizations need to identify and leverage their business-critical knowledge. Business capability and modeling methods are qualitative methods that leverage qualitative sources. Quantitative methods may provide partial support, but additional qualitative machine-based methods are needed. Organizations have vast stores of qualitative sources (e.g., text, speech, audio in a variety of languages) which hold important knowledge and insights that can help us to achieve Bollinger and Smith's goals. To derive meaning from these sources, we must be having a working understanding of language.

Whether you are a forward-looking or a traditional organization, your use of analytics should be guided by what you are trying to achieve, what good models or practices you have to work with, the level of risk you are willing to accept, the level of investment you are willing to make, and the scale and scope of the application.

What Are You Trying to Achieve? What Is Your Research or Business Goal?

To prosper, an organization must leverage all of the resources available to it. In an increasingly competitive world, all organizations—universities, research institutes, global private companies, non-governmental organizations—understand that it is no longer acceptable for individuals to make business decisions or define research projects based on "gut feelings," "knee jerk reactions," or personal experience.

Business and research goals define the purpose and reason a company, organization or institution exists.

What Good Practices and Good Practice Models Exist?

When a new tool or method is developed, we try to fit it into the existing repertoire. More often than not a new method or tool complements or supplements what exists. It is rarely the case—except in fundamental theoretical research—that an entire way of thinking or seeing is replaced. By definition, complementary and supplementary methods and processes add to or support what exists. Where the focus is on new machine-based methods, in all probability we have good human-based practical and theoretical models. The new process or method gives us the opportunity to test what exists on a broader scale, to expand the scope of analysis, to achieve a comparable result in a much more cost-effective or cost-efficient manner. The availability of new tools does not replace or eliminate existing models and methods. In all cases, the tool is the last decision or choice we make. We follow basic research and business practices to define our goals, lay out our questions, and consider what knowledge already exists. We need to build on, enable, or enhance existing knowledge wherever it exists.

The important lesson learned here is that an effective model applied manually can and should serve as the basis for a machine-based model. If there is a well-formulated theoretical model, the first question should be how that model can be adapted to machine processing. This can often be a critical success factor. In fact, in some cases, the machine transformation of a model that has been applied manually for years can provide a rigorous test of that model. It does not make good business or research sense to disregard a good theoretical model because it is not in machine-processable form. Just because you have a new technology or tool to work with does not mean you can disregard well-formulated and long-standing business rules, theory, or reference models.

How Will You Measure the Results of Your Analysis?

Thinking about measurement forces us to transform our thoughts and ideas into quantitative results and to express our observations numerically. Measurement applies to both quantitative and qualitative analyses. The goal of measurement should be the reduction of uncertainty and bias. Measurement of quantitative is supported by well-formed and time-tested statistical and mathematical methods. Measurement of qualitative sources is more challenging because it has traditionally involved human review and interpretation of results. Even the most seasoned researcher or decision maker finds it challenging to hold back their experiences, beliefs, and perspectives. The new tools do not preempt or negate the need to measure results, to define potential errors, or to dismiss concerns of validity and reliability. Rather, new tools will require new measures. Measurement is a core element of ensuring that we achieve our business and research goals. Errors occur in degrees. Defining our tolerance for errors is an import part of measurement. Two considerations that help us to manage errors are reliability and validity. Both reliability and validity errors are possible with the new analytical tools. In fact, both may increase if we chose tools and sources incorrectly.

Reliability speaks to the consistency in measure and considers whether you would obtain the same result if you repeated the analysis with all factors remaining constant (Babbie, 1989; Carmines & Zeller, 1979; Kirk & Miller, 1986). A simple example of a reliability error occurs when we apply quantitative analytical tools to dynamic text or language. The reliability and generalizability of results is compromised on each time the corpus of text changes. Validity is understood as truth in measurement and tells us whether we measured what we intended to measure (Adcock, 2001; Guion, 1980; King, Keohane, & Verba, 1994). Validity is easier to measure in quantitative analyses than it is in qualitative analyses (Atkinson, 2005; Reason and Bradbury, 2001; Eriksson & Kovalainen, 2015; Seale, 1999). A simple example of a validity error occurs when we apply quantitative analytical tools to text or language without first developing a deep understanding of the linguistics or the knowledge structures inherent to the text. The simplest examples of errors, though, are type 1 and type 2 errors.

What Level of Risk Are You Willing to Assume?

Analytical tools are not free of risk. It is important to understand the nature and extent of risk associated with your approach, regardless of whether you are analyzing evidence manually or using machines. Risk is related to uncertainty. Uncertainty is a potential, unpredictable, and uncontrollable outcome; risk is a consequence of action taken in spite of uncertainty. The risks we encounter in working with analytics include risks of project failure and the risk of a failed or poor business decision. These risks can be immediate or may have long-term consequences. The risks may fall to the researchers, the business analyst, or decision maker, or they may fall to the population that is impacted by the research project or the decision maker. We are just beginning to understand the nature of the risks associated with using analytical methods, particularly the application of quantitative methods applied to qualitative sources, and the transformation of qualitative methods from manual to machine-based. As we gain more experience, it is important to build questions of risk and uncertainty into our measurement models.

What Level of Investment Are You Willing to Make?

The wisdom of business intelligence methods applies to analytics and particularly to the new analytical tools. Business intelligence reminds us that there is a trade-off of time, quality and resources. The new analytical tools require investments of time and resources, if quality results are to be achieved. One of the reasons business analysts and researchers often choose these tools is the belief they will save them time (e.g., just apply the tools and the solution appears) and require fewer resources (e.g., the tools will solve all the complex problems). In fact, the tools can require more computing resources and an extended set of competencies. The tools that have the greatest value and the most analytical power also come with a high price tag and maintenance costs. Quality results do not result from lower cost solutions or lower level resources.

Is This a Project or Enterprise-Level Goal?

We should be clear before we begin a project whether the goal is a one-time effort or an on-going, enterprise-level operation. In the author's experience, it is important to manage expectations. If you never intend to replicate the exercise, if there is no expectation that the solution has value beyond the immediate application, be clear about that with your stakeholders. If the solution will be an enterprise application or one that support continuous operations, it is critical to plan for scale up and integration. One-time, one-off solutions generally do not integrate well into enterprise architecture. They also tend to lack ongoing support from the technology or design teams. On the research side, it is important for researchers to explain their methodology and the tools underlying engines and business rules and details for replicability and reuse.

Understanding Why in Context—Use Case Scenarios

In this section we will walk through the seven use cases and explain how we used the key questions to guide each of the projects (Tables 12.2 through 12.8). The use cases are real examples that have been completed over the past 20 years. Each example illustrates a different type of goal, a different methodology, different sources of evidence, and different outcomes. Every use case demonstrates the value of combining both knowledge management and business analysis strategies. In this section, we introduce each of the use cases, their goals, the traditional approach to the problem, measurement strategies, reliability and validity issues, and the risk of errors.

Table 12.2   Use Case 1: Causal Relationship between Arab Spring Revolutions and the Release of Wiki-leaked Cables

Dimension	Use Case Context
Primary goals	Test the media's characterization of the relationship between the content of the Wiki-leaked diplomatic cables and the motivation for the Arab Spring.
Conceptual models	Conceptual models of the language of diplomacy and the nature of intelligence that is sent from field offices and diplomatic posts to the Department of State. There are few theoretical models available given the classification of the information. The knowledge of subject matter experts was used to develop models.
Sources of evidence	Two primary sources: (1) media reports describing the Arab uprising and (2) the leaked diplomatic cables. Cables for two countries were collected—Tunisia and Egypt—for two years.
Measurement strategy	Measurement focused first on the categorization of the language used in the cables by topics and the organizations and geographical locations referenced. Semantic analysis provides the factors for comparison of diplomatic cables and the media's characterization of the causes of the uprisings. Measurement demonstrated that there was little in little in the cables that was not already known to citizens of those countries.
Probability of errors	The greatest risk was human perspective and subjective interpretation of the source materials, or the injection of personal opinion into the analysis. This risk was reduced by the construction of objective control models of diplomatic intelligence, and the interpretation of those models by a neutral machine. Had a control model not been available, there is a high probability that the full extent of coverage of intelligence would not have been exposed.
Level of acceptable risk	The initial analysis was research-based and carried little risk to either the Department of State or the media. The researcher carried the majority of the risk in the event that the analysis provided no clear results or produced results that were not reproducible or which could not be validated by subject matter experts.

(Continued)

Table 12.2 (*Continued*)   Use Case 1: Causal Relationship between Arab Spring Revolutions and the Release of Wiki-leaked Cables

Dimension	Use Case Context
Reliability and validity issues	Because the conceptual model was designed around a full view of diplomatic intelligence, it can be reliably applied to any other country for analysis. Because a conceptual model was available for control purposes, internal validity was also strengthened.
Follow-on goals	The initial goal was achieved—the research team was able to demonstrate that a cause-effect relationship of the leaked diplomatic cables and the uprising did not exist. Rather, the deep analysis of language and content of the diplomatic cables suggested that the amount of unique or novel information in the cables was low when compared to media reports and social media communication in the country. This lead to a second research question comparing the content of diplomatic cables with in-country media reports.

Table 12.3   Use Case 2: Emotional and Social Tone of the Discourse of the 2012 U.S. Presidential Campaign

Dimension	Use Case Context
Primary goal	Characterize the discourse of the 2012 U.S. Presidential election in terms of its emotional and social nature, to profile the extensional and intentional nature of the candidate's language, and to gauge the density of the information content.
Existing models	Two strong conceptual models existed, including the Gottschalk-Gleser Model of Emotional Stability and McLaughlin's characterization of intentional and extensional language. The model is available in manual and automated formats. It gauges levels of emotional stability, depression, social inclusion, and alienation. McLaughlin's model was developed in the 1940s and has been available only as a manual instrument. In this project it was transformed into a machine-understandable profile for semantic analysis. McLaughlin's model characterized four indicators of intentional language and four indicators of extensional language.

(Continued)

Table 12.3 (*Continued*)   Use Case 2: Emotional and Social Tone of the Discourse of the 2012 U.S. Presidential Campaign

Dimension	Use Case Context
Sources of evidence	Primary sources were speeches and interviews of the Republican and Democratic candidates for President.
Measurement strategy	Semantic analyses produced the information on factors for analysis. Measurement focused on the variations (e.g., central tendencies and dispersion) of factors across political candidates. The language of the text was first characterized according to the factors, and variations across factors were analyzed using quantitative statistics.
Probability of errors	The greatest risk of error would have come from a non-expert characterization of the factors, and in the subjective interpretation of those factors found in the text or speech. Risk was reduced by using authoritative conceptual models. The human subjectivity risk was reduced through the machine-level interpretation of those models to the text.
Level of acceptable risk	The research goal was best served by a low level of acceptable risk, given the sensitive nature of the political process. Risk was managed by identifying and using or transforming a well-tested and trusted conceptual model. Both models are accepted by experts in the fields of psychology, psychiatry, and linguistics.
Reliability and validity issues	The reliability challenge was in the discovery of sufficiently rich texts and speeches from the candidates. The validity challenge was the possibility that the candidates had not written the speeches or text, but were delivering thoughts expressed by others. This was mitigated to a degree by the inclusion of interview transcripts for each of the candidates. Another validity challenge that surfaced through the analysis was the low density of content in most political speeches and texts.
Follow-on goals	The success of the initial semantic analysis of the political candidates prompted the research team to conduct the same analysis on media persona from three major networks. The media texts were contemporaneous to the candidate's texts. The follow-on research goal tested the same factors for media persona and found that in most cases, the media persona were more extreme than the political candidates.

Table 12.4   Use Case 3: Language and Knowledge Structures of Mathematics

Dimension	Use Case Context
Goals	Develop insights into how the language used in mathematical textbooks might contribute to the success or failure of students learning.
Existing models	Conceptual models for this project were simple—simple linguistic analysis of the use of language in mathematics, and the characterization of types of knowledge structures (e.g., noun phrases) used in each area of mathematics (e.g., algebra, geometry, trigonometry, calculus, etc.). This intent was to develop conceptual models of mathematical language.
Sources of evidence	Representative sets of textbooks for each area of mathematics were selected, deconstructed, OCR'd, and semantically processed.
Measurement strategy	The language of mathematics was characterized through natural language processing. Grammatical variations were observed across areas. These variations were offered for review to subject matter experts.
Probability of errors	The primitive use of natural language processing generated descriptive information about the noun and verb phrases used in areas of practice. The large volume of text analyzed for each area was a protection against either type 1 or type II errors. The analyses generated clear patterns of variations.
Level of acceptable risk	As this was exploratory research, we allowed for a high level of risk. This research was understood to be the first of several phases. We expected that new hypotheses would be produced by the findings, and more rigorously tested.
Reliability and validity issues	Reliability was high for the characterization of language and knowledge structures because each text is grounded in basic principles, assumptions, functions, and symbols. How language is used to explain concepts or functions may vary, but the nature and frequency of use of that language is consistent across texts.
Follow-on goals	A new hypothesis resulted from the exploratory research related to the need to interpret abstract concepts and to explain mathematical operations in everyday language. We discovered that while the language of mathematics is intended to be universal among experts, a layer of interpretation and translation is required for novices learning the field.

Table 12.5   Use Case 4: Knowledge Transfer Practices among Beekeepers

Dimension	Use Case Context
Primary goals	Understand and characterize knowledge sharing and transfer among beekeeping communities. Because beekeepers are one of the most effective and long-lived communities of practice, we assumed they would be a good source for study. The intent was to characterize knowledge sharing in the community (e.g., simple broadcasts, questions and answers, in-depth discussions, and disagreements).
Existing models	We began by knowledge modeling of beekeepers interactions, and validated the models with expert members of that community. We used linguistic registers of actions where they existed.
Sources of evidence	Messages from three online beekeeper Listservs were captured during the most active seasons of the year. Listserv transcripts were downloaded, linguistically characterized, and categorized in terms of type of sharing.
Measurement strategy	We provided simple measurements of factors that were semantically modeled to establish a baseline of understanding.
Probability of errors	Because of the primitive linguistic nature of the analysis, the probability of errors is very low.
Level of acceptable risk	This was exploratory research intended to test whether we could develop a language-based model of knowledge sharing activities. Risk was borne by the research team. There was no intervention into a community or its discourse.
Reliability and validity issues	While the source sample was statistically reliable, the conceptual model was applied to only three communities of beekeepers. The reliability of the research can only be established for the beekeeping communities. Reliability of the semantic model of knowledge sharing can be tested in other communities. Validity of the results is high because it is grounded on natural language processing and rule-based linguistics characterizations.
Follow-on goals	The language model that was developed is being developed to other types of communities of practice (e.g., craft communities). In addition, a new hypothesis has surfaced for testing—whether the behavior of individuals to share knowledge and expose errors in an informal setting changes in a business context.

Table 12.6   Use Case 5: Precision of Automated Geographical Classification

Dimension	Use Case Context
Goals	To determine which approach—concept extraction or categorization—is more effective for determining with a high level of accuracy what country a text is about.
Existing models	Several authoritative models of countries including the Getty Thesaurus of Geographical Terms, the U.S.G.S. characterization of geographical entities, and the ISO country names list. We chose two models—the Getty model and the ISO country names and synonyms list for testing. Both models were transformed to machine-based profiles.
Sources of evidence	Control set of documents that had been manually indexed for geographical "aboutness" by human indexers. Same set of documents were then machine-processed using two different analytical methods.
Measurement strategy	Three way comparison of the accuracy of automated categorization using the Getty profile, the accuracy of concept extraction using the ISO model, and with the manually indexed document as the control point.
Probability of errors	Both type 1 and type 2 errors were noted in the results. However, these errors are not errors of research but errors of machine-based processes. The fact that the errors were identifiable and explainable was a significant research result.
Level of acceptable risk	Level of risk depends on the context in which the profile is used. If used in the intelligence community, the risk of missing or mischaracterizing a geographical entity can be significant. If used for general search and discovery, the risk may take the form of frustration or mistrust—a risk that may be acceptable.
Reliability and validity issues	Internal validity of the research was high because of the availability and use of the control set of documents. Reliability may vary, though, depending on the nature of documents analyzed and of the regional focus of the documents. The geographical descriptions of some countries were found to generate higher errors rates than of other countries.
Follow-on research	In this case, the follow-on research focused on developing deeper understanding of the geographical characteristics of selected countries.

Table 12.7   Use Case 6: Analysis of Physician-Patient Communications

Dimension	Use Case Context
Goals	Provide a more cost effective and reliable method for characterizing doctor-patient communication. The ultimate goal is to provide guidance to doctors on how to communicate for better health outcomes. Intended to create a machine-processable tool that can emulate the human coder at an equally effective or more effective level in a shorter time.
Existing models	Six conceptual models representing the aspects of doctor-patient communication. All of them require manual application—human review, interpretation, and coding. Transformed the factors to machine-based profiles. Validated the profile with subject matter experts, remove the subjectivity of application.
Sources of evidence	Transcribed text of videotaped conversations between doctors and patients and doctors and interns.
Measurement strategy	Comparison of machine-processable coding to human coding work. Intent is to continuously improve until we achieve human level of quality.
Probability of errors	Errors are primarily those of interpretation. Reference manuals describe the nature of the factor in very general terms. The reduction of errors is dependent upon our ability to objectively represent the factors and to represent them in sufficient detail to meet the expectations or standards of the subject matter experts.
Level of acceptable risk	Low level of risk. There is no interference with the actual medical conversation.
Reliability and validity issues	Internal validity is an existing problem that we're trying to address through the development of the machine-based profile. We believe there is inter-coder and possibly intra-coder variation. Reliability can only be achieved after the profile has been validated, and tested on a larger set of interactions.
Follow-on research	Research to date suggests that there are variations in communications among doctors and between doctors and patients. New factors have surfaced in the focused analysis of doctor-doctor communication.

Table 12.8　Use Case 7: Trafficking in Persons Semantic Discovery

Dimension	Use Case Context
Goals	Identify and model all of the facets of trafficking in persons, and create a semantic profile that would enable discovery across the organization, repositories, locations, and stakeholders.
Existing models	Authoritative model exists within the Department of State but there are variations to be incorporated from outside the department and by areas of practice.
Sources of evidence	Trafficking in persons reports, primary source materials, formal and informal sources of information, media reports, law enforcement reports, etc.
Measurement strategy	Intent is to improve discovery by creating a single virtual structure that is easily accessible. Solution must be comprehensive and inclusive.
Probability of errors	Manageable if we work at the facet level—there are many dimensions of trafficking in persons, each having its own caveats.
Level of acceptable risk	Low because we are focusing only on access and discovery. If the profile moves beyond this to tracking communications and discovery actual activities, the risk would increase significantly.
Reliability and validity issues	Challenge is to reliability and gaining access to primary source materials which are highly classified.
Follow-on research	Each facet of the profile can be broken off and used independently. We would expect further refinement of pieces of the profile.

How We Use the Tools—Analysis as a Process

After we have defined our business or research goals, levels of investment, measurement strategies, and tolerance for risks and errors, we need to choose an analytical method. At this stage, it is important that we begin with a good understanding of analysis. Analysis is a systematic thinking process—a detailed examination of the elements or structure of something in preparation for discussion, interpretation, problem solving, or decision making. Historically, there are three basic approaches to analyzing a situation or thing—quantitative analysis, qualitative analysis, and a combination of both qualitative–quantitative (e.g., mixed methods). We have a longer tradition of quantitative methods. Qualitative methods have evolved over the past half-century, and mixed methods are the most recent of the three. Quantitative methods have the longest history and are grounded in mathematical and statistical methods. Qualitative methods are more recent and align more with social sciences and humanities. Mixed methods are the youngest of all. The majority of the analytical tools available today on the commercial market and as open source tools are statistically or stochastically based. These technologies derive from and align with quantitative analytical methods. This makes sense because they are "computer or machine-based" applications.

What Analytical Method Is Best Suited to the Goals?

It is important to understand these analytical methods and how they work. Applying a good method to an inappropriate or unsuitable source will result in suboptimal or compromised results. Table 12.9 describes four common analytical approaches, their functionality and the best-fit sources. The answer to this question is—it depends on what you're trying to achieve, what kinds of questions you're trying to answer, and what type of source evidence you have to work with.

Quantitative Analysis and Data Analytics

Quantitative analysis is a systematic approach that involves measuring or counting attributes (i.e. quantities). Quantitative analysis makes use of mathematical measurements and calculations, statistical modeling and research to either describe the evidence or to infer something from the evidence. Quantitative methods are most often associated with machine-based strategies because of the complexity of the calculations and of the volume of data being analyzed. It is simply more efficient and effective to take a quantitative approach because of the rich foundation of quantitative methods. In the context of this chapter, quantitative methods translate well to data analytics. Data analytics is supported by a robust methodological

Table 12.9 Domain, Analytical Methods and Source Types

Domain	Analytical Method	Type of Source
Data analytics	Quantitative analysis	Quantitative sources
Text analytics	Quantitative analysis	Qualitative sources
Computational linguistics	Quantitative and qualitative analyses	Qualitative sources
Language-based semantics	Qualitative analysis	Qualitative sources

foundation from mathematics and statistics (Han, Pei, & Kamber, 2011; Hand, Mannila, & Smyth, 2001; Witten, Frank, Hall, & Pai, 2016). There is a robust market of tools—in the hundreds—to support data analysis, and this market has been stable for close to two decades. The open source and the commercial markets are mature so there are options available for organizations of all sizes—from the simple functions embedded into Excel to the more sophisticated capabilities of Informatica and SAS. These tools have been used by organizations for decades. The change in use is in the scale and scope of the evidence they can process and the fact that they can process live streams of transactional data.

Another type of quantitative analysis that is relevant to this chapter is text analytics. Text analytics is sometimes also referred to as text mining, text analysis, text processing, text mining, or text data mining. This type of analysis involves the derivation of information from text—it attempts to infer things from text based on patterns. This approach dates back to the late 1950s and early 1960s and is grounded in the application of quantitative methods to text (Abbasi & Younis, 2007; Abilhoa & De Castro, 2014; Arya, Mount, Netanyahu, Silverman, & Wu, 1998; Bejou, Wrap, & Ingram, 1996; Bengio, 2009; Carletta, 1996; Debortoli, Muller, Junglas, & vom Brooke, 2016; Deerwester, Dumais, Furnas, Landauer, & Harshman, 1990; Fernandez et al., 2015; Hofmann, 1999; Lehnert, Soderland, Aronow, Feng, & Shmueli, 1995; Luhn, 1958a, 1958b; Maarek, Berry, & Kaiser, 1991; Miner, 2012; Pereira, Tishby, & Lee, 1993; Salton, 1970, 1986; Shmueli-Scheuer, Roitman, Carmel, Mass, & Konopnicki, 2010; Srinivas & Patnaik, 1994; Teufel & Moens, 2002; Tischer, 2009). Organizations often apply text analytics to a large corpus of text in an attempt to identify new or interesting patterns or areas of convergence and similarity. While these methods have continued to evolve over the past 50 years, the growth has largely been focused on improving the relevance and precision of the categories or the coverage and recall of targeted entities extracted. In general, these methods have evolved parallel to, but not with, the evolution of language-based methods. Text is processed to remove formatting, structure, and language variations prior to applying quantitative algorithms. Some text analytics tools have natural language processing (NLP) components that are used to reduce language variations and surface lemmas and roots. Natural language components are not a function that is exposed to decision makers or researchers for direct use. The majority of the tools available on the market today support text analytics solutions.

Qualitative Analysis and Language Based Analytics

Qualitative methods are most frequently associated with manual or human analysis. There are many types of qualitative research methods and many ways to understand and apply them. For the purpose of this chapter, we refer to Tesch's hierarchical categorization of qualitative research methods because they focus on the elements that are most closely aligned with knowledge and knowledge management. Tesch (2013) identified 27 types of qualitative research and organized them into three major substantive questions: (1) What are the characteristics of the language itself? (2) Can we discover regularities in human experience? (3) Can we comprehend the meaning of a text or action? Tesch correctly observes that most qualitative analysis is done with words and focuses on language. Words and language is a very important starting point for selecting an analytical method and analytical tools. Typically, qualitative research tools are designed to facilitate working with information, such as the conversion or analysis of interviews; the analysis of online surveys; the interpretation and analysis of focus groups, video recordings, audio records; or the organization of images. These tools still require human intelligence and effort to support qualitative analysis. They have little embedded human intelligence. The challenge is that until recently there were no robust or affordable machine-based tools that automated the process of analyzing the language of qualitative sources.

One thing we have learned over the years is that a machine can only do high performance analysis when it has a human-equivalent level of understanding of that language. Since the 1960s, we've devoted considerable research to understanding how to embed linguistic competence into machine-based applications. Since the 1970s, considerable work has been devoted to building a machine-based understanding of language. Computational linguistics is focused on the use of machines and technologies to support deeper and broader understanding of linguistics and language. It is a young and interdisciplinary field concerned with the statistical or rule-based modeling of natural language from a computational perspective (Anwar, Wang, & Wang, 2006; Berman, 2014; Brandow, Mitze, & Rau, 1995; Church & Hanks, 1990; Church & Mercer, 1993; Croft, Coupland, Shell, & Brown, 2013; Eggins, 2004; Hatzivassiloglou & McKeown, 1997; Hearst, 1992; Hindle & Rooth, 1993; Kaplan & Berman, 2015; Marcus, Marcinkiewicz, & Santonini, 1993; Mitkov, 2005; Nir & Berman, 2010; Pustejovsky, 1991; Shieber, Schabes, & Pereira, 1995; Sproat et al., 2001; Van Gijsel, Geeraerts, & Speelman, 2004; Yarowsky, 1995). The theoretical foundations of computational linguistics are in theoretical linguistics and cognitive science. Applied computational linguistics focuses on modeling human language use. Since the 1980s, computational linguistics has been a part of the field of artificial intelligence. Its initial development was in the field of machine translation. The simple automated approach to machine translation was revisited in favor or more complex development of language and domain-specific vocabularies, extensive morphological rules engines, and patterns that reflect the actual use of language in different contexts. Today, computational linguistics focuses on the modeling of discourse and language patterns. Traditionally, computational linguistics was performed by computer scientists focused on applying stochastic methods to natural language. In the twenty-first century, though, computer programmers and linguists work together as interdisciplinary teams. Linguists provide critical knowledge of language structure, meaning, and use that guide the use of analytical tools. Because these tools have an academic or research focus, computational tools are often open source or laboratory-based. By and large, they are not available on the commercial market.

Natural language programming is a critical component of today's qualitative analysis methods and tools. It focuses on the interactions between human language and machines. It is an interdisciplinary field that lives primarily in computer science but includes major contributions from linguistics, artificial intelligence, and cognitive science (Ballard & Biermann, 1979; Biermann, 1981; Biermann, Ballard, & Sigmon, 1983; Chowdhury, 2003; Dahl & Saint-Dizier, 1985; Dijkstra, 1979; Halpern, 1996; Heidom, 1971; Jurafsky, 1996; Liddy, 2001; Loper & Bird, 2002; Manning & Schutze, 1999; Marcu, 1997; Mihalcea, Liu, & Lieberman, 2006; Spyns, 1996; Tomita, 2013; Weizenbaum, 1986). Over the past 50 years, computer programmers, linguists, and information scientists have worked together to create these core components. We have constructed machine-based components to support all of the levels of language. NLP applications must be developed for each language, and for geographical variations of languages. The most important element of an NLP application for analysts and knowledge scientists is the part of speech (POS) tagger. In order for a POS tagger to work, we need two things: an extensive characterization of the target language and a framework or set of tags. The tag set should include all possible combinations of category values for a given language. A tag set is generally represented by a string of letters or digits. As with computational linguistics tools, natural language programming and processing tools are generally used in academia by researchers. NLP is a key tool used by computational linguists, and computational linguists have been critical to the development of these methods.

Language-based semantic methods and tools is the youngest of all and perhaps the least understood of all the terms. Language-based semantic methods are a combination of all of the above concepts and capabilities. Language and knowledge-based applications typically leverage these components but in a designer or configurable way. These applications are like toolboxes—they contain core language and knowledge modeling tools and allow users to build solutions to suit different business or research problems. Among the tools in a semantic analysis toolbox we would find NLP and natural language programming, POS taggers, concept and entity extraction tools, grammar and rule-based categorization construction tools, indexing tools, abstracting rule builders, and clustering engines. The capabilities are often beyond the current competencies of business, so they do not receive a significant treatment in the popular media or even in the scholarly press (Edmundson, 1969; Goddard, 2011). These methods are rarely all found in commercial tools, though there are a few examples.

Mixed Methods Analysis and Variant Sources

Mixed methods analysis involves collecting, analyzing, and integrating quantitative and qualitative methods and sources. In essence, this type of analysis takes advantage of using multiple ways to explore a research problem. We may begin by using linguistic registers and knowledge modeling to understand the content of text corpora, then switch to quantitative classification models to model what we have learned, and switch back to semantic methods to classify the full corpus. The mixed methods approach is a fairly young discipline—such methods have only been accepted by the research community for the last decade. When using a mixed methods approach, the rigor of the design is critical. Mixed methods are not the simple application of quantitative methods to qualitative sources.

When Is a Quantitative Analysis Approach Warranted?

Quantitative analysis is appropriate when the source evidence is quantitative or structured in nature, where the volume of structured data is significant, or when the quantitative variables are complex, and when the intent is to generalize the results to a larger base or population. We generally know something about the data under analysis when using quantitative methods. We would not apply a quantitative tool to "messy" or unmanaged data. Using standard methods means that the research can be replicated, analyzed, and compared with similar studies (Kruger, 2003). One of the attractions of quantitative methods is that they minimize the personal bias that is inherent to qualitative methods—the machine is doing the analysis following predictable rules.

When Is a Qualitative Analysis Approach Warranted?

Qualitative analysis is appropriate when we are working with language or text, when we are interested in learning the characteristics of the language, when we are interested in discovering regularities in human experience, and when we are trying to discern meaning of text, language, or action. Qualitative methods are closely aligned with the focus and interests of knowledge sciences—sense-making, understanding, deriving meaning. They are solidly anchored in the use of language and the analysis of text and speech.

When Should We Choose a Mixed Methods Approach?

A mixed methods analysis is particularly well-suited for contexts where we are trying to develop a better understanding of a research or a business problem, where interpretation of the problem is continuously evolving, where sample sizes may vary depending on what we are trying to learn, and where we may need to leverage more than one technique to collect source evidence. We used mixed methods to explain and interpret, to explore, to strengthen a single design strategy, to overcome the weakness of a particular design, and to generally look at a research problem from multiple levels.

Which of These Analytical Methods Are Supported by Tools?

The challenge with the market today is that there is no clear description of the functions that the tools support—the types of analyses that we can perform. Tools are described with words like "analytics," "semantics," "data to text," "discovery," and "language tools" without any real explanations of what they are and what they do. The only way to learn what you're getting is to have a good understanding of what you need, and a set of sample test cases to use to determine what any given tool does. Ideally, we should be able to identify tools that will align well with quantitative analyses and quantitative sources, tools that align well with qualitative analyses and qualitative sources, and a variety of tools that support different stages of a more complex mixed methods analysis. We need to understand how the tool treats words (e.g., are they reduced to data points or are the words treated with language tools?), how the tool prepares data for analysis (e.g., is there a set of routines that removes formatting, stop words, etc.?), what the analytical engine does (e.g., does it look for patterns, determine frequencies and associations, or are there multiple tools that can be "designed into a solution?"), is there an underlying dictionary and how is it used (e.g., what is the source—Wall Street Journal vocabulary, Oxford English Dictionary, etc.?).

In addition to the basic functionality of the tool, we need to understand whether the tools bundle products (e.g., they generate a search index but do not expose the results to developers, they create and assign categories but do not enable direct definition of categories, etc.), whether the tool is designed to work with particular content (e.g., social media, text messages, words in spreadsheets or databases, or large text documents), and products that intentionally blend the analysis of data and text without treating either according to its basic characteristics or behaviors. It is critical to know which analytical methods are supported and how they align with your business needs.

Which of These Analytical Methods Are Not Supported by Tools?

It is not surprising that the majority of the analytical tools today are quantitative because it has taken many more years to develop the machine readable and understandable linguistic foundations that are needed to support qualitative source analyses. And, this foundation is more complex than simple statistical or mathematical methods. In fact, analyzing language will generally require the use of multiple qualitative tools—POS taggers, linguistic register generators, knowledge modeling tools, entity extraction tools, categorization scheme tool builders, automated indexing tools, and tools that allow us to specify abstracting and indexing rules. There is no one tool that supports all of the steps in a qualitative analysis simply because each of these qualitative analyses methods works from a different set of processes and rules. Consider how a human approaches and thinks about each of these tasks, and what we need to teach a machine before it can perform the same functions. Other key tasks involved in qualitative analysis such as knowledge elicitation, knowledge modeling, and knowledge representation are inherently human.

What Opportunities Are There for Mixing Analytical Methods?

The opportunities for supporting mixed analytical methods are significant but they are dependent upon good design. Understanding where and when we can use quantitative and qualitative methods in our design is the key question. In fact, every qualitative analysis will be a mixed methods solution. We may begin with an analysis and characterization of language, extract knowledge structures, and then use quantitative methods to characterize the use of those structures in a corpus.

How in Context—Use Cases

In this section, we consider these questions as they apply to the seven use cases (Tables 12.10 through 12.16). Choices made at this point can mean the difference between success and failure.

Table 12.10 Use Case 1: Exploring the Causal Relationship between Arab Spring Revolutions and Wiki-leaked Cables

Dimension	Use Case Context
Design choices	Multistep design process that focused on the topical categorization of the content, references to geographical and organizational entities, and concept extraction of media texts.
Quantitative methods	Quantitative methods could be applied to the entities and categories generated through semantic analysis. Quantitative data analytics, though, were not appropriate for characterizing the language content. Research studies which took this approach reported achieving no significant results.
Qualitative methods	Majority of the initial preparation work was qualitative in nature—focusing on the topical description of the corpus, and the identification of targeted entities. The qualitative analysis was more rigorous through the intentional and objective development of semantic profiles based on the knowledge models of experts.
Mixed methods	Mixed methods were the preferred approach. Neither qualitative nor quantitative methods would have sufficed.

Table 12.11 Use Case 2: Emotional and Social Tone of the Discourse of the 2012 U.S. Presidential Campaign

Dimension	Use Case Context
Design choices	Because this project had so many different dimensions and two distinct populations—media and political candidates—a multi-step design process was required. This approach worked well because it allowed us to pause and check the results of each step.
Quantitative methods	Quantitative methods were used for the final analysis—to formulate scales of impact and to determine the significance of variations.
Qualitative methods	Knowledge models of factors were critical to this project. Expert models of psychological stability, reference models of language use, and concept identification and modeling were all critical to the analysis.
Mixed methods	The entire project should be seen as a mixed methods analysis.

Table 12.12 Use Case 3: Language and Knowledge Structures of Mathematics

Dimension	Use Case Context
Design choices	Primary focus was on the nature of language (linguistic register) and the knowledge structures. It was clear that there was variation in both of these by area of mathematics—it is not one universal language. Stepwise design was also critical here.
Quantitative methods	Quantitative methods were only used to describe the results and to analyze the variations across areas of practice.
Qualitative methods	Primary focus was on profiling language and extracting knowledge structures.
Mixed methods	Primarily a qualitative research analysis, but with some use of quantitative methods in the final stage.

Table 12.13   Use Case 4: Knowledge Transfer Practices among Beekeepers

Dimension	Use Case Context
Design choices	Primary focus was on the use of language in beekeeper discourse, so we began by building a linguistic register. Factors pertaining to knowledge transfer were identified, modeled and tested.
Quantitative methods	Quantitative methods were only used to describe the basic population and the nature of the interactions (i.e., communication networks), and to analyze the significance of the factors surfaced through semantic analysis.
Qualitative methods	Knowledge structures were identified from authoritative sources; models were constructed and built into profiles. Profiles were applied to each thread and each list entry.
Mixed methods	Quantitative methods were used to characterize and validate the sample population and to perform the final quantitative analysis on the results of the semantic analysis. This was a mixed methods analysis.

Table 12.14   Use Case 5: Precision of Automated Geographical Classification

Dimension	Use Case Context
Design choices	Two strategies were built into the design, one focusing on categorization profiles and the other focusing on text analytics. Two parallel designs were run simultaneously and then quantitatively tested in the final stage.
Quantitative methods	Quantitative methods were used to support the text analytics strategy, and they were used in the final stage to compare the results of both design strategies..
Qualitative methods	Qualitative methods were used to model and construct the country based categorization profile. Qualitative methods were also used to align the model with the expert profile developed by the Getty Institute.

(*Continued*)

Table 12.14 (*Continued*)   Use Case 5: Precision of Automated Geographical Classification

Dimension	Use Case Context
Mixed methods	The categorization profile leveraged mixed methods, though the text analysis strategy leveraged only quantitative methods. One strategy was not an example of mixed methods, but the entire project as a whole was.

Table 12.15   Use Case 6: Analysis of Physician-Patient Communications

Dimension	Use Case Context
Design choices	The design is a machine-transformation of existing manually applied profiles. Six profiles were found in the literature, and each was interpreted and transformed for machine processing. This required first generating knowledge models of each factor from each profile, then developing the rules to represent that factor.
Quantitative methods	The only quantitative methods are those that measure the thresholds established for categorization. We have profiles to work with but we do not have access to highly confidential manually coded conversations from other researchers. There is not a control group for us to work with.
Qualitative methods	The majority of the analysis is grounded on qualitative methods used to create profiles of medical and health language.
Mixed methods	The majority of the analysis here was qualitative rather than quantitative. Quantitative methods would have introduced additional risk, and risk would not have been acceptable for tagging future conversations with a machine-based profile.

Table 12.16   Use Case 7: Trafficking in Persons Semantic Discovery

Dimension	Use Case Context
Design choices	Two methods were used—text analytics and semantic categorization. The intent was to compare and contrast the results, and to identify the best fit role for each in the organization's information management and discovery strategy. Two parallel analyses were implemented and compared in the final step.
Quantitative methods	Quantitative methods in the form of stochastic clustering were used for one track. The results were not well suited to integration and discovery in a dynamic environment and across different types of language and domains.
Qualitative methods	The semantic profile was constructed using concept extraction and knowledge modeling methods. The profile was labor and consultation intense to construct but it is sustainable over time.
Mixed methods	Semantic categorization profile strategy was mixed methods. Text analytics strategy was largely quantitative. In total, the project was mixed methods.

What Sources Are We Analyzing?

It is important to distinguish between quantitative data or evidence (represented as numerical or structured and interpreted) and qualitative data or evidence (expressed through human language). Quantitative sources and qualitative sources are fundamentally different in their structures, interpretation, meaning, and behaviors. Reducing language to data or digits removes context and meaning. Adding meaning back into the results after the data processing and cleansing step of text analytics introduces additional subjectivity that can either compromise the interpretation or render the results invalid. Human knowledge and understanding is still the best source of evidence we have to support analysis. This is the area of greatest potential and opportunity for knowledge organizations. It is also the dimension with the greatest threat of failure. How will the machine treat or process the language? It is not sufficient to build common language dictionaries and stemming algorithms into a product and assume that this means the machine will understand the language. We need to stop and ask what components need to be designed into the application for a machine to process language as well as a human. These components are available today but they are not always present in a "text analytics" tool.

What Is the Nature of Language, Information, and Knowledge We're Using as Source Evidence?

Text, speech, and knowledge are all qualitative sources of evidence. Qualitative sources of evidence are expressed in words. Words have meaning and are embedded into linguistic structures that support the language of expression. Words cannot and should not be treated as "data points" without a full understanding of the language. Reducing language to make it palatable for quantitative analysis does not surface or expose meaning—it masks and buries it. Language is essential to understanding meaning, to knowledge transfer, to teaching machines how to communicate, understand, and manipulate knowledge. Any analysis of text, speech or expressed thoughts must begin with an understanding of the underlying language patterns and variations.

Not all languages have the same linguistic structures. Not all uses of language are comparable even within the same language. The context in which language is used varies for a single individual, across communities, across disciplines, and across social classes. Humans intuitively understand these variations and bring that understanding to any manually performed qualitative analysis. When we ask a machine to perform that same type of analysis, the machine must be grounded on a foundation of language understanding. No computer-based analysis of text, speech, or expressed thoughts can be valid unless it takes into consideration the basic elements of language. To humans, language looks simple because after we have learned it we no longer think about its construction and use. We just use it. But language is complex. Humans doing qualitative analysis understand language—machines doing qualitative analysis must be able to attain a similar level of understanding if we are to trust their results. There are five levels (Figure 12.1) that explicitly and formally describe a language. Machines must be able to understand and process all five levels. The five levels are phonetics, phonology, morphology, syntax, and semantics.

> *Level 1*: Phonetics represents the most primitive signal level—whether expressed as text or acoustic signals. Phonetics is generally represented in a language's phonetic alphabet. This is the level at which the machine begins to recognize language—it understands symbols. Without this level of recognition there is nothing on which to build. An important system level issue here is representation of symbols in Unicode. If the tool you are using does not require text in Unicode, you are not working with a language-based tool.

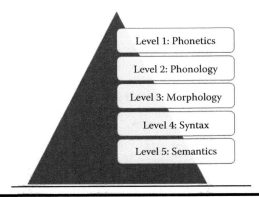

Figure 12.1   Five levels of language.

Level 2: Phonology is a pattern or sequence of phones (or sounds) or text patterns sequence of letters. This is the first level at which we have formal patterns of sounds and symbol patterns. This level of understanding is also essential for audio and speech recognition, for translating sound into text patterns, and for smart stemming.

Level 3: Morphology deals with the building of word forms from phonemes and their lemmas. Morphology helps us with dictionary lookup by identifying dictionary headwords, and with and understanding its grammars. Every language has morphological rules and behaviors—such as inflection, derivation, compounding, and so forth. This level of language understanding helps the computer to understand some of the rules for building and also deconstructing words in a particular language. It is not simply a matter of the machine identifying nouns and verbs, but determining the tense of the verb, whether the noun is singular or plural, and the nature of the adjective. All of this knowledge—which humans bring to a manual qualitative analysis task—must be engineered into computer-level understanding.

Level 4: Syntax is the acceptable structure of a sentence in a language. Syntax is important because it is the first level at which the machine begins to understand the context of the language we wish to analyze. Syntax also is the level at which the complexity of language becomes evident. We think of syntax in terms of open and closed POS. Closed categories typically do not change—they do not decline. Closed categories include preposition, conjunction, article, interjection, and particles. This is a limited (finite) set of words. It is easier to build these into a language base. Prepositions (of, without, by, to), coordinating conjunctions (and, but, or, however), subordinating conjunctions (that, if, because, before, after, although, as), articles or determiners (a, an, the), interjection (wow, eh, hello), and particles (yes, no, not). Open POS typically can change—through inflection, declension, newly created words, transforming words. Open categories include verbs, nouns, pronouns, adjectives, and adverbs—each important to analyzing language. For example, the investigation of conditional statements will rely heavily on adverbs. The exploration of hyperbolic speech or superlatives will focus on adjectives. Verbs form the key representation of actions or operations—for example, we would focus on verbs to study medical advice or a team's activities. If we cannot isolate and explore verbs, we cannot effectively portray how a call center is advising callers. It is not sufficient to work simply with the roots or lemmas of these words—in some cases the tense or the declension is a critical aspect of the analysis. Working with a tool that simply stems, matches letters, and looks for patterns among words does not support this level of exploration. Reducing language to "data points" may simplify language for processing, but in so doing we lose the meaning and context—understanding.

What Meaning or Understanding Are We Deriving from the Language?

Semantics is level 5 in the characterization of language. This is the highest level of language understanding and representation. Computer applications that purport to understand and interpret "text" or "speech" must have the building blocks of levels 1 through 4. There can be no derivation of meaning from text, speech, or any other expression of language without morphological (level 3) and syntactical (level 4) analyses. It is at level 5 that humans begin to teach the machine what we understand and what we know. We do this by modeling knowledge and translating what we know intuitively into linguistic registers and knowledge structures, and by constructing knowledge-based processes.

What Linguistic Registers Are Represented in the Source Evidence?

Language varies by social context, by economic class, and by subject discipline. Most of the work on linguistic registers is systematic qualitative research by linguists. A linguistic register is a set of specialized vocabulary and preferred (or unpreferred) syntactic and rhetorical structures used by specific socioprofessional groups for special purposes. Bar-Hillel (1966), Barton and Hamilton (2012), Berman (2014), Berman and Nir (2009), Berman and Verhoeven (2002), Childton (2004), Cope and Kalantzis (2000), Eggins (2004), Eggins and Slade (2005), Fairclough (2003, 2013), Ferguson (1983), Finegan and Biber (1994), Gee (2015), Halliday, McIntosh, and Strevens (1964), Henzl (1973), Kaplan and Berman (2015), Kress (2009), Neumann (2014), Ravid and Berrman (2009, 2010), Schiffman (1997), Spigelman (1999), and Ure (1982) were among the first to address the idea of registers. Linguistic registers represent a person's ability to adapt their use of language to conform to standards or traditions in a given professional or social situation. The human-equivalent level of understanding of a context or domain looks exactly like a fully-elaborated linguistic register. Over their lifetime, humans develop and use a repertoire of linguistic registers to use in social contexts, in work environments, when talking with others in a subject domain, and when expressing emotions or spiritual thoughts. Before we can undertake an analysis of text or a corpus of text, we must understand the embedded linguistic registers.

Machines will not be able to distinguish these variations with their limited understanding of language unless we can teach them how to distinguish. To build out linguistic registers for major domains may take us as long as it did to build some of the semantic foundations of our current applications and technologies. It is well worth the effort, though, because this is the key to really building more intelligent semantic applications. The register of law is different from the register of medicine, which is different from the language of engineering and from the language of mathematics.

What Knowledge Structures Are Represented in the Source Evidence?

Critical components of any linguistic register are the structures used to represent knowledge. Each person who uses language to communicate, to share ideas or to record information is drawing upon a rich repertoire of knowledge structures. Analyzing text without understanding the linguistic registers or knowledge structures is comparable to taking uncontrolled sources of data, throwing them all into a repository without any data management effort, and expecting statistical applications to sort out all of the data variations and discrepancies for us. At the highest level, we develop models that represent knowledge structures, including declarative knowledge, procedural knowledge, and structural knowledge. Declarative knowledge tells us how and why things work the way they do, names, locations, concepts, elements, and relationships among concepts. Procedural knowledge describes the steps or activities required to perform a task, processes, and actions. Structural knowledge is used for plans, strategies, problem solving, critical thinking, and so on. At the lowest or most primitive level, though, we are talking about POS. All of these are important to creating a linguistic register for a context or domain. We begin to identify knowledge structures, though, by looking for tendencies and patters at a very primitive level; by identifying all of the noun and verb phrases that are commonly used. Elements of knowledge structures include concept names and variations on those names, definitions and concepts, related or associated concepts, and elaborated descriptions of the relationships among concepts.

What Types of Semantic Methods Are Available for Us to Work With?

Almost every analytical tool on the market today is described as "semantic." In fact, these products may not be semantic as we've defined it but data analytics, text analytics, or simple NLP applications. The important issue here is to know what you need and what the tool you've chosen is capable of doing.

What Does the Analytical Workflow Look Like for Each of These Methods?

Let's compare and contrast the workflow models for text analytics and for language-based applications. Text analytics generally follows a three step process: (1) text acquisition and preparation—identifying the source evidence and performing preliminary formatting of the text; (2) making the text readable at the computer level by producing optical character recognition (OCR)-ready text, removing stop words, parsing and chunking text into meaningful "data points"; and (3) applying statistical methods such as Bayesian statistics or k-nearest neighbor to identify patterns within the text. In contrast, linguistic or language-focused analysis begins with a statement of goals and requirements, considers the type of solution needed, proceeds to testing and ultimately to deployment. This strategy focuses on understanding the source evidence and aligning the design with the goals. Knowledge modeling and representation, as well as consultation with subject matter experts, and existing theoretical models are important elements of this analytical approach. The goal is to teach the computer what we are looking for and how a human would think about it. This is in contrast to focusing on the text, preparing the text for data analysis, and then applying the analytical methods.

Do These Methods Offer the Possibility of Reducing or Eliminating the Subjective Interpretation Element?

Reducing subjectivity and increasing objectivity should be a primary goal whenever we are working with qualitative sources of evidence. Our ability to achieve this goal, though, is dependent upon our ability to objectively model the problem we are exploring and the knowledge we are using as source evidence. Any new reference or conceptual models we construct of factors should be derived from the work of experts in those fields—wherever it exists. In addition, any models we construct should be validated with subject matter experts before they are applied to text. We reduce subjectivity by exposing our assumptions and constructions at each step of the analytical process.

What Is the Return on Investment for the Analysis?

The return should be greater than the investment. This may sound simple and logical, but it is easy to become enamored with the tools and to overinvest in an analysis. While the marketing and promotional materials lead us to believe that every application will be cost effective, these materials do not take into consideration the setup, configuration, modeling, training, and operational support the tools require. Where there is a good understanding of the level of investment required and where the return is greater than the investment, we have a good opportunity.

What Level of Effort Is Involved in Doing a Rigorous Analysis?

The answer to this question is—it depends. Some applications can take up to six months and require one or more fulltime employees. The scale of the effort depends on the complexity of the design model, the complexity of the language, and the scale of the application. It is important to understand the organization's tolerance for or expectation of resources required. Many organizations may expect little or no investment based on the promotional materials. A general rule of thumb is that if effort and thought are invested in the development of the model and methods, the level of effort to apply will be reduced.

What Types of Competencies Are Needed to Support the Analysis?

The competencies required to work with qualitative sources, to model and characterize language, and to develop machine-based solutions include linguistic methods, linguistic registers, knowledge modeling and structure methods, knowledge engineering, knowledge elicitation, natural language programming, statistics, general research methods, subject matter expertise, and knowledge and information organization methods. Any successful application clearly requires a team effort.

What Is the Feasibility of the Analysis Without Semantic Methods?

None of the seven use cases we have presented would have been feasible or economically practical without semantic methods. While semantic methods may not have been the only method we used, it played a critical role in each use case. It will be the case that any analysis that leverages language-based source evidence requires the use of semantic methods at some point in the design.

Understanding Sources in Context—Use Case Examples

As the use case explanations suggest, the most important thinking point is the nature of the source evidence and our characterization and representation of language. The key questions pertaining to sources are considered in the context of the seven use cases (Tables 12.17 through 12.24).

Table 12.17   Use Case 1: Causal Relationship between Arab Spring Revolutions and the Release of Wiki-leaked Cables

Dimension	Use Case Context
Nature of information and knowledge	Diplomatic cables from Tunisia and Egypt, media reports of the Arab spring, other treatments and discussions of the nature of diplomatic language.
Linguistic register(s)	Yes, linguistic registers were critical to understanding and validating the nature of diplomatic language, as well as the language of media reports.
Knowledge structures	It was important to expose and describe the types of knowledge structures included in diplomatic cables. We found people, organizations, locations, events, and topics.
Semantic methods used	Concept extraction, entity extraction, categorization
Design workflow	Language-focused workflow
Competencies needed	Knowledge of diplomacy, knowledge of the foreign policy interests in the countries, linguistic competencies, understanding of media reporting and language.
Return on investment	Very high in the longer term if the profile can be used to model and assess the intelligence information that is being sent back to headquarters. If the incoming intelligence is equivalent to media reports from the country, there is a need to improve foreign intelligence gathering work.
Qualitative risk reduction	Because of the nature of diplomatic language, there is little emotion or personal perspective expressed. This language carries a lower risk for analysis.

(Continued)

Table 12.17 (*Continued*)   Use Case 1: Causal Relationship between Arab Spring Revolutions and the Release of Wiki-leaked Cables

Dimension	Use Case Context
Feasibility without semantic tools	The multifaceted nature of the semantic profile suggests it would take a very long time to complete the analysis. Also suggests that manually operationalizing the analysis in the future is not practical.

Table 12.18   Use Case 2: Emotional and Social Tone of the Discourse of the 2012 U.S. Presidential Campaign

Dimension	Use Case Context
Nature of information and knowledge	Speech transcripts from candidates, media transcripts of television programs.
Linguistic register(s)	Linguistic registers were generated for individual test populations, and for the full set.
Knowledge structures	Knowledge structures were developed around the expertise of psychologists and psychiatrists. Entity based knowledge structures were used to examine the density of information in the language of both groups.
Semantic methods used	Concept extraction, categorization
Design workflow	Language based workflow was used.
Competencies needed	Subject matter expertise in area of emotional and social stability defined by psychiatrists and psychologists; linguistic competencies, knowledge modeling, categorization scheme construction, entity extraction.
Return on investment	The return on investment was significant because it clearly explained and supported the depressive nature of media and political discourse. It also provides a profile for analyzing speech as it is delivered.
Qualitative risk reduction	Because the project leveraged semantic profiles developed by psychologists and psychiatrists the qualitative control points are strong.

(*Continued*)

Table 12.18 (*Continued*)   Use Case 2: Emotional and Social Tone of the Discourse of the 2012 U.S. Presidential Campaign

Dimension	Use Case Context
Feasibility without semantic tools	Without the machine-based tools this project would not have been possible or feasible. The complexity and labor intensive nature of the analysis would have been cost prohibitive. In addition, without the expert profiles, the degree of subjective interpretation would have been significant.

Table 12.19   Use Case 3: Language and Knowledge Structures of Mathematics

Dimension	Use Case Context
Nature of information and knowledge	Mathematical textbooks, mathematical instruction materials, mathematical research papers in algebra, geometry, trigonometry, advanced algebra, calculus, discrete mathematics, industrial mathematics, and general advanced research topics.
Linguistic register(s)	Linguistic registers were generated to understand the use of verbs in particular as indications of how operations and functions are explained.
Knowledge structures	Knowledge structures were built around noun phrases to represent the basic concepts and objects addressed in each area of mathematics.
Semantic methods used	Simple part of speech tagging, simple concept extraction.
Design workflow	Language focused workflow was followed.
Competencies needed	Basic competencies such as language modeling, and interpretation of verb phrases. Knowledge of mathematics was also important.
Return on investment	In the longer term, the return can be significant if the conclusions and findings are adopted by teachers of mathematics and writers of mathematical textbooks.

(*Continued*)

Table 12.19 (*Continued*)   Use Case 3: Language and Knowledge Structures of Mathematics

Dimension	Use Case Context
Qualitative risk reduction	Because there is a high level of fear of mathematics among the general publication, managing the risk of subjective interpretation was important. The fundamental nature of the analysis kept this risk to a minimum.
Feasibility without semantic tools	The intensity and the simple density of the language of mathematical textbooks make this project cost prohibitive and labor intensive without machine-based methods.

Table 12.20   Use Case 4: Knowledge Transfer Practices among Beekeepers

Dimension	Use Case Context
Nature of information and knowledge	Listserv conversations captured as threads over a 3-month period. Three different Listservs were monitored during the most active beekeeping season of the year.
Linguistic register(s)	Linguistic registers were created to understand the language of beekeeping. The register was compared to the language used in beekeeping training materials, beekeeping textbooks, and beekeeping course materials. Also analyzed was the language of knowledge transfer, sharing, and failure exchange.
Knowledge structures	Knowledge structures were extracted and compared to those used in primary and secondary source materials.
Semantic methods used	Language modeling, part of speech tagging, semantic categorization of knowledge transfer, and failure exchange.
Design workflow	Language modeling workflow was followed.
Competencies needed	Practical knowledge of beekeeping, linguistic knowledge, semantic analysis strategies.

(Continued)

Table 12.20 (*Continued*)   Use Case 4: Knowledge Transfer Practices among Beekeepers

Dimension	Use Case Context
Return on investment	The return will be realized by organizations as the research moves forward. The study explores the knowledge sharing behavior outside of formal organizations. The next step in this research is to determine how those behaviors change when the context changes and individuals no longer have a direct stake in success.
Qualitative risk reduction	There is a risk of subjective interpretation inherent to all knowledge sharing studies. By focusing on language and neutralizing the subjective elements through profile development, there is an opportunity to reduce risks.
Feasibility without semantic tools	The volume of data could have been managed manually. However, the subjective interpretation of discussion of errors and knowledge sharing exchanges would have presented challenges for the validity of the research.

Table 12.21   Use Case 5: Precision of Automated Geographical Classification

Dimension	Use Case Context
Nature of information and knowledge	Internal documents of various types from an international development organization. Documents had been country classified manually to provide a robust control group. Documents and their metadata were used in the research analysis.
Linguistic register(s)	Registers were not created because we focused on a single entity—country.
Knowledge structures	Two knowledge structures were developed—one a simple representation of country names and variations of those names, and the other a fully elaborated country classification that included all geographical features, the nationalities and ethnic groups in the country, names of all cities, and administrative jurisdictions.

(Continued)

Table 12.21 (*Continued*)   Use Case 5: Precision of Automated Geographical Classification

Dimension	Use Case Context
Semantic methods used	Simple entity extraction, rule-based categorization.
Design workflow	Language-focused workflow was followed for the second strategy, and simple text analytics workflow was followed for the first strategy.
Competencies needed	Knowledge of geographical descriptions, categorization, knowledge of authoritative sources.
Return on investment	Potential return is significant because this is currently a process which is prone to high error rates and to high costs from manual work.
Qualitative risk reduction	The risk is of type 1 and type 2 errors, but not associated with subjective human interpretation of source materials.
Feasibility without semantic tools	Without machine-based methods, it would have been cost prohibitive to test the categorization profile. The complexity and number of match points in the profile is too large for humans to compare against individual documents.

Table 12.22   Use Case 6: Analysis of Physician-Patient Communications

Dimension	Use Case Context
Nature of information and knowledge	Videotaped conversations of doctors, interns, and patients which were manually transcribed by graduate students.
Linguistic register(s)	Linguistic registers were created to understand the language of communication between these different stakeholders. Part of speech tagging was critical to this project.
Knowledge structures	Knowledge structures were defined through reference to the six expert profiles found in the literature. Concept extraction was used to identify factors highlighted in the profiles.
Semantic methods used	Part of speech tagging, concept extraction, categorization.

(Continued)

Table 12.22 (*Continued*) Use Case 6: Analysis of Physician-Patient Communications

Dimension	Use Case Context
Design workflow	Language-focused workflow.
Competencies needed	Communications expertise, medical and health language expertise, semantic analysis methods.
Return on investment	Return could be significant in terms of scaling up the profiling of existing communications, and transforming what we have learned to teaching methods for medical students and doctors.
Qualitative risk reduction	Subject risk for this area is very high depending on one's role and perspective. An objective analysis of the discourse can identify ways that communication can be improved. There is a potential to impact health outcomes by improving communication and trust between doctor and patient.
Feasibility without semantic tools	The risk of subjective interpretation is high, and this risk can impact the trust and validity of any research based on manual coding.

Table 12.23 Use Case 7: Trafficking in Persons Semantic Profile

Dimension	Use Case Context
Nature of information and knowledge	Department of State trafficking in persons annual reports, discussions of facets of trafficking in research papers, in law enforcement, in health services, and in economic contexts. Leveraged reports and language from other organizations also working in this area such as the UN, ILO, WHO, USAID, and so on.
Linguistic register(s)	Linguistics registers were generated to better understand the variations in language used across stakeholder groups.
Knowledge structures	Knowledge structures are primarily focused on noun phrases; however, the profile also exposes attitudes toward trafficking in persons expressed through the use of adverbs and adjectives.

(*Continued*)

Table 12.23 (*Continued*)   Use Case 7: Trafficking in Persons Semantic Profile

Dimension	Use Case Context
Semantic methods used	Part of speech tagging, concept extraction, categorization.
Design workflow	Language workflow is followed.
Competencies needed	Knowledge of trafficking from multiple perspectives, the language of trafficking in a diplomatic context, knowledge of extensional and intentional language practices, and general semantic methods.
Return on investment	Potentially significant if the profile can be used to identify and integrated all treatments of trafficking across the organization. At this time, that information is scattered and challenging to find.
Qualitative risk reduction	This is a highly sensitive and emotional topic. Subjective interpretation is an inherent challenge. In addition, the information on this topic is hard to find—it is not a comfortable topic for most to talk about or read about.
Feasibility without semantic tools	Not economically feasible across the organization without machine-based methods. Manual approach is also at high risk for subjective interpretation.

Table 12.24   Successes Achieved in the Sample Use Cases

Use Case Number	Success Factors
1	Topical characterization of content of diplomatic cables. Discovery of intelligence content in the cables.
2	Use of Gottschalk–Gleser semantic profile to characterize political discourse and media coverage.
3	Initial focus on linguistic registers and knowledge structures which exposed the abstract and symbolic nature of mathematical language and its relation to young students' ability to learn concepts.
4	Development of a profile for discussing and exposing mistakes and failures. This profile can be reused in other contexts in the future.

(Continued)

Table 12.24 (*Continued*)   Successes Achieved in the Sample Use Cases

Use Case Number	Success Factors
5	Both approaches produce high levels of type 1 and type 2 errors which had heretofore not been investigated in depth.
6	Discovery of existing doctor-patient communication profiles by experts in the field.
7	Having an authoritative reference model and nationally recognized experts to work with in constructing and testing the profile.

Fitting Tools to Methods and Sources

Organizations in the industrial and the knowledge economy should use every analytical method and every source available to achieve their goals. It is easier to select and apply quantitative analytics than qualitative analytics, and there are more quantitative analysis products on the market today. Qualitative analytics, though, holds great promise—particularly for knowledge organizations and for knowledge management researchers. Regardless of the type of organization, the business or research focus, it is important to choose a tool that fits the methods and the sources of evidence. The 27 questions can serve as a guide ensuring a good fit.

The challenge we face when selecting a tool is that not all of the commercially or open-source tools described as "semantic" are constructed on top of such a foundation. The technologies we need to do qualitative analysis should be built on a strong foundational understanding of language—at least one language. They must be able to interpret any content we give them at a primitive language level. They must also be able to take further instructions from us on what to do with their understanding of the content. Knowledge scientists and business analysts must be able to instruct a computer to extract and categorize the adjectives used in a corpus of text or speech, or to describe the use of pronouns, or conditional verbs. These can be important analytical elements. In this chapter, we distinguish between truly semantic technologies—which have these capabilities—and stochastic technologies—which do not have these capabilities. A stochastic application will apply quantitative or statistical methods to text with the intent of determining frequencies or descriptive statistics of the use of words, or infer categories from the patterns of usage it finds in the text. These types of technologies may provide some level of information, but they are not comparable to the level of qualitative analysis that is performed by humans based on human understanding.

While the market appears to be rich and robust, it is also volatile and difficult to navigate. A review of the products on the market over the past ten years might show that over a 3- year period close to 40% of the tools will have disappeared or have been integrated into another product. The volatility is more often found on among language and linguistics focused tools because they require a significant investment to create and to sustain. These are also the tools that have the greatest value for enterprise applications, so they are often purchased and integrated as a component of another transactional or content management system. As consumers and producers of these tools, we need to lobby for a more accurate portrayal of the functionality of the tools.

Reviewing Your Tool Choices

If you have addressed all of the key questions to this point, you're ready to select a tool or multiple tools to support your project. You should be prepared to develop functional requirements, evaluate and test products before your purchase or acquire. While acquiring a "free" or open source tool might seem attractive, it could be a poor choice that suboptimizes your business or research goals. You should be prepared to explain your choice based on evidence, not on marketing materials, a sales promotion, or a conference presentation. You should be able to explain which tools you considered and how you evaluated the tools. And you should consider a proof of concept test as a condition of the procurement.

Reviewing Your Successes

The field of analytics and the field of knowledge management are emerging, even though both can be traced back to the 1960s. It is important for everyone who is working or experimenting in this area to share what they have learned, both successes and failures. Did you achieve your goal? Did the tools behave as expected? Was the solution affordable? What other costs did you incur? Did your project generate new research projects? Did it change the way we think about a business issue? Were you able to explore a topic that would not have been possible without these tools? Table 12.24 describes how we would describe our successes if they were applied to the use cases. We note that none of the success factors is related to a tool selection or to the functionality of the tool. This is because the tool selection was the last decision.

Exposing Your Failures

It is equally important to share your failures and lessons learned. Where did your solution fall short? How would you change the strategy if you were to reuse it in the future? Were the results valid? Were they reliable? Based on what you learned, would you not undertake the project? Were your expectations reasonable or unreasonable? What was the most significant thing you learned? How would you explain your failure to achieve your goals? Table 12.25 describes the lessons learned and failures experienced in the use cases. We note that none of the lessons learned were related to or a result of a tool selection or to the functionality of the tool. Again, this is because the tool selection was the last decision.

Table 12.25 Lessons Learned from the Sample Use Cases

Use Case Number	Lessons Learned
1	Late discovery that Wikileaks was credited with causation of Arab Spring by only a few, but highly visible, media outlets. The more accurate characterization was the increased use of social media by those involved in the protests to chronicle the activities.
2	Difficulty of discovering and obtaining the political transcripts and the variability of quality of the texts. Need for four distinct tools to support he research.
3	Rather than generating registers of full textbooks and other materials, we could have been more selective of portions of textbooks. Having generated the linguistic register, we can now adapt our future strategy.
4	We learned that the three months of communications were representative of the most active periods of the year. A more representative sample would have covered a full calendar year.
5	The extensive development of the country categorization profile introduced some errors particularly across countries in a region—where there were common naming practices.
6	Labor intensive effort to transcribe the videotaped interactions. Next round of research we might test audio-to-text transcription methods.
7	The size of the authoritative reports and the need to break them into facet-focused chunks prior to doing any linguistic analysis or extracting any knowledge structures.

Lessons Learned and Future Work

Working from the guiding questions, the use cases and other practical experience, we offer 15 general lessons learned.

Lesson learned 1: Knowledge organizations have much to gain from wise and considered use of the full range of analytical tools.

Lesson learned 2: Using analytics wisely means understanding the difference between data and text analytics, good practice business intelligence models and methods, and the nature and representation of information and knowledge sources.

Lesson learned 3: Managing expectations is critical—promotional materials may lead administrators to think they have found a silver bullet solution to qualitative analysis. There is a need to explain how the tools work, what they will produce, and what resources are required to achieve those results.

Lesson learned 4: The availability of new smart tools does not override the need for good analytical design. Fitting the right methods to the source data and aligning with business or research goals will likely involve multiple tools and methods. There are no silver bullet solutions on the analytics or the knowledge management side. As the use cases suggest, the challenges and risks associated with poor choices and poor designs are high. Wise design choices can generate meaningful insights.

Lesson learned 5: Knowledge management methods such as knowledge modeling, knowledge elicitation, and knowledge representation are critical to the wise use of analytical tools. The profession needs to be more deliberate in their use with analytics.

Lesson learned 6: Validity, reliability, and trust risks arise when applying data analytics tools to knowledge expressed in language. These risks are mitigated where we approach data analytics with a deep understanding of language and qualitative sources, and apply the appropriate analytical methods.

Lesson learned 7: Mixing methods and sources does not create a new analytical method and it certainly does not create a better analytical strategy. The application of quantitative methods to qualitative data introduces a high risk of error.

Lesson learned 8: Text analytics, business analytics and semantic analysis are not WYSIWYG. Smart choices are made by researchers and decision makers who understand methods and sources and who are willing to critically examine the tools.

Lesson learned 9: The burden of bridging the gaps between business analytics and knowledge management—to create the common ground—lies primarily with the knowledge management discipline because these professionals have a deeper understanding of knowledge, language, and qualitative methods.

Lesson learned 10: Academics and practitioners should work together to expand the body of knowledge around language-based semantic applications. Academic-business partnerships and academic-developer partnerships can go a long way to growing a more robust and stable market. Most universities or institutes are not set up to undertake this level of investigation, the best tools are too expensive for academic researcher to use, and open source tools are often unusable by anyone other than the developer.

Lesson learned 11: The tool developers and promoters should be encouraged to describe the functionality of their products and services accurately, and consumers must be willing and able to critique and provide constructive feedback to developers.

Lesson learned 12: Text analytics and semantic analysis methods are not silver bullets. They require investment to set up, configure, and sustain. They have significant costs so most businesses get one chance to make the right choice.

Lesson learned 13: Research papers should describe in detail the tools used and their capabilities. Peer reviewers should be willing to critique tool choices and to question their use.

Lesson learned 14: The two most significant challenges for selecting a tool are understanding the nature of the products that are available on the market today, and working towards a stable and growing market for tools.

Lesson learned 15: The volatility of the market is related to the level of vendor investment required to sustain these tools and the level of organizational investment required to maintain and operationalize them. As a result, the "quick-fix" tools tend to dominate the market. A greater presence in the market, though, does not signal a long product life span—the text analytics tools disappear from the commercial market at the same rate as the language-based applications. Language-based products may have a longer life span where they are integrated into a larger enterprise application (e.g., SAP, SAS). The effect for the consumer is the same, though—they disappear from the open market. Tools that support academic research—computational linguistic tools—tend to have a longer life span. However, these tools are generally not business or designer-friendly.

Conclusions

We know from experience that we can improve the effectiveness of data and text analytics by leveraging knowledge management methods such as knowledge modeling, knowledge representation, and knowledge engineering. The opportunities and rewards of applying analytics to knowledge management challenges hold equal opportunities. The opportunities for knowledge management professionals lie primarily in the transformation of qualitative methods to machine-based methods. The challenge, though, is the amount of design, thinking, and investment required to achieve those results. What we have learned over the past 45 years is that this is possible if we have dedicated academic and development resources to support the transformation. The research questions we can pose and explore have increased exponentially. These questions would have been too manually intensive or too time consuming to investigate in the past. Business decisions that have had to rely on experience and "gut instincts" can now be tested with machine-based qualitative solutions.

We need to learn to think more expansively and creatively in defining our research agendas and setting our business goals. Creativity and discovery begins with the first strategic dimension—setting expectations and the framework for analysis and solution. It means modeling knowledge and thought processes, applying appreciative inquiry process at every step of the way, expecting to do discovery at a micro level with macro level impacts. It means having the opportunity to increase our understanding of human knowledge and remapping knowledge, and achieving greater exposure for tacit knowledge. Knowledge management practitioners also need to deepen the language-focused components of the discipline. This means collaborating with developers of analytics tools, and with bridging the gaps that exist across disciplines.

References

Abbasi, A. A., & Younis, M. (2007). A survey on clustering algorithms for wireless sensor networks. *Computer Communications, 30*(14), 2826–2841.

Abilhoa, W. D., & De Castro, L. N. (2014). A keyword extraction method from Twitter messages represented as graphs. *Applied Mathematics and Computation, 240*, 308–325.

Adcock, R. (2001). Measurement validity: A shared standard for qualitative and quantitative research. *American Political Science Review, 95*(3), 529–546.

Anwar, W., Wang, X., & Wang, X. L. (2006). A survey of automatic Urdu language processing. In *Machine learning and cybernetics, 2006 international conference on* (pp. 4489–4494). IEEE.

Arya, S., Mount, D. M., Netanyahu, N. S., Silverman, R., & Wu, A. Y. (1998). An optimal algorithm for approximate nearest neighbor searching fixed dimensions. *Journal of the ACM (JACM), 45*(6), 891–923.

Atkinson, P. (2005). Qualitative research—Unity and diversity. In *Forum qualitative sozialforschung/forum: Qualitative social research* (Vol. 6, No. 3). London: Routledge.

Babbie, E. R. (1989). *The practice of social research*. Belmont, CA: Wadsworth Publishing Company.

Ballard, B. W., & Biermann, A. W. (1979, January). Programming in natural language: "NLC" as a prototype. In *Proceedings of the 1979 annual conference* (pp. 228–237). New York, NY: ACM.

Bar-Hillel, Y. (1966). *Language and information; selected essays on their theory and application*. Reading, MA: Addison-Wesley.

Barton, D., & Hamilton, M. (2012). *Local literacies: Reading and writing in one community*. London: Routledge.

Bedford, D., & Platt, C. (2013, September 16). *Mathematical languages—Barriers to knowledge transfer and consumption*. Scientific Information Policies in the Digital Age: Enabling Factors and Barriers to Knowledge Sharing and Transfer, Aula Marconi, Consiglio Nazionale delle Ricerche, Rome, Italy.

Bedford, D. A. D., Bekbalaeva, J., & Ballard, K. (2017, November 27–31). *Global human trafficking seen through the lens of semantics and text analytics*. American Society for Information Science and Technology Annual Conference, Crystal City, VA.

Bedford, D. A. D., Turner, J., Norton, T., Sabatiuk, L., & Nassery, H. (2017, November 27–31). *Knowledge translation in health sciences*. American Society for Information Science and Technology Annual Conference, Crystal City, VA.

Bedford, D. A. D. (2012a). *Enhancing the precision of geographical tagging—Embedding gazetteers in semantic analysis technologies*. Poster session presented at the TKE 2012, Madrid, Spain.

Bedford, D. A. D. (2012b). *Semantic analysis of the political discourse in the presidential and congressional campaigns of 2012*. Paper presented at the Text Analytics World Conference, San Francisco, CA.

Bedford, D. A. D., & Neville, L. (2016, September). Knowledge sharing and valuation in beekeeping communities. In *Proceedings of the international conference on intellectual capital knowledge management and organizational learning*. Ithaca, NY: Ithaca College.

Bejou, D., Wray, B., & Ingram, T. N. (1996). Determinants of relationship quality: An artificial neural network analysis. *Journal of Business Research, 36*(2), 137–143. Oxford: Oxford University Press

Bengio, Y. (2009). Learning deep architectures for AI. *Foundations and Trends® in Machine Learning, 2*(1), 1–127.

Berman, R. A. (2014). Linguistic perspectives on writing development. In B. Arfe, J. Dockrell, & V. Berninger (Eds.), *Writing development in children with hearing loss, dyslexia or oral language problems: Implications for assessment and instruction* (pp. 176–186). Oxford: Oxford University Press.

Berman, R., & Nir, B. (2009). Cognitive and linguistic factors in evaluating text quality: Global versus local. In V. Evans & S. Pourcel (Eds.), *New directions in cognitive linguistics* (pp. 421–440). Amsterdam, the Netherlands: John Benjamins.

Berman, R., & Verhoeven, L. (2002). Cross-linguistic perspectives on the development of text-production abilities: Speech and writing. *Written Language and Literacy, 5*(1), 1–43.

Biermann, A. W. (1981). Natural language programming. In A. Biermann & G. Guiho (Eds.), *Computer program synthesis methodologies* (pp. 335–368). Amsterdam, the Netherlands: Springer.

Biermann, A. W., Ballard, B. W., & Sigmon, A. H. (1983). An experimental study of natural language programming. *International Journal of Man-machine Studies, 18*(1), 71–87.

Bollinger, A. S., & Smith, R. D. (2001). Managing organizational knowledge as a strategic asset. *Journal of Knowledge Management, 5*(1), 8–18.

Brandow, R., Mitze, K., & Rau, L. F. (1995). Automatic condensation of electronic publications by sentence selection. *Information Processing and Management, 31*(5), 675–685.

Carletta, J. (1996). Assessing agreement on classification tasks: The kappa statistic. *Computational Linguistics, 22*(2), 249–254.

Carmines, E. G., & Zeller, R. A. (1979). *Reliability and validity assessment* (Vol. 17). Newbury Park, CA: Sage Publications.

Chilton, P. (2004). *Analysing political discourse: Theory and practice.* London: Routledge.

Chowdhury, G. G. (2003). Natural language processing. *Annual Review of Information Science and Technology, 37*(1), 51–89.

Church, K. W., & Hanks, P. (1990). Word association norms, mutual information, and lexicography. *Computational Linguistics, 16*(1), 22–29.

Church, K. W., & Mercer, R. L. (1993). Introduction to the special issue on computational linguistics using large corpora. *Computational Linguistics, 19*(1), 1–24.

Cope, B., & Kalantzis, M. (Eds.). (2000). *Multiliteracies: Literacy learning and the design of social futures.* London: Psychology Press.

Croft, D., Coupland, S., Shell, J., & Brown, S. (2013). A fast and efficient semantic short text similarity metric. *Computational Intelligence (UKCI), 2013 13th UK Workshop on* (pp. 221–227). IEEE.

Dahl, V., & Saint-Dizier, P. (1985). *Natural language understanding and logic programming.* New York, NY: Elsevier.

Debortoli, S., Müller, O., Junglas, I. A., & vom Brocke, J. (2016). Text mining for information systems researchers: An annotated topic modeling tutorial. *CAIS, 39*, 7.

Deerwester, S., Dumais, S. T., Furnas, G. W., Landauer, T. K., & Harshman, R. (1990). Indexing by latent semantic analysis. *Journal of the American Society for Information Science, 41*(6), 391.

Dijkstra, E. W. (1979). On the foolishness of "natural language programming". In F. L. Bauer, E. W. Dijkstra, S. L. Gerhart, & D. Gries (Eds.), *Program construction* (pp. 51–53). Berlin: Springer.

Edmundson, H. P. (1969). New methods in automatic extracting. *Journal of the ACM (JACM), 16*(2), 264–285.

Eggins, S. (2004). *Introduction to systemic functional linguistics.* London: A&C Black.

Eggins, S., & Slade, D. (2005). *Analysing casual conversation.* London: Equinox Publishing.

Eriksson, P., & Kovalainen, A. (2015). *Qualitative methods in business research: A practical guide to social research.* London: Sage Publications.

Fairclough, N. (2003). *Analysing discourse: Textual analysis for social research.* London: Psychology Press.

Fairclough, N. (2013). *Critical discourse analysis: The critical study of language.* London: Routledge.

Ferguson, C. A. (1983). Sports announcer talk: Syntactic aspects of register variation. *Language in Society, 12*(2), 153–172.

Fernández, A., Gómez, Á., Lecumberry, F., Pardo, Á., & Ramírez, I. (2015). Pattern recognition in Latin America in the "Big Data" era. *Pattern Recognition, 48*(4), 1185–1196.

Finegan, E., & Biber, D. (1994). Register and social dialect variation: An integrated approach. In D. Biber & E. Finegan (Eds.), *Sociolinguistic perspectives on register* (pp. 315–347). Oxford: Oxford University Press.

Gee, J. (2015). *Social linguistics and literacies: Ideology in discourses.* New York, NY: Routledge.

Goddard, C. (2011). *Semantic analysis: A practical introduction.* Oxford: Oxford University Press.

Guion, R. M. (1980). On Trinitarian doctrines of validity. *Professional Psychology, 11*(3), 385.

Halliday, M. A. K., McIntosh, A., & Strevens, P. (1964). *The linguistic sciences and language teaching.* London: Longmans.

Halpern, M. (1966). Foundations of the case for natural-language programming. In *Proceedings of the November 7–10, 1966, all joint computer conference* (pp. 639–649). San Francisco, CA: ACM.

Han, J., Pei, J., & Kamber, M. (2011). *Data mining: Concepts and techniques.* San Diego, CA: Elsevier.

Hand, D. J., Mannila, H., & Smyth, P. (2001). *Principles of data mining.* Cambridge, MA: MIT Press.

Hatzivassiloglou, V., & McKeown, K. R. (1997, July). Predicting the semantic orientation of adjectives. In *Proceedings of the eighth conference on European chapter of the association for computational linguistics* (pp. 174–181). Morristown, NJ: Association for Computational Linguistics.

Hearst, M. A. (1992, August). Automatic acquisition of hyponyms from large text corpora. In *Proceedings of the 14th conference on computational linguistics* (pp. 539–545, Vol. 2). Nantes: Association for Computational Linguistics.

Heidorn, G. E. (1971). *Natural language inputs to a simulation programming system: An introduction.* Monterey, CA: Naval Postgraduate School.

Henzl, V. M. (1973). Linguistic register of foreign language instruction. *Language Learning, 23*(2), 207–222.

Hindle, D., & Rooth, M. (1993). Structural ambiguity and lexical relations. *Computational Linguistics, 19*(1), 103–120.

Hofmann, T. (1999). Probabilistic latent semantic indexing. In *Proceedings of the 22nd annual international ACM SIGIR conference on research and development in information retrieval* (pp. 50–57). Berkeley, CA: ACM.

Jurafsky, D. (1996). A probabilistic model of lexical and syntactic access and disambiguation. *Cognitive Science, 20*(2), 137–194.

Kaplan, D., & Berman, R. (2015). Developing linguistic flexibility across the school years. *First Language, 35*(1), 27–53.

King, G., Keohane, R. O., & Verba, S. (1994). *Designing social inquiry: Scientific inference in qualitative research.* Princeton, NJ: Princeton University Press.

Kirk, J., & Miller, M. L. (1986). *Reliability and validity in qualitative research.* Beverly Hills, CA: Sage Publications.

Kress, G. (2009). *Multimodality: A social semiotic approach to contemporary communication.* London: Routledge.

Kruger, D. J. (2003). Integrating quantitative and qualitative methods in community research. *The Community Psychologist, 36*(2), 18–19.

Lehnert, W., Soderland, S., Aronow, D., Feng, F., & Shmueli, A. (1995). Inductive text classification for medical applications. *Journal of Experimental and Theoretical Artificial Intelligence, 7*(1), 49–80.

Liddy, E. D. (2001). *Natural language processing.* Syracuse, NY: Syracuse University.

Loper, E., & Bird, S. (2002, July). NLTK: The natural language toolkit. In *Proceedings of the ACL-02 workshop on effective tools and methodologies for teaching natural language processing and computational linguistics* (pp. 63–70, Vol. 1). Philadelphia, PA: Association for Computational Linguistics.

Luhn, H. P. (1958a). A business intelligence system. *IBM Journal of Research and Development, 2*(4), 314–319.

Luhn, H. P. (1958b). The automatic creation of literature abstracts. *IBM Journal of Research and Development, 2*(2), 159–165.

Maarek, Y. S., Berry, D. M., & Kaiser, G. E. (1991). An information retrieval approach for automatically constructing software libraries. *IEEE Transactions on Software Engineering, 17*(8), 800–813.

Manning, C. D., & Schütze, H. (1999). *Foundations of statistical natural language processing* (Vol. 999). Cambridge, MA: MIT Press.

Marcu, D. (1997, July). The rhetorical parsing of natural language texts. In *Proceedings of the 35th annual meeting of the association for computational linguistics and eighth conference of the European chapter of the association for computational linguistics* (pp. 96–103). Madrid: Association for Computational Linguistics.

Marcus, M. P., Marcinkiewicz, M. A., & Santorini, B. (1993). Building a large annotated corpus of english: The penn treebank. *Computational Linguistics, 19*(2), 313–330. Berlin, Heidelberg: Springer.

Mihalcea, R., Liu, H., & Lieberman, H. (2006, January). NLP (natural language processing) for NLP (natural language programming). In A. Belbukh (Ed.), *International Conference on intelligent text processing and computational linguistics* (pp. 319–330). Berlin, Heidelberg: Springer.

Miner, G. (2012). *Practical text mining and statistical analysis for non-structured text data applications.* Amsterdam, the Netherlands: Academic Press.

Mitkov, R. (Ed.). (2005). *The Oxford handbook of computational linguistics.* Oxford: Oxford University Press.

Neumann, S. (2014). Cross-linguistic register studies: Theoretical and methodological considerations. *Languages in Contrast, 14*(1), 35–57.

Nir, B., & Berman, R. (2010). Parts of speech as constructions: The case of Hebrew "adverbs". *Constructions and Frames, 2*(2), 242–274.

Pereira, F., Tishby, N., & Lee, L. (1993, June). Distributional clustering of English words. In *Proceedings of the 31st annual meeting on association for computational linguistics* (pp. 183–190). Montreal: Association for Computational Linguistics.

Pustejovsky, J. (1991). The generative lexicon. *Computational Linguistics, 17*(4), 409–441.

Ravid, D., & Berman, R. (2009). Developing linguistic register across text types: The case of Modern Hebrew. *Pragmatics and Cognition, 17*(1), 108–145.

Ravid, D., & Berman, R. A. (2010). Developing noun phrase complexity at school age: A text-embedded cross-linguistic analysis. *First Language, 30*(1), 3–26.

Reason, P., & Bradbury, H. (Eds.). (2001). *Handbook of action research: Participative inquiry and practice.* London: Sage Publications.

Salton, G. (1970). Automatic text analysis. *Science, 168*(3929), 335–343.

Salton, G. (1986). Another look at automatic text-retrieval systems. *Communications of the ACM, 29*(7), 648–656.

Schiffman, H. (1997). *The study of language attitudes.* Philadelphia, PA: University of Pennsylvania.

Seale, C. (1999). Quality in qualitative research. *Qualitative Inquiry, 5*(4), 465–478.

Shieber, S. M., Schabes, Y., & Pereira, F. C. (1995). Principles and implementation of deductive parsing. *The Journal of Logic Programming, 24*(1), 3–36.

Shmueli-Scheuer, M., Roitman, H., Carmel, D., Mass, Y., & Konopnicki, D. (2010). Extracting user profiles from large scale data. In *Proceedings of the 2010 workshop on massive data analytics on the cloud* (p. 4). New York, NY: ACM.

Spigelman, J. J. (1999). Statutory interpretation: Identifying the linguistic register. *Newcastle Law Review, 4*, 1.

Sproat, R., Black, A. W., Chen, S., Kumar, S., Ostendorf, M., & Richards, C. (2001). Normalization of non-standard words. *Computer Speech and Language, 15*(3), 287–333.

Spyns, P. (1996). Natural language processing. *Methods of Information in Medicine, 35*(4), 285–301.

Srinivas, M., & Patnaik, L. M. (1994). Genetic algorithms: A survey. *Computer, 27*(6), 17–26.

Tesch, R. (2013). *Qualitative research: Analysis types and software.* London: Routledge.

Teufel, S., & Moens, M. (2002). Summarizing scientific articles: Experiments with relevance and rhetorical status. *Computational Linguistics, 28*(4), 409–445.

Tischer, S. (2009). U.S. Patent No. 7,483,832. Washington, DC: U.S. Patent and Trademark Office.

Tomita, M. (2013). *Efficient parsing for natural language: A fast algorithm for practical systems* (Vol. 8). New York, NY: Springer Science and Business Media.

Ure, J. (1982). Introduction: Approaches to the study of register range. *International Journal of the Sociology of Language, 1982*(35), 5–24.

Van Gijsel, S., Geeraerts, D., & Speelman, D. (2004). A functional analysis of the linguistic variation in Flemish spoken commercials. In G. Purnelle, C. Fairon, & A. Dister (Eds.), *Le poids des mots. Proceedings of the 7th international conference on the statistical analyses of textual data* (pp. 1136–1144). Louvain-la-neuve: Presses Universitaires de Louvain.

Weizenbaum, J. (1966). ELIZA—A computer program for the study of natural language communication between man and machine. *Communications of the ACM, 9*(1), 36–45.

Witten, I. H., Frank, E., Hall, M. A., & Pal, C. J. (2016). *Data mining: Practical machine learning tools and techniques.* Burlington, VT: Morgan Kaufmann.

Yarowsky, D. (1995, June). Unsupervised word sense disambiguation rivaling supervised methods. In *Proceedings of the 33rd annual meeting on association for computational linguistics* (pp. 189–196). Stroudsburg, PA: Association for Computational Linguistics.

Chapter 13

Data Analytics for Cyber Threat Intelligence

Hongmei Chi, Angela R. Martin, and Carol Y. Scarlett

Contents

Introduction

Cybersecurity nowadays is a national priority due to the widespread application of information technology and the growing incidence of cybercrime. Data science and big data analysis is gaining popularity due to widespread utility in many subfields of cybersecurity, such as Intrusion detection system (IDS), social networks, insider threat, and wireless body network. The use of advanced analytic techniques, computational methods, and traditional components that have become representative of "data science" has been at the center of cybersecurity solutions focused on identification and prevention of cyberattacks.

Most companies are capable of hardening their external defenses using various information security methods available. However, companies often fail in protecting against attacks that originate from within the company. These are known as insider threats that are difficult to guard against, because employees are given a certain level of access to internal company networks that normally bypass the external security measures put in place to protect the company. One way to identify insider threats is through the use of predictive analysis monitoring social networks. Social networking is a form of communication over virtual spaces (Castells, 2007). Monitoring such networks requires tools that can readily analyze linguistic patterns.

Analyzing the huge amount of content in social networks needs linguistic tools such as linguistic inquiry word count (LIWC), Stanford's Core NLP Suite, or SentiStrength. Natural language processing is the ability of computers to understand what users are writing using frequently used human writing patterns. This means the computer must be capable of identifying various dialects, slang, and even properly identify homonyms to better interpret human language. Most NLP algorithms are based on statistical machine learning techniques, which use statistical inferences. These inferences come from rules gained from analyzing large sets of documents known as corpora which contain the correct values that need to be studied (Chi et al., 2016). An example of using NLP is the auto selection in the Google search bar. As the user types a query into the search bar, Google's search algorithm considers the words that the user has already typed and compares them to the words most often typed by other users. This allows Google to provide users with options and the ability to auto-complete the remaining statement based on these similar searches. Such tools could aid in detecting insider threats associated through identifying conversations that show some agenda or malice.

In this chapter, we focus on the potential use of computational methods in intelligence work, insider security threats, and mobile health. The standard bearer for mobile computing is the smartphone, which connects people anytime and anywhere. Smartphones also store large amounts of personal information and run applications that may legitimately, inadvertently, or maliciously manipulate personal information. The relatively weak security models for smartphone applications, coupled with ineffective security verification and testing practices, have made smartphones an ideal target for security attacks. Advancing the science, technology

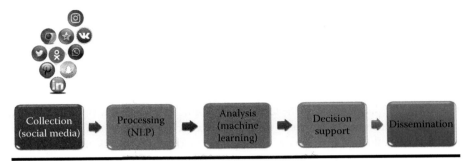

Figure 13.1 Cyber intelligence processing.

and practices for securing mobile computing are essential to support the inevitable use of mobile computing in areas with even requirements for privacy, security, and resistance to tampering (Figure 13.1).

Cyber Threat Intelligence

Cyber threat intelligence (CTI) has become a primary focus in the intelligence community as cyber threats invade almost every branch of national security. On February 23, 2000, the Central Intelligence Agency (CIA) briefed the Senate Select Committee on Intelligence that foreign cyber threat was one of the key challenges to the nation acknowledging the issue as a developing form of warfare for weaker nations unable to compete with the United States in terms of raw power. In this briefing, the CIA emphasized that both China and Russia saw cyber warfare as a viable tool of war. The briefing also pointed out that cyber capabilities in the hands of terrorists posed an uncontrollable threat as non-state actors are devoid of normal deterrents that would prevent powerful nations from conducting cyberattacks (CIA, 2000). In 2002, the Federal Information Security Management Act (FISMA), was passed into federal law acknowledging cybersecurity as inherent to the nation's security and mandating agency wide program development (U.S. Congress, H.R. 3844, 2002).

Since then, the United States has experienced spectacular, costly, and embarrassing cyberattacks, the exposure of which has had major implications for U.S. security. Titan Rain is the code name for Chinese-based cyberattacks against U.S. government computers that began in 2003 and continued for over several years stealing terabytes of government data (Thornburgh, 2005). The Department of Homeland Security (DHS) computers were hacked over a period of three months in 2006 (Nakashima and Krebs, 2007). The Secretary of Defense was hacked to exploit the Pentagon in 2007 (Sevastopulo, 2007). The National Security Agency (NSA) revealed in a high-level report that the private emails of top officials in Obama's administration were hacked in 2010 (Windrem, 2015). Iran, in 2011, claimed to have hacked a U.S. drone (Rawnsley, 2011). Although the United States never publicly acknowledged the hack, Obama did acknowledge the capture of the drone (CNN Wire Staff, 2011). In response to these events, the DOD Defense Science Board (2013) released a report detailing their assessment of the cybersecurity threat to national security with the recommendation that the U.S. government take cyber threat seriously and rethink the way they operate (2013). Shortly thereafter, in March of 2014, the Office of Personnel Management (OPM) systems and two government contractors were hacked affecting over 4.2 million employees (Koerner, 2016). The Senate, in response to this attack, passed the Federal Information Security Management Act of 2014 which redirected the security oversight of the OPM systems to the Department of Homeland Security and established a Federal Information Security Incident Center (U.S. Congress, S-2521, 2014).

Despite the government's best efforts to develop and improve the nation's cybersecurity posture, the implementation plan did not happen fast enough. In 2015, the Pentagon's joint staff unclassified email system was taken down for two weeks due to an attack suspected to originate from Russia (Bennett, 2015). The same OPM hacker, in a similar attack, stole personal information from 21.5 million people all related to government service (Davis, 2015). Events such as these brought to the forefront the scale of vulnerability within the intelligence community. In March of 2017, Wikileaks published a treasure trove of professed CIA hacking capabilities, the implications of which have yet to be fully understood (Lapowsky and Newman, 2017). In April of 2017, an NSA hacking tool called EternalBlue was stolen and released online (Larson, 2017). This tool was then used in the WannaCry hack in May 2017 that reportedly affected 99 countries (BBC Europe, 2017). These events showed that even the most secure organizations are vulnerable and the threat is moving at a faster rate than CTI. The government quickly realized that advanced, efficient, and rapid solutions are needed to secure intelligence information used to safeguard the nation.

Related Work in Cyber Threat Intelligence

The systemization of the CTI field is in the development stage but preliminary works to provide structure and organization are evident (Lee, 2014). A patented research study, completed in 2002, systematized CTI into a series of standardized steps that included data collection, processing, analysis, record creation, knowledge base creation, product creation, and client information collection (Edwards et al., 2002). Cloppert, from the SANS institute, categorized indicators of attacks into three categories: atomic (things that cannot be broken down like IP addresses and emails), computed (things that can be broken down like code and hash tags) and behavioral (attack patterns) (2009). Lockheed Martin formalized what is known as the Cyber Kill Chain to describe the stages of attack that include reconnaissance, weaponization, delivery, exploitation, installation, command and control, and actions on objectives. (Hutchins et al., 2011).

Since then, there have been several related works and attempts to categorize indicators of intrusion like the Cyber Observable Expression (CybOX), the Open Indicators of Compromise (OpenIOC), and the Incident Object Description Exchange Format (IODEF) frameworks (Gragido, 2012). Around the same time, the Cyber Intelligence Sharing and Protection Act was passed which allows for government to share intelligence with private corporations for the purpose of securing networks and systems (112 Congress, 2012). SANS provides an in-depth white paper on some major CTI works titled "Tools and Standards for Cyber Threat Intelligence Projects" (Farham, 2017). Shortly after these developments, the cybersecurity and communications section under the Department of Homeland Security led a collaborative environment to establish a structured language for CTI sharing titled Structured Threat Information eXpression (STIX) which uses the Trusted Automated eXchange of Indicator Information (TAXII™) to exchange CTI information between the government and private sectors. The structure consists of eight core constructs (Barnum, 2014). The eight core constructs of STIX are campaign, course of action, exploit target, incident, indicator, observable, threat actor, and tactics, techniques, and procedures (TTP). Many of the earlier frameworks were incorporated into STIX and the effort is now being led by the OASIS Cyber Threat Intelligence Technical Committee (OASIS, 2017).

Computational Methods in Cyber Treat Intelligence

CTI has developed at amazing speed some impressive analytical visualization tools to meet the growing demands for cyber threat indication. Table 13.1 demonstrates the wealth of tools available to conduct CTI. Most of these tools have the primary function of assisting the CTI analyst with collecting, categorizing, organizing, and visualizing data quickly. Each tool may have its own special utility that makes it stand out from the rest but there are some basic functions or tools that most of these platforms should have to be competitive (Department of Justice, 2007). Some common functions found in these platforms include data mining tools like word cloud, link analysis, geographic information system (GIS) capabilities, and statistical visualization tools like charts and plots.

Ultimately, the success of these platforms will be defined by their utility to the security analyst. Word cloud is a quantitative tool that provides visual displays of frequently used words. Word clouds are useful to the security analyst in narrowing the scope of study by allowing the security analyst to select reports that have words of interest. They also help the security analyst understand the quantity of reports they are dealing with. Word clouds can be useful when needing to separate relevant nouns or action words from nonrelevant ones and can assist in finding a word in a report like finding a needle in a haystack (this is usually a person). Link analysis helps the security analyst establish associations particularly as they relate to discovering central nodes in a network of interest or outliers that do not belong; an example is Thycotic's use of link analysis to recognize unauthorized access

Table 13.1 Computational Platforms for Cyber Threat Intelligence

Brand	Name	Use	Web Address
	Anomali	Threat intelligence platform	https://www.anomali.com/
	ArcSight	Security intelligence platform	http://www.ndm.net/arcsight/
	Cyber4Sight	Predictive threat intelligence platform	https://sso.cyber4sight.com/idp/SSO.saml2
	Haystack	Constellation analytics security platform	https://haystax.com/
	Maltego	Open source data mining tool	https://www.paterva.com/web7/index.php
	Metasploit	Penetration testing software	https://www.metasploit.com/
	Qualys	Cloud based security management tool	https://www.qualys.com/
	Palantir	Big data analysis tool	https://www.palantir.com/
	Recorded future	Threat intelligence platform	https://www.recordedfuture.com/
	Securonix	Security analytics platform	https://www.securonix.com/
	Spectorsoft	User behavior monitoring software	http://www.veriato.com/
	Splunk	Operational intelligence platform	https://www.splunk.com/
	Open stack	Open source cloud computing software	https://www.openstack.org/
	Tableau	Big data visualization tool	https://www.tableau.com/
	Threat connect	Security operations and analysis platform	https://www.threatconnect.com/
	Threat quotient	Threat intelligence platform	https://www.threatq.com/

(Thycotic, 2017). GIS capabilities help security analysts visualize attack patterns in terms of geography. Geography plays a strong role in cyber activity as the actions of a hacker correlate with the hacker's environment even if the hacker operates in the digital world (Bronk, 2016). An example of this is the MalwareTech botnet tracker that uses GIS capabilities to track and visualize botnets throughout the globe (MalwareInt, 2017). The Department of Defense realizes the relevancy of geography to cyber activity demonstrated by the strategic objective to build and maintain strong global alliances to combat cyber threats (The Department of Defense Cyber Strategy, 2015). GIS capabilities can also assist with predicting future attack patterns through the visualization of previous patterns. Data visualization tools help conduct comparative or trend analysis such as heat signatures for identifying high activity concentrations or graphs to identify escalation of events.

Regardless of the analytical capability, security analysts require tools that can talk and communicate with each other for quick exchange of data sharing. Once a threat is identified, time is of the essence which means the tool should facilitate not only thorough and in-depth analysis but also be able to share information with other security analysts quickly. Thus, the data should flow fluidly from one system to another to give others adequate time to respond and secure their systems. As demonstrated in Table 13.1, CTI tools are developed from a variety of sources in response to the competitive and lucrative field but are not necessarily able to cross-talk with other intelligence tools. The more the tool can communicate with the other tools out there, the more utility it has. The usability of the tool will also be a factor, as the security analyst is not necessarily going to be a subject matter expert in computer science even though the intelligence community is certainly trying to change this. Until then, the tool should be relatively easy for a security analyst to use, as the most common security analyst tasks include the ability to read, interpret, and apply that data to world events. As of right now, the more user friendly the CTI tool, the more it will be used (Tableau, 2017).

Challenges in Cyber Threat Intelligence

The challenges in CTI are for the most part the same as any intelligence field. They revolve around the ability to pull, store, find, and share reliable data. SANS institute completed a study on CTI to establish a baseline for CTI. The results of the study demonstrated that a demand exists to enable the aggregation of information from all available sources, visualize that data in an expedient and comprehensible manner, and verify the reliability of the data via multiple reporting sources (Shackleford, 2015). This is especially true in the intelligence community, as tools that pull from unclassified source data, like social media sites, do not directly feed into classified intelligence channels so the transfer of data is particularly challenging.

The ability to selectively collect and filter relevant information from the wealth of data out there presents another challenge. Konkel reports that even though the intelligence community is one of the largest collectors of Big Data, the information is fragmented and assembling the information is the real challenge (2014). CTI currently relies on the collection of data to find signatures and anomalies to identify hackers but these techniques have overwhelmed security analysts. A research study from Enterprise Strategy Group found that 74% of security operations teams reported that security events and alerts are ignored due to volumes of data and not enough staff (Davis, 2016). With so much data being collected, expiration dates on the information become an issue as the data expires faster than the security analyst can come to conclusions and act on those conclusions (Dickson, 2016). Expiration dates are also an issue because, as one Big Data platform pointed out, threat analysis is only released to the public after the fact to avoid tipping off the hacker that someone is tracking the hack (SecureWorks, 2017).

One of the main attractions to cybercrime is the anonymity of the act. Experts in the field are frustrated by the fact that origins of attack are difficult to trace (J.G., 2014). Although the CTI professional can remove malware from a system, the malware itself is not eradicated. Once the malware is identified and extracted, a hacker can still change small portions of the code and then use the same malware again. One piece of malware derived from one person can spread to thousands of computers and then be manipulated in seconds to spread again (Koerner, 2016). This highlights the futility of trying to track down a hacker from the trace, signature, design, or timestamp. One alternative response would be to identify the source from its point of origin rather than from a trace. This is where monitoring social media in conjunction with IT systems has the potential to lessen the time it takes to identify a threat.

Identity management is another challenge in the intelligence field as the internet offers the element of anonymity. For example, one report on social media sites estimates that approximately 31 million Facebook user accounts are fake (Hayes, 2016). Regardless of how a tool tracks an identity, determining identities will always be a challenge when dealing with the digital world as the slightest mistake in reporting can change the identity of a person. Biometric tools have this challenge despite the capability of establishing unique identities, such as fingerprints. The data that is assigned to the fingerprint still must be tagged. Should a fingerprint have similar biometric data to another individual, this information can easily become scrambled; many fingerprint readers use only selected points along the fingerprint ridges for identity, which results in up to 3% misidentification or overlapping fingerprints between individuals. The intelligence world is thus turning to machine learning and predictive analysis to remedy these issues.

The solution to these challenges is an obvious one: Track the hacker by developing easy to use tools that can collect data from a variety of independent sources in a selective manner. These tools should be able to communicate with other tools in an easy and timely fashion. Although the solution maybe is obvious, the implementation of the solution is not. Further development of machine learning programs is required to selectively pull relevant information. This paper views cyber intelligence as the computational ability to identify what information to collect to successfully identify the threat without collecting irrelevant data. With the advancement of technology and the speed in which the intelligence world is moving towards cybersecurity, the ability to develop these types of computational methods is what will define one Big Data analytic tool over another. This is also where algorithms of Big Data have a place in cyber threat intelligence.

Prevalence of Social Media Analysis in Cyber Threat Intelligence

Social media analysis has its place in the field of CTI in multiple ways, but for the future of CTI, the attraction is primarily due to its predictive nature, because it has the potential to establish more than just an identity. It can also establish moods, tones, and feelings prior to acts. Text analysis through social media has been useful in other areas of intelligence to establish security levels in an environment and to detect when that level is changing. Establishing security levels is important when one is concerned about the safety of those who operate in that space, be they government officials, military, law enforcement, intelligence professionals, or civilians. Text analysis through social media provides insight into the status of populations, such as levels of satisfaction using quantitative and qualitative analytics. The significance of this is that social media analysis has the potential to discover the source of a cyberattack, prior to the attack taking place, because of its ability to establish human motivations and moods.

Data analytics, as it relates to social media, appears to be an underdeveloped and underutilized source in the CTI field. In a threat intelligence capability briefing, establishing motivations is listed under strategic threat intelligence but when looking at the sources available to pull from, the human factor is surprisingly absent with the internal list including "logs, networks, endpoints, malware, phishing e-mails, and past incidents" and the external list including "industry sharing groups, government, organization partnerships, vendors, and open source (Nelson, 2016)." Granted social media can be categorized under open source but the necessity for it to become its own entity is becoming increasingly apparent. SANS survey results (Shackleford, 2015) show that tools were used by CTI professionals to "aggregate, analyze, and present CTI" and that 76% of intelligence information gathers from the security community. The results highlight that social media is still an untapped resource although the demand exists.

Attempts to understand, dissuade, deter, and stop a hacker via social media intelligence gathering could present an alternative solution to wading through the wealth of attacks that lead to endless mazes of data. The study of hacker motivations behind attacks may be limited because CTI falls directly into the realm of information technology and computer science whereas the human criminal factor typically falls under the umbrella of law enforcement or military intelligence. Human intelligence (HUMINT), for example, is a field that is not necessarily focused on computers or advanced technology. Rather, HUMINT typically focuses on who is behind the latest national threat. Cybercrime is just now coming into mainstream focus for national threats. HUMINT means deriving intelligence from people rather than computers and an investigation usually begins with an event. It is the event that drives the intelligence professional to discover the five Ws of who, what, when, where, and why with the "why" usually being the least of concern due to the fast-paced environment of threats. There is a relatively new field that focuses more on the "why" and human behavior called sociocultural research. However, the focus of this research, in general, is societies and the objective of the study is to understand how and why society allows the threat to operate in that space. Social media analysis could fall under the umbrella of all source intelligence (OSINT) but it is still not designed specifically to look at those behind computer-related activity and this type of analysis would be limited if not used in conjunction with CTI. In summation, the study of social media to predict cyber threats by analyzing human motivations is really a field or area of expertise that has yet to be established.

Benefits of Social Media Analysis in Cyber Threat Intelligence

Tracking hackers via regular every day activity, country environments, history, education, and social contact has the potential to narrow the scope of investigation compared to tracking from the end of the attack. To put this in terms of scale, there are approximately 7.5 billion people in the world, 3.7 billion of whom are internet users (Internet World Stats, 2017). The average ratio of criminal or crime type per capita is under 10% (Nation Master, 2002). This means the plausible number of potential criminals among internet users is less than 3.7 million users. To establish capability, the ratio of user to technician is estimated to be anywhere from 50:1 to about 80:1 (National Center for Education Statistics, 2000; Rumberg 2012), which means out of 3.7 million internet users with the possible intent to conduct criminal activity, there is a good chance that only 46,000–74,000 have the actual capability of being a hacker. These numbers can be further diminished by narrowing the scope to countries of interest as one research study showed that most hacks came from the United States and China (Kumar and Carley, 2016). The global threat intelligence report from Dimension Data states that 63% of attacks based on IP addresses originate from the United States and the top five countries for attack origin account for 75% of all attacks in 2016 (Thomas, 2017). Many of these attacks will be state-sponsored so the concern for a state would be either an out of state hacker or a nonsanctioned in-state hacker which narrows the field of potentials even further. Although these are broad estimations, these numbers are a lot more manageable than tracing hackers from attack end points. For example, a single threat tracker found almost 4 million attacks in one day, on July 2, 2017 (ThreatCloud, 2017). The benefit of tracking humans as opposed to attacks is the number of possibilities is substantially smaller and will not change as drastically the way cyberattacks morph by the day. Just these estimates create a good incentive to track computer operators as a form of threat mitigation.

Cyber threat intelligence (CTI) is a growing area in the intelligence community which is occurring at unprecedented rates to secure technology vulnerabilities that have taken the lead ahead of those responsible for securing it. Establishing a structured and defined trade in CTI has proven to be a colossal task as the very basics for defining the field, like qualifications standards, common taxonomies for technical language, and formalized trade techniques, are unstructured. As if transitioning the intelligence tradecraft to CTI is not daunting enough, the use of Big Data analytics is still in its infancy in the intelligence world. Meanwhile, incorporating social media into that CTI analysis is an even more untapped resource. The *New York Times* recently reported that hackers are now embedding attacks into social media recognizing it as an easier gateway into secured systems (Frenkel, 2017). As this field develops, the role of algorithms that support text analysis will grow in demand since the overwhelming amount of data will require simplified solutions to sort through the maze.

Establishing Motivations in Cyber Threat Intelligence through Social Media Analysis

Indications that establishing hacker motivations is growing in relevance is seen in CTI analysis but is usually portrayed from the perspective of cyberattacks. The image in Figure 13.2 shows an example of how hacker motivations (Hackmageddon, 2017) are typically displayed or understood. Humans, however, are complex individuals and motivations are often much more nuanced than what appears at face value. For example, kidnappings can be a tool to generate income. However, kidnappings, although classified as crime, are often used as a functional tool to meet a variety of objectives. Motivations driving this tool can vary greatly from love, desire, revenge, control, religious beliefs, cultural practices, political influence, or economics (ThreatRate Risk Management, 2017).

To clearly understand the motivations behind an attack, the mind of the attacker should be analyzed and understood. This can be done by looking at human behavior. Considering the operator exists in the cyber world, insight into

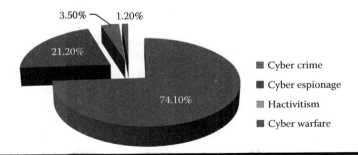

Figure 13.2 Motivations behind cyberattacks.

the minds of the operator using Big Data analysis and text analysis is a logical method of CTI collection. These operators, like all humans, have a basic human desire for social interaction. Granted, there may be anomalies to this but it is safe to assume that if an operator lives in the world of cybercrime, then the operator's social interaction would likely take place in the cyber world as well. Using text analysis of social media could thus be an effective strategy for narrowing and intercepting the threat.

Understanding the motivations behind the actions of an intruder also offers the potential to provide alternative means for mitigating threat as each human motivation is satisfied in different ways. There have been numerous studies regarding the complexity of motivations (Marsden, 2015) and how varied human motivations can be. The Department of Defense is realizing the advantages of behavioral analysis in detecting intrusion, which is why it is pushing for a new acquisition model that would allow them to invest in private sector machine learning technologies (Owens, 2017). As strategic politics and war move into the digital era and away from conventional techniques, the development of this type of intelligence may become more and more relevant, especially as governments begin to lose traditional forms of power and control with the proliferation of information dissemination.

Role of Behavioral and Predicative Analysis in Cyber Threat Intelligence

Scales are often used in behavioral analysis to diagnose human and animal behavior. For example, Bech, Hamilton, and Zung developed a social dysfunction and aggression scale, which allows a person's level of aggression to be rated from irritability to physical acts of violence (Bech, 1996). This type of scale can be beneficial in establishing a threat level in a work environment. Freedom House developed the political terror scale seen in Table 13.2, which rates the level of political terror across a country. This type of scale assists in identifying dictatorial or corrupt rule and assists in decision making for international intervention or sanctions but more importantly it can be used as a warning for potential government collapse or civil war. Scales like these are useful in predictive analysis in a multitude of other fields from economic trends to the development of social issues.

Scales in behavioral analysis can also be used to assign weight to words or actions. Weights help the researcher identify the development or trajectory of behavior. A classic example is the scale used by canine behaviorist to monitor the transition of aggression in canines: (biting, snapping, growling, stiffening up, lying down, standing crouched, creeping, walking away, turning body away, turning head away, and yawning). In this ladder, the development of canine aggressive behavior is demonstrated in the form of a ladder. The bottom of the ladder begins with subliminal signs of being uncomfortable that are obvious to dogs but are often ignored by humans. The top of the ladder ends with overt signs of aggression that

Table 13.2   The Political Terror Scale

Level	Interpretation
1	Countries under a secure rule of law, people are not imprisoned for their views. Political murders are extremely rare.
2	There is a limited amount of imprisonment for nonviolent political activity. However, few people are affected and torture and beatings are the exception. Political murder is rare.
3	There is extensive political imprisonment, or a recent history of such imprisonment. Execution or other political murders and brutality may be common. Unlimited detention, with or without a trial, for political views is accepted.
4	Civil and political rights violations have expanded to a large number of the population. Murders, disappearances, and torture are a common part of life. In spite of its generality, on this level terror affects those who interest themselves in politics or ideas.
5	Terror has expanded to the whole population. The leaders of these societies place no limits on the means or thoroughness with which they pursue personal or ideological goals.

Source: The Political Terror Scale, *Documentation: Coding Rules,* http://www.politicalterrorscale.org/Data/Documentation.html#PTS-Levels, 2017.

are more familiar to humans but once displayed are often also ignored and usually end in a bite. Canine behaviorists understand that if the problem can be identified at its onset, mitigating action can take place to prevent the aggressive behavior.

Scales like the example in the canine ladder of aggression (Horwitz and Mills, 2009) can be used in a similar way for threat identification and prevention using social media. Once the scale or trajectory of behavior is understood, the analyst can data mine for these key words, assign weight to the words like scales and then focus on the words that carry the most weight to prevent a hack. For example, if the objective is to save lives by preventing dog bites, a canine analyst would monitor behaviors in the middle of the aggression scale. Likewise, with human behavior, if the goal is to diminish potential hacks, an analyst can search for words found in the middle of the identified hacking "scale of aggression." Both will narrow the search for subjects that may commit the unwanted behaviors. The scope can be narrowed even further by focusing on indicators that the behavior has yet to occur, but the subject being reviewed has the potential or drive to commit the offense. In the case of hackers, priority of intervention would then be placed on text indicating the individual is in a current state of danger of performing a hack. Logically, this method will be extended to CTI using the information available from social media.

The same technique can be applied to hackers using the language of a hacker. Hypothetically, if each phase of the Cyber Kill Chain requires a separate set of tasks, the words assigned to these tasks could potentially indicate where in the Cyber Kill Chain the hacker is postured. The weight of the words used could provide warning or indication of movement to the next stage and thus narrow the scope of intervention. The Office of the Director of National Intelligence developed a common cyber threat framework that defines the actions of hackers in each stage of the Cyber Kill Chain (Sweigert, 2017). What is missing and could be developed is a hacker dictionary with the words associated to the most logical stage of the framework. There is already a preliminary work titled "The Hacker's Dictionary" that tracks the lexicon of hackers (Raymond, 2017). After the words are assigned to the stages, they could then be given weight and priority based on where the words fall in the Cyber Kill Chain.

Text Analysis Tools

The field of text analysis is a rapidly growing type of research. The reason for this recent surge in growth is because an increase in computing power allows for the analysis of extremely large sets of data. Linguistic analysis is now becoming a computer science topic just as much as it is a social science. By using linguistic analysis, it is possible to develop machine-learning algorithms that do tasks such as auto completion or even smarter machines. Another use for language analysis is that it can give insight into the thinking of the writer. It is for this reason that many researchers are using social media outlets as a source of data. By analyzing the text on social media, it is possible for researchers to get a feel of how large groups of people act regarding certain topics like politics.

Textual analysis is the act of making an educated guess at the meaning of that text. By analyzing text from various aspects of life such as television, films, and magazines it is possible to get an understanding of the way certain cultures view and interact with the world around them. Currently the largest repository of text can be found on the internet, which means that it is useful to have tools that allow us to use and manipulate digital text. The use of these text analysis tools is referred to as computer assisted text analysis (Gee, 2015).

Analyzing written language is called text analysis, also referred to as text mining, and involves analyzing text-based content using natural language processing (NLP). Natural language processing is the ability of computers to understand what users are writing in human writing patterns. This means that the computer must be capable of identifying various dialects, slang, and even properly identify homonyms.

To allow computers to better understand human language, most NLP algorithms are based on statistical machine learning, using statistical inferences. These inferences come from rules gained from analyzing large sets of documents know as corpora, that contain the correct values that need to be learned.

Linguistic Inquiry Word Count

Linguistic analysis is the scientific analysis of a language sample and is a form of textual analysis. A language sample can be spoken or written language. By analyzing the language sample, it is possible to gain an understanding of the writer. Linguistic analysis has 5 branches which are phonology, morphology, syntax, semantics, and pragmatics. Phonology is concerned with determining which speech sounds contrast with one another to produce differences in meaning and which speech sounds are in complementary distribution and never produce meaningful contrasts. Morphology is the identification and study of morphemes, a minimal part of a word that cannot be further subdivided. Syntax is the structure that sentences must follow to make sense in that given language. Semantics is the study of the meaning of a word, including its connotation and its denotation. Pragmatics is how humans use linguistic cues in actual situations.

Linguistic inquiry word count (LIWC) is text analysis software. It is used to gain insights about the mood of the writer. The research for it began from a study done by Pennebaker and Beall in 1986. The study discovered that when people were asked to write about emotional moments in their life, there was evident improvement in their physical health. To determine the correlation between these two things, a panel of judges was assembled and ask to read emotional essays and rate them among categories such as how coherent it was, how well it was organized, emotional words used, and others. The issue that was found with this approach is that most judges could not agree on the ratings of the papers in these categories. Also, it was noticed that when the judges read emotional essays it had an effect on their moods, which could impact on their ratings of the work.

It is for this reason that LIWC was created. LIWC has two primary components: the processing component and the dictionary component. The processing component uses natural language processing to read in the text file. The dictionary component is a user-defined dictionary that is filled with words. Each word in the dictionary is categorized. There are 64 categories overall and each word can fall into more than one category. These categories can be either subjective, such as happy and sad, or objective, such as nouns and verbs.

As LIWC goes through the text file, it compares the words from the text file to the words in the dictionary file to determine their category. If a word in the text file matches a word in the dictionary file, all the categories that the word belongs to are incremented.

Even though LIWC currently has more than 80 categories, when it was first developed it started with two broad categories, which are still important within the program. These two categories are style words and content words. Content words are words such as nouns and regular verbs. The purpose of these words is to convey the meaning of the communication. Style words are words such as pronouns or prepositions, which reflect how people communicate, as opposed to content words, which reflect what people are saying.

Since LIWC is a major tool used in the text analysis field there have been experiments done to determine how effective the program is. At the University of Illinois (Kahn et al., 2007) three experiments where done to test the effectiveness of LIWC. In the first experiment, users where asked to write an amusing essay and an emotional essay. LIWC was then used to analyze these essays and determine the emotions the author was attempting to convey in them. LIWC was successful at this and was able to identify the positive word usage in the amusing essay as well as the negative emotions being conveyed in the emotional essay. During the second experiment, users' emotions were manipulated using film clips. After the users were shown these clips, they were asked to give an oral presentation about how they felt. This oral presentation was transcribed into text and analyzed using LIWC. The results were accurate and they demonstrated that LIWC could correctly pick up various traits from one text. The third experiment was similar to the second experiment only this time the users were asked in-depth questions that helped determine more about their actual personality. They were then once again shown clips similar to those in the second experiment. The purpose of this experiment was to determine if LIWC could actually tell the difference in users' personalities based on the words they used. The results of this application, that LIWC could not tell the difference, reflected a more challenging scenario.

There have been several ways in which people have sought to use sentiment analysis. One of the most popular ways has been in prediction of elections. Twitter is a medium allowing for microblogs up to 140 words in length, known as "Tweets," to be posted by users. Other users on the site then read these Tweets and have the option of retweeting them. Tumasjan et al. (2010) analyzed over 10,000 messages that contained references to a political party of politician during the German federal election. The purpose of this analysis was to determine whether you could use Twitter as a valid medium for determining the offline political landscape in regards to the public's opinion.

This study was based off the fact that many believe that the current United States president's victory was due to his ability to effectively use social networks in his campaigning. The president not only used social networking but he even created his own website to increase his online presence. This led the researchers to wonder to what extent social networks play roles in politics and whether it was an effective place for gathering information on what the public currently felt about the political parties and the politicians in those parties. The researchers also wanted to determine whether Twitter was a platform for political deliberation, whether it was an accurate reflection of political sentiment, and whether it could accurately predict election results.

After examining 104,003 political Tweets regarding the 6 political parties that were part of the election or Tweets that mention prominent politicians in those parties, the researchers were able to answer these questions. Regarding whether Twitter is a platform for political deliberation, it was found that there was a high amount of political debate on Twitter; however, it was coming from a small group of heavy users, which may lead to an unequal representation overall. Regarding the second question, it was found that the examination of users' Tweets about the political parties and politicians did show that users could accurately identify certain traits about the moderate politicians and parties unless they were completely veered towered one side of the political spectrum. Regarding the last question, there was a correlation found between the number of Tweets where the party was mentioned and the winner of the election.

Sentiment Analysis

Sentiment analysis is the use of NPL and text analysis to determine the attitude of the speaker on the topic. The primary purpose of sentiment analysis is to determine if a sentence or even a document is positive or negative; this is called polarity sentiment analysis. A common use for polarity sentiment analysis is on product reviews. By analyzing the text of the review, it can be determined whether the review is positive or negative.

There are numerous ways that have been proposed to conduct sentiment analysis. Pang and Lee (2004) proposed a machine learning method that categorized the subjective portions of the document. This method differs from previous methods of classifying documents which was based on selecting lexical features of the document such as the word good being present. Instead, the author proposed to only use subjective sentences and not the objective sentences, which can be considered to be misleading.

Web 2.0 has the chance for businesses to use sentiment analysis in more business efficient matters. Web 2.0 is characterized by its increase in interactions between users and web pages. Facebook posts, Twitter posts, and even product reviews allow for analysis of large datasets, which can be used to create predication models. These predication models can be used to help companies determine how successful their next product would be or what features they should include in their future products. Another use for this analysis is to attempt outcomes of governmental elections.

When attempting to determine the sentiment of writing one of the issues that may cause skewed results is the misreading of the text into the analysis software. One of the foundations of sentiment analysis is polarity. The polarity is either negative or positive. An example of this can be seen in application reviews, which can either have a positive review (thumbs up) or a negative review (thumbs down).

When attempting to determine the polarity of a document there have been many approaches purposed. One of the most common methods is to select lexical features like indicative words such as "good." Upon seeing "the word good" in the text of a review of a novel, many analysis tools may identify that as a positive polarity which may skew the overall results even though the sentence where it was used had nothing to do with the reviewers feeling about the actual novel. For example, the reviewer may write "The lead character attempted to protect his good name."

It is for this purpose that Pang has proposed the use of a minimum cuts method, which separates the subjective and objective sentences from text and only considers subjective sentences in the analysis phase. They propose a document level subjectivity detector, which will allow for subjectivity detection on the sentence level. Only sentences that passed through the subjectivity detector will be allowed in to the polarity classifier for document analysis.

The results of applying this subjectivity detector are that they gained a statistically significant improvement of 4% in the accuracy of the sentiment analysis of the given documents.

The reason that Twitter is such a popular place for sentiment analysis is due to the high value of text that is present there as well, how easy it is to gather the text, the wide variety of users it has, and the possibility to collect data from other cultures. This offers a plethora of data for companies such as what is the best demographic for their product and how the public reacts to certain forms of marketing.

To demonstrate how effective Twitter can be for a corpus of text for sentiment analysis, Pak and Paroubek (2010) collected 300,000 texts from Twitter, which they then separated into three groups evenly. The categories were: text that conveyed positive emotions, text that conveyed negative emotions, and objective text. Linguistic analyses were then collected on this corpus and a sentiment classifier was then developed using the corpus as training data.

The corpus was collected using the Twitter API that allows you to actually query a collection of Tweets from the website. In the query, happy emoticons and sad emoticons were searched. Tweets containing sad emoticons where considered negative text and tweets using positive emoticons where considered positive text. Objective text was pulled from Tweets of magazines and newspaper such as *The New York Times*. The classifier was built using a multinomial naive Bayes classifier with binary *n*-grams from the text as the input.

SentiStrength

SentiStrength is a tool for sentiment analysis. Textual sentiment analysis tools focus largely on determining positive, negative, or lack of, sentiments within the given text. To accomplish this, tools either use a machine-learning or lexical approach to analyzing the text. The machine-learning method works by using *n*-grams. The software is trained to mark certain *n*-grams as positive and others as negative. The lexical approach is similar to that of LIWC where the software is given a dictionary file that identifies certain words as negative or positive, and then assigns a value during the analysis of the text. The lexical approach also uses pseudo *n*-gram styles such as the word "not"; it is read as a negation of the following word, so the words "not happy" would be read as a negative statement (Thelwall, 2013).

SentiStrength is a sentiment analysis tool that uses the lexical approach for analysis. There are three versions of SentiStrength: the online version, the free Java version which is made available for research and educational purposes, and the paid version which is also written in Java, but made for commercial use. All the versions contain similar functions but the commercial version offers the ability to scan and organize much larger datasets.

Unlike LIWC, which is popular due to its dictionary, the main draw to SentiStrength is the algorithm it uses to determine the sentiment in a document. As mentioned earlier, SentiStrength is a lexical sentiment analysis tool; this means that it uses a dictionary to determine the sentiment of words in the given text. The dictionary file is made up of a combination of the LIWC dictionary but has been updated periodically through testing. In the dictionary file for SentiStrength, each word is given a positive sentiment score of 1 to 5 and a negative sentiment score of −1 to −5. A score of 1 means no positive found and a score of −1 means no negative emotion was found. The sentiment of the overall sentence is then given the sentiment score of the highest positive and negative sentiment in that statement

Textual analysis is the analysis of text to gain useful information from it. Computer-assisted textual analysis done by computer programs using natural language processing which allows computers to convert high-level human language into a computer-readable format. Many NLP programs also use statistical analysis within their algorithm to be predictive of text that is being read in.

The difference between LIWC and sentiment analysis is that linguistic analysis is a form of textual analysis that is done by using certain language structures found in most languages to get an understanding of the text. There are a variety of linguistic analysis tools available, but one of the most well used tools is LIWC. The two main parts of LIWC are the processing component and the dictionary. The processing component is what reads in the text file and it is then compared to the dictionary component and sorted based on the category of the word in the dictionary. By doing this, LIWC can then find patterns within those categories that may give information about the meaning behind the writing, such as whether it is joyous or sad. Sentiment analysis is similar to linguistic analysis in what it seeks to accomplish. The notable difference between the two is that linguistic analysis is attempting to understand the language that is being used in the writing whereas sentiment analysis is attempting to determine the overall polarity of the document. With the rise of social media, sentiment analysis is being used to analyze large sets of text to determine certain trends found in it.

Case Study Using Linguistic Inquiry Word Count

To better combat cyber threats, many researchers (Chi et al., 2016; SANS, 2017) have begun to focus more on the psychology of insiders who become threats and what causes them to do so. Collecting data from targeted personnel, such as social networking and emails, is the best way to prevent or identify an insider threat early. Text analysis will analyze that data and give policy makers the resilient decision support.

Text data is analyzed using LIWC to determine if a given actor is a threat. The difference between the past methods of determining an insider threat is that there has never been a study that used linguistic analysis to bridge the gap between an employee's text and the demonstration that they could possess the characteristics of the dark triad model. The data is tested for agreeableness, neuroticism, conscientiousness, and extraversion. Since both agreeableness and conscientiousness are traits consistent with all three of the dark triad traits, this is the first pair that is looked for in the actor. The LIWC scores in the required categories are tested against the average of known insider threat cases to determine if the actor possesses these traits.

If these traits are identified, the actor's text is tested against the LIWC categories for neuroticism and extroversion. Depending on the scores for these categories it becomes evident if the actor does or does not possess the characteristics of these mental diseases. If the actor does possess the traits for one of the dark triad personality traits, then they are considered to be a possible insider threat.

Of the three dark triad traits, this is the first pair that is looked for in the actor. The LIWC scores in the required categories are tested against the average of known insider threat cases to determine if the actor possesses these traits. Figure 13.3 represents the method by which threat values will be assigned to users.

The underlying problem, when determining whether or not the psychology of a person makes them more likely to pose a threat to the company, is if their actions make them less trustworthy. Trust is a concept, which is multidimensional and effects relationships on both small and large scales. When an organization hires an individual, it is important that they trust that individual. It is for this reason that many companies give new employees personality evaluations. These personality questions are

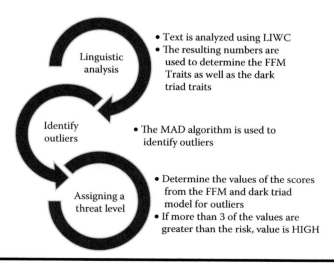

Figure 13.3   Threat assessment method.

based on the traits identified in the five factor model (FFM) (Judge and Bono 2000). FFM is a widely accepted model that is used to identify certain personality traits in people based on certain mental cues. The problem with this personality test is that many people lie to get hired by companies. Response distortion is common among many applicants; therefore, many companies may not get an accurate understanding of an employee's true personality.

The outliers for social networking after LIWC analysis can be identified using median absolute deviation (MAD). MAD is a method to identify outliers. The MAD (Leys et al., 2013) was calculated as shown below. Here M is the median and x the values:

$$\text{MAD} = 1.4826 * M_i(\mid x_i - M_j(x_j) \mid)$$

The decision criteria for the outliers are as shown below. We chose mi = 2.5. A score of 3 is very conservative, 2.5 is moderately conservative, and 2 is poorly conservative.

Conclusions

This chapter gives an overview of data analysis uses in the field of CTI and insider threat. All data collection resources are related to mobile devices. This chapter examined the process of collecting and organizing data, various tools for text analysis, and several different analytic scenarios and techniques. Applying data analytics into the field of CTI is still in its infancy. Meanwhile, incorporating social media into CTI analysis is an even more untapped resource (Clark, 2016). Dealing with a collection of very huge datasets with a great diversity of types from social networks, data analytic provides enormous novel approaches for CTI professionals. How to use Big Data analytic technique to achieve valuable information is still big challenge.

References

Barnum, S. (2014). *Standardizing Cyber Threat Intelligence Information with the Structured Threat Information eXpression* (*STIX™*), Mitre. February 20. Version 1.1, Revision 1. http://stixproject.github.io/about/STIX_Whitepaper_v1.1.pdf. Accessed June 26, 2017.

BBC Europe (2017). Cyber-attack: Europol says it was unprecedented in scale. *BBC News*, May 13. http://www.bbc.com/news/world-europe-39907965. Accessed June 23, 2017.

Bech, P. (1996). *The Bech P., Hamilton and Zung Scales for Mood Disorders: Screening and Listening.* Springer-Verlag: Berlin, Germany.

Bennett, C. (2015). Russian hackers crack pentagon email system. *The Hill*, August 6. http://thehill.com/policy/cybersecurity/250461-russian-hackers-crack-pentagon-email-system. Accessed July 8, 2017.

Bronk, C. (2016). *Cyber Threat: The Rise of Information Geopolitics in U.S. National Security.* Praeger: Santa Barbara, CA.

Castells, M. (2007). Communication, power and counter-power in the network society. *International Journal of Communication* 1(1): 29.

Chi, H., Scarlett, C., Prodanoff, Z.G., and Hubbard, D. (2016, December). Determining predisposition to insider threat activities by using text analysis. *Future Technologies Conference (FTC)* (pp. 985–990). IEEE.

CIA (2000). *Statement for the Record Before the Joint Economic Committee on Cyber threats and the US Economy by John A. Serabian, Jr. Information Operations Issue Manager, CIA.* Central Intelligence Agency News and Information: Cyber Threats and the U.S. Economy. https://www.cia.gov/news-information/speeches-testimony/2000/cyberthreats_022300. html. Accessed June 21, 2017.

Clark, R.M. (2016). *Intelligence Analysis: A Target-centric Approach.* CQ Press: Washington, DC.

CNN Wire Staff (2011). Obama says U.S. has asked Iran to return drone aircraft. *CNN,* December 13. http://edition.cnn.com/2011/12/12/world/meast/iran-us-drone/index. html. Accessed June 23, 2017.

Davis, J. (2015). Hacking of government computers exposed 21.5 million people. *New York Times,* July 9. https://www.nytimes.com/2015/07/10/us/office-of-personnel-management-hackers-got-data-of-millions.html. Accessed June 23, 2017.

Davis, K. (2016). Phantom and ESG research finds companies ignore majority of security alerts. *Business Wire Inc.,* March 15. http://www.businesswire.com/news/home/20160315005555/en/Phantom-ESG-Research-Finds-Companies-Ignore-Majority. Accessed July 9, 2017.

The Department of Defense Cyber Strategy (2015). https://www.defense.gov/Portals/1/features/2015/0415_cyber-strategy/Final_2015_DoD_CYBER_STRATEGY_for_web.pdf. Accessed July 24, 2017.

Department of Justice (2007). Analyst toolbox: A toolbox for the intelligence analyst. *Global Justice Information Sharing Initiative.* https://it.ojp.gov/documents/analyst_toolbox.pdf. Accessed July 20, 2017.

Dickson, B. (2016). How predictive analytics discovers a data breach before it happens. *Crunch Network,* July 25. https://techcrunch.com/2016/07/25/how-predictive-analytics-discovers-a-data-breach-before-it-happens/. Accessed July 9, 2017.

DOD Defense Science Board (2013). *Task Force Report: Resilient Military Systems and the Advanced Cyber Threat.* Office of the Under Secretary of Defense for Acquisition, Technology and Logistics: Washington, DC, January. http://nsarchive.gwu.edu/NSAEBB/NSAEBB424/docs/Cyber-081.pdf.

Edwards, C., Migues, S., Nebel, and Owen, D. (2002). System and method of data collection, processing, analysis, and annotation for monitoring cyber-threats and the notification thereof to subscribers. https://www.google.com/patents/US20020038430. Accessed June 23, 2017.

Farham, G. (2017). Tools and standards for cyber threat intelligence projects. *SANS Institute InfoSec Reading Room,* October 14. https://www.sans.org/reading-room/whitepapers/warfare/tools-standards-cyber-threat-intelligence-projects-34375.

Frenkel, S. (2017). Hackers hide cyberattacks in social media posts. *New York Times,* May 28. https://www.nytimes.com/2017/05/28/technology/hackers-hide-cyberattacks-in-social-media-posts.html. Accessed July 2, 2017.

Gee, J. (2015). *Social Linguistics and Literacies: Ideology in Discourses.* Routledge: New York.

Gragido, W. (2012). Understanding Indicators of Compromise (IOC) Part I, October 3, 2012. https://blogs.rsa.com/understanding-indicators-of-compromise-ioc-part-i/. Accessed July 7, 2017

Hackmageddon (2017). Motivations behind attacks. *Information Security Timeline and Statistics.* http://www.hackmageddon.com/2017/06/09/april-2017-cyber-attacks-statistics/.

Hayes, N. (2016). Why social media sites are the new cyber weapons of choice. *Dark Reading,* September 6. http://www.darkreading.com/attacks-breaches/why-social-media-sites-are-the-new-cyber-weapons-of-choice/a/d-id/1326802. Accessed July 9, 2017.

Horwitz, D., and Mills, D. (2009). *BSAVA Manual of Canine and Feline Behavioral Medicine.* BSAVA: Gloucester, UK.

Hutchins, E., Cloppert, M., and Amin, R. (2011). *Intelligence-Driven Computer Network Defense Informed by Analysis of Adversary Campaigns and Intrusion Kill Chains.* Lockheed Martin Corporation. http://www.lockheedmartin.com/content/dam/lockheed/data/corporate/documents/LM-White-Paper-Intel-Driven-Defense.pdf. Accessed July 8, 2017.

Internet World Stats (2017). http://www.internetworldstats.com/stats.htm. Accessed July 24, 2017.

Judge, T.A., and Bono, J.E. (2000). Five-factor model of personality and transformational-leadership. *Journal of Applied Psychology* 85(5): 751.

Kahn, J.H., Tobin, R.M., Massey, A.E., and Anderson, J.A. (2007). Measuring emotional expression with the linguistic inquiry and word count. *The American Journal of Psychology* 120: 263–286.

Koerner, B. (2016). Inside the cyberattack that shocked the US government. *Wired,* October 23. Condé Nast: New York, NY. https://www.wired.com/2016/10/inside-cyberattack-shocked-us-government/. Accessed June 22, 2017.

Kumar, S., and Carley, K. (2016). *Approaches to Understanding the Motivations Behind Cyber Attacks.* Carnegie Mellon University. http://www.casos.cs.cmu.edu/publications/papers/2016ApproachestoUnderstanding.pdf. Accessed July 6, 2016.

Lapowsky, I., and Newman, L.H. (2017). Wikileaks CIA dump gives Russian hacking deniers the perfect ammo. https://www.wired.com/2017/03/wikileaks-cia-dump-gives-russian-hacking-deniers-perfect-ammo/. Accessed July 24, 2017.

Larson, S. (2017). NSA's powerful Windows hacking tools leaked online. *CNN Technology,* April 15. http://money.cnn.com/2017/04/14/technology/windows-exploits-shadow-brokers/index.html. Accessed June 23, 2017.

Lee, R. (2014). Cyber threat intelligence. *Tripwire,* October 2. https://www.tripwire.com/stateof-security/security-data-protection/cyber-threat-intelligence/. Accessed July 7, 2017.

Leys, C., Ley, C., Klein, O., Bernard, P., and Licata, L. (2013). Detecting outliers: Do not use standard deviation around the mean, use absolute deviation around the median. *Journal of Experimental Social Psychology* 49(4): 764–766.

MalwareInt (2017). Malware Tech Botnet tracker. https://intel.malwaretech.com/. Accessed July 11, 2017.

Marsden, P. (2015). The science of why. *Brand Genetics,* June 22. http://brandgenetics.com/the-science-of-why-speed-summary/. Accessed July 12, 2017.

Nakashima, E., and Krebs, B. (2007). Contractor blamed in DHS data breaches. *The Washington Post,* September 24. http://www.washingtonpost.com/wp-dyn/content/article/2007/09/23/AR2007092301471.html?hpid=sec-tech. Accessed June 22, 2017.

National Center for Education Statistics (2000). Technology in schools. *Chapter 5: Maintenance and Support, Technology in Schools: Suggestions, Tools, and Guidelines for Assessing Technology in Elementary and Secondary Education.* https://nces.ed.gov/pubs2003/tech_schools/chapter5.asp. Accessed July 11, 2017.

NationMaster (2002). http://www.nationmaster.com/countryinfo/compare/Ireland/United-States/Crime. Accessed July 25, 2017.

Nelson, M. (2016). *Threat Intelligence Capability*. North Dakota State Capability. https://www.ndsu.edu/fileadmin/conferences/cybersecurity/Slides/Nelson-Matt-Threat_Intel_Capability_Kick_Start_.pptx. Accessed July 14, 2017.

OASIS (2017). OASIS cyber threat intelligence (CTI) TC. https://www.oasis-open.org/committees/tc_home.php?wg_abbrev=cti. Accessed June 26, 2017.

Owens, K. (2017). Army uses behavioral analytics to detect cyberspace invaders. https://defensesystems.com/articles/2017/06/29/army-cyber.aspx. Accessed June 29, 2017.

Pak, A., and Paroubek, P. (2010, May). Twitter as a corpus for sentiment analysis and opinion mining. In *LREc* (Vol. 10). Université de Paris-Sud: France.

Pang, B., and Lee, L. (2004, July). A sentimental education: Sentiment analysis using subjectivity summarization based on minimum cuts. In *Proceedings of the 42nd annual meeting on Association for Computational Linguistics* (p. 271). Association for Computational Linguistics: Barcelona, Spain.

The Political Terror Scale (2017). Documentation: Coding rules. http://www.politicalterrorscale.org/Data/Documentation.html#PTS-Levels. Accessed July 12, 2017.

Rawnsley, A. (2011). Iran's alleged drone hack: Tough, but possible. *Wired*. https://www.wired.com/2011/12/iran-drone-hack-gps/. Accessed June 23, 2017.

Raymond, E. (2017). *The New Hacker's Dictionary*, 3rd ed. The MIT Press. https://mitpress.mit.edu/books/new-hackers-dictionary. Accessed July 11, 2017.

Rumberg, J. (2012). Metric of the month: Tickets per technician per month. *MetricNet*. http://www.thinkhdi.com/~/media/HDICorp/Files/Library-Archive/Insider%20Articles/tickets-per-technician.pdf. Accessed July 11, 2017.

SANS (2017). FOR578: Cyber threat intelligence. https://www.sans.org/course/cyber-threat-intelligence. Accessed July 14, 2017.

SecureWorks (2017). Cyber threat basics, types of threats, intelligence & best practices. https://www.secureworks.com/blog/cyber-threat-basics. Accessed July 9, 2017.

Sevastopulo, D. (2007). Chinese hacked into Pentagon. *Financial Times*, September 3. https://www.ft.com/content/9dba9ba2-5a3b-11dc-9bcd-0000779fd2ac?mhq5j=e2. Accessed June 23, 2017.

Shackleford, D. (2015). *Who's Using Cyberthreat Intelligence and How?* Sans Institute InfoSec Reading Room. https://www.sans.org/reading-room/whitepapers/analyst/cyberthreat-intelligence-how-35767. Accessed June 21, 2017.

Sweigert, D. (2017). *A Common Cyber Threat Framework: A Foundation for Communication*. Cyber Threat Intelligence Integration Center—ONDI. https://www.slideshare.net/dgsweigert/cyber-threat-intelligence-integration-center-ondi?qid=4d6055a5-ead1-40d5-84b8-60439c570852&v=&b=&from_search=6. Accessed July 14, 2017.

Tableau (2017). Top 10 big data trends for 2017. https://www.tableau.com/sites/default/files/media/Whitepapers/whitepaper_top_10_big_data_trends_2017.pdf?ref=lp&signin=66d590c2106b8d532405eb0294a4a9f1. Accessed July 6, 2017.

Thelwall, M. (2013). Heart and soul: Sentiment strength detection in the social web with sentistrength. *Proceedings of the CyberEmotions* 5: 1–14.

Thomas, M. (2017). 2017 global threat intelligence report. *Dimension Data*, May 4. http://blog.dimensiondata.com/2017/05/2017-global-threat-intelligence-report/. Accessed July 9, 2017.

Thornburgh, N. (2005). Inside the Chinese hack attack. *TIME Magazine*, August 25. http://content.time.com/time/nation/article/0,8599,1098371,00.html. Accessed June 23, 2017.

ThreatCloud (2017). *Live Cyber Attack Threat Map.* Check Point Software Technologies Inc. https://threatmap.checkpoint.com/ThreatPortal/livemap.html. Accessed June 3, 2017.

ThreatRate Risk Management (2017). Types of kidnappings. http://www.threatrate.com/pages/47-types-of-kidnappings. Accessed July 6, 2017.

Thycotic (2017). PBA access use case. https://vimeo.com/209209431. Accessed July 11, 2017.

Tumasjan, A., Sprenger, T.O., Sandner, P.G., and Welpe, I.M. (2010). Predicting elections with Twitter: What 140 characters reveal about political sentiment. *ICWSM* 10(1): 178–185.

U.S. Cong. Senate—Homeland Security and Governmental Affairs 113 Cong. (2014). *Federal Information Security Management Act of 2014.* 113 Cong. S.2521. Washington, DC. https://www.congress.gov/bill/113th-congress/senate-bill/2521. Accessed June 22, 2017.

U.S. Cong. House-Government Reform; Science. 107 Cong. (2002) *Federal Information Security Management Act of 2002.* 107 Cong. 2nd Sess. H. R. 3844. Washington, DC. https://www.congress.gov/bill/107th-congress/house-bill/3844. Accessed June 22, 2017.

Windrem, R. (2015). China read emails of top U.S. officials. *NBC News*, August 10. http://www.nbcnews.com/news/us-news/china-read-emails-top-us-officials-n406046.

Index